Short Wavelength Lasers and their Applications

SHORT WAVELENGTH LASERS AND THEIR APPLICATIONS

PROCEEDINGS OF INTERNATIONAL SYMPOSIUM,
SAMARKAND, USSR, MAY 14–18, 1990

Edited by
V.V. KOROBKIN AND M.YU. ROMANOVSKY

Nova Science Publishers, Inc.
New York

Nova Science Publishers, Inc.
6080 Jericho Turnpike, Suite 207
Commack, New York 11725

Library of Congress Cataloging-in-Publication Data
available upon request

ISBN-1-56072-020-9

Book Production Manager Janet Glanzman White
Graphics Elenor Kallberg and Debbie Lippus

Printed in the United States of America

CONTENTS

PART IV: X-RAY OPTICS AND SPECTROSCOPY

PART V: EXCIMER LASER RESEARCH

Preface

International Symposium "Short Wavelength Lasers and Their Application" was held in May, 14-18, 1990 in Samarkand, USSR. It was the third meeting on this topic, the first was held in Aussois, France at 1986 and the second in Osaka, Japan at 1987. This third Symposium was organized because of a currently growing interest in the research of X-ray and VUV lasers, novel excimer lasers, X-ray and VUV optics , and their application. The Symposium was sponsored by the General Physics Institute of the Academy of Sciences of the USSR, International Commission for Optics - USSR Territorial Committee, and Samarkand State University. There were about 120 attendees. Approximately one third of them were from 10 countries other than the USSR.

The Proceedings of the Symposium include both invited papers (mostly of a review character) and contributed ones - totally 44. It should be noted that the book has been published within a reasonable interval after the Symposium and that the papers have not been refereed. We let the reader determine the quality of the individual papers.

The book includes five parts. The first one - "X-ray and VUV lasers. Theory and experiment" - is the largest of them and gives a good impression of the investigation spectrum in this field of physics. Selected papers in the second part - "Laser plasma and Application" - demonstrate the main directions of the research. We would like to note here the paper of Prof. C.Yamanaka from Osaka University. The third part - "X-ray nonlinear generation" - shows non-traditional approaches to problems of the X-ray generation and physical processes in the superbright optical laser pumping field. The fourth part - "X-ray optics and spectroscopy" - includes new results in the field of X-ray optical devices and new theoretical calculations and databases of X-ray transitions in high-stripped ions. The last part - "Excimer laser research" - presents new investigations in the excimer laser physics. The paper of Prof. C.K.Rhodes and coworkers from the University of Illinois at Chicago has an interdisciplinary character and is published out of frames of the book parts. Also the reports from the General Physics Institute of the Academy of Sciences of the USSR can be noted.

October 1990

V.V.Korobkin

M.Yu.Romanovsky

MOLECULAR X-RAY LASER RESEARCH

T.S. Luk, A. McPherson, G.N. Gibson, K. Boyer, and C.K. Rhodes
Laboratory for Atomic, Molecular, and Radiation Physics
Department of Physics, University of Illinois at Chicago
P.O. Box 4348, Chicago, Illinois 60680

ABSTRACT

Recent experimental results indicate that a new technique for the generation of amplification at x-ray wavelengths may be feasible. It involves the combination of (1) a new ultrahigh brightness subpicosecond laser technology, (2) a recently discovered mode of strong-field excitation, particularly applicable to molecules, and (3) a new channeled mode of propagation. Concerning the production of electronically excited states by strong-field processes, a concept of molecular x-ray laser design emerges which involves the matching of the conditions of excitation to the molecular structure. The molecular approach enables the generation of a vigorous level of electronic excitation to be achieved in an environment characteristic of dense cold matter, a situation exceptionally conducive to x-ray amplification. Both high efficiency and wavelength tunability are intrinsic features of this method.

I. INTRODUCTION

Amplification in the x-ray region requires prodigious energy deposition rates spatially organized in a high-aspect-ratio volume of material[1]. Therefore, the fundamental question centers on the controlled deposition of energy at high specific powers. We show that the use of extremely intense (10^{18} – 10^{21} W/cm^2) short pulse (~ 100 fs) radiation, which can be produced by high brightness laser technology[2,3], may be able to produce both (1) the necessary deposition rates and (2) the required spatial control. This outcome results from combining the energy deposition [4,5] arising from high-order multiphoton processes with a new mode of channeled propagation involving a relativistic charge-displacement mechanism. Experimental evidence also suggests that a new class of molecular strong-field multiphoton processes may be able to selectively generate excited ions to serve as the amplifying systems. Certain preliminary aspects of this molecular hypothesis will be examined below. A significant point that will emerge is that the conditions needed for the strong-field multiphoton excitation are largely identical to those found required for the confined propagation. Roughly speaking, we are interested in the physical regime involving intensities $\gtrless$ 10^{18} W/cm^2 so that the peak field strength ε is significantly above an atomic unit (e/a_0^2). The experimental picture[2,3], is now such that this regime can be systematically studied.

It has been suggested[4–8] that, for sufficiently strong coupling, fundamentally different nonlinear modes of direct electronic excitation may occur. The experimental results of two recent studies with 248–nm radiation, one involving ion production[9] and the other concerning fluorescence[10], have produced strong confirming evidence for this hypothesis with the unambiguous observation of the direct generation of $2\sigma_g$ holes in N_2. Further preliminary results concerning the pattern of ionic fragmentation of N_2O, which are presented below, give additional information on the nature of these nonlinear molecular amplitudes. Indeed, the main conclusions for molecules that emerge from the consideration of these results are (1) that inner electrons can be directly excited and (2) that the strong–field coupling is qualitatively different from those that can be generated by other forms of interaction, particularly charged particle collisions, even though the force field can be of comparable peak strength in the collisional case. Consequently, the strong–field interaction characterizes a unique regime of molecular interaction enabling the efficient production of new forms of highly excited matter. It is the ability to produce these unusual states of excited matter, under conditions consistent with the appropriate spatial control of the deposited energy[11], which gives promise for the achievement of stimulated emission at wavelengths in the x–ray range. This approach differs fundamentally in its dynamics and scaling from those involving recombination in expanding plasmas[12–14].

II. EXCITED STATE PRODUCTION IN MOLECULES

A. Ion Studies of N_2

Experimental results now exist which clearly show the excitation of inner electrons by a direct multiquantum process. The specific system illustrating this process is N_2 in which the production of $2\sigma_g$ holes has been detected in two different experimental studies. The initial hint came from the measurement[9] of kinetic energy distributions of ionic fragments produced in N_2 with subpicosecond 248–nm radiation which indicated the presence of the charge asymmetric channel

$$N_2^{2+} \longrightarrow N^{2+} + N \tag{1}$$

resulting from multiphoton ionization. Interestingly, in experiments performed with subpicosecond 600–nm radiation, although some differences in the branching of the fragment charge distributions appear[9,15], additional evidence was also indicated[16] for reaction (1).

Since other experiments[17] examining soft x–ray produced fragmentation of N_2 showed that process (1) corresponds to the production of a two–hole state having an excitation energy of ~ 70 eV, the obvious implication was that the laser excitation produced a similar configuration of molecular excitation. The states[18,19] implicated in the x–ray studies[17] of N_2, which has the ground state configuration $(1\sigma_g)^2(1\sigma_u)^2(2\sigma_g)^2(2\sigma_u)^2(1\pi_u)^4(3\sigma_g)^2$, were $^1\Sigma_g[2\sigma_g^{-1}, 3\sigma_g^{-1}]$, $^1\Sigma_u[2\sigma_g^{-1}, 2\sigma_u^{-1}]$, and $^1\Pi_u[2\sigma_g^{-1}, 1\pi_u^{-1}]$, all involving a $2\sigma_g$ hole. We note that the $2\sigma_g$ orbital has a binding energy of 37.7 eV, a value considerably higher than the $2\sigma_u$ (18.7eV), $1\pi_u$ (15.6 eV), and $3\sigma_g$ (16.7eV) orbitals. Therefore, the N_2^{2+} states having a $2\sigma_g$ vacancy and correlating to the N^{2+} + N asymptote have an energy substantially above the lowest manifold of states[20] in N_2^{2+} which are associated with the $N^+ + N^+$ limit. Experiments indicate that the dynamics of the multiphoton coupling favor a parallel alignment of the field with the molecular axis[9,15,16].

Furthermore, it has been noted[21] that the $2\sigma_g$ hole is a highly correlated state and does not fit a quasi-particle picture, an aspect that makes these states of particular interest.

B. Fluorescence Studies of N_2

Measurements of fluorescence point toward new insights into the mechanisms of multiphoton excitation that are directly relevant to the implementation of x-ray lasers. We describe below the observation[10] of a very intense band of radiation at ~ 55.8 nm which appears to be fundamentally related to the mode of excitation producing the charge asymmetric fragmentation represented by process (1). These experiments were performed at an intensity of ~ 10^{16} W/cm^2 with a pulse width of ~ 600 fs, conditions close to those pertaining to the ion studies.

The 55.8-nm emission is, at the outset, highly unusual in one obvious respect. Although the spectrum of N_2 has been studied for nearly a century[19] with many techniques of excitation, it was not possible to find any previous[19,22,23] report of such a characteristic emission originating from material composed solely of nitrogen atoms or molecules in either neutral or ionic form. From this fact alone, it appears that the observation of the 55.8-nm band originates from the excitation of a new class of states and that the strong production of these levels is intimately connected with the basic nature of the multiquantum mechanism involved in the coupling to the molecule. This line of reasoning and its implications in regard to the generation of coherent x-rays are explored in the following sections.

The hypothesis that emerged from the analysis of the 55.8-nm radiation is that it originates from a quasi-bound $^3\Pi_u$ level of N_2^{2+} having a $(2\sigma_g^{-1},1\pi_u^{-1})$ configuration which undergoes a bound-bound transition terminating on the D $^3\Pi_g$ level associated with the $N^+ + N^+$ asymptote[10]. In addition, the previous studies of ionic fragmentation[9], combined with the comparison of the results under comparable experimental conditions with N_2O, indicate that the $2\sigma_g$ vacancy is produced by a direct multiphoton process in N_2 leading to an excited state of the type (N^{2+} + N), namely, a level with evident charge-transfer character. Therefore, the findings concerning the fluorescence are fully consistent with the results independently derived from the ion studies[9].

Overall, several transitions were assigned arising from the production of the $2\sigma_g$ vacancy. Qualitatively, the relative strengths of the molecular features could be accounted for by the overlap of the wavefunctions involved in the transitions. Basically, the transition producing the 55.8-nm radiation is

$$2\sigma_g^{-1} \longrightarrow 2\sigma_u^{-1} + \gamma, \qquad (2)$$

a case which has a very favorable overlap, a fact that can be seen directly by visualization[24,25] of the wavefunctions of those two orbitals.

C. Molecular Site-Specific Energy Deposition

An electronic motion of the type suggested by the excitations observed in N_2 has general implications for the behavior of molecular systems under strong-field short-pulse irradiation. One of them is a <u>site-selective</u> energy deposition. This could be observed, for example, with the nitrogen atoms in N_2O, a linear molecule with the structure NNO. The two nitrogen atom sites in N_2O are nonequivalent; one resides on the end of the molecule while the other is located at the central position. Furthermore, since the bond energy associated with the binding of the oxygen is low (~ 1.6 eV), very roughly, N_2O can be considered as $(N_2)O$. In this picture, the terminal nitrogen atom might be expected to behave as in N_2, while the response of the central nitrogen atom is expected to be different. Given the physical picture of the electronic excitation implied in Section II.B, it is expected that the system would exhibit a site–dependent probability of ionization that would be revealed in the dissociated ionic fragments. Such a process would represent the multiphoton analogue of the site–specific excitation[26–28] that can be achieved with the absorption of x–rays.

In preliminary experiments[29], the comparison of the fluorescence spectra exhibited by N_2 and N_2O in the 70–nm to 80–nm range, in relation to the behavior of certain N^{2+} lines involving $1s^22p^3$ doubly excited levels, has furnished evidence suggesting this type of behavior. In order to verify this hypothesis, isotopic studies of ionic fragmentation can be performed to more fully examine these processes. For example, since $^{15}N^{14}NO$ and $^{14}N^{15}NO$ can be readily synthesized[30] from isotopically substituted NH_4NO_3, the phenomenon of site–selective energy deposition can be sensitively and easily detected by comparative studies of ion production in $^{14}N^{14}NO$ and the isotopic variants. Using the apparatus available[9] for ion studies, it is possible to measure both the ion yields and the kinetic energy distributions characteristic of the fragments[9,31,32].

Initial experiments using $^{14}N^{15}N^{16}O$, under conditions comparable to those involved in the study of fluorescence[10] from N_2 and N_2O, have confirmed the expected site dependence from measurements of the charge state spectrum of the nitrogen ion fragments. A clear demonstration of this dependence is illustrated in Fig. (1). For the range of flight times involving the singly charged species N^+ and O^+ ($13 \leq M/q \leq 17$), normal ($^{14}N^{14}N^{16}O$) material produces the signal shown in Fig. (1a). In this case, it is impossible to distinguish the N^+ ions coming from the two different molecular sites.

The shapes of the ion signals are determined by the instrumental response and the kinetic energies of the fragments. The individual ion signals are actually composed of two components which are not fully resolved in the data. One component arises from ions initial formed with velocities parallel to the time–of–flight direction while the other arises from ions whose initial velocity is anti–parallel to that vector. The two peaks are of unequal height and width due to the detailed nature of the trajectories which, in turn, are determined by the extraction and acceleration voltages used in the time–of–flight apparatus[9]. The separation of the two components provides a measure of the initial kinetic energy of the fragment, which, for the data shown in Fig. (1a), is estimated to be several electron–volts for the N^+ and O^+ signals. Naturally, the total ion yield is given by the integral of the signal.

The same region of the time–of–flight spectrum is shown in Fig. (1b) for $^{14}N^{15}N^{16}O$. In this case, a small quantity of ^{40}Ar was present and the $^{40}Ar^{3+}$ signal is clearly seen. The argon peak can be used to calibrate the time–of–flight axis and give a direct measure of the instrumental width determined by the apparatus. In this regard, note the narrow and symmetric shape of the $^{40}Ar^{3+}$ signal.

Three distinct signals arising from the $^{14}N^{15}N^{16}O$ are clearly represented in Fig. (1b). Plainly, the total yield of the $^{15}N^+$ ions is approximately one half of the $^{14}N^+$ yield. Therefore, the total probability of producing singly charged

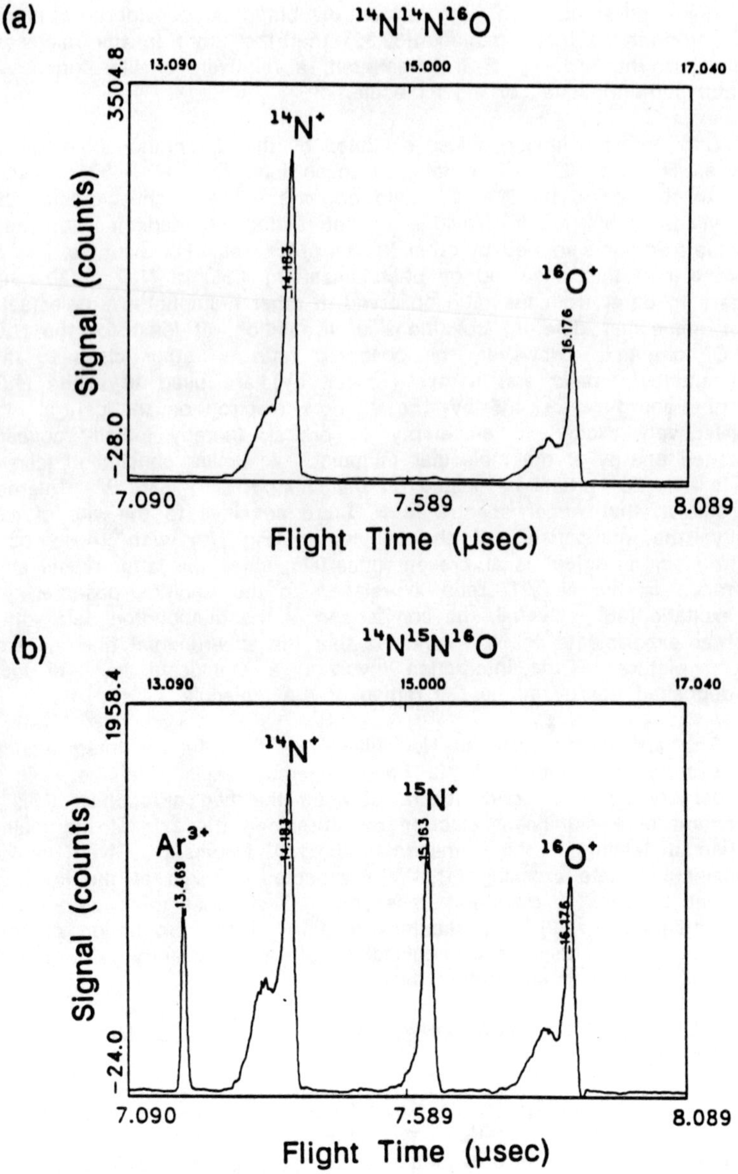

Fig. (1): Charge state fragments observed from N_2O in the range $13 \leq M/q \leq 17$. (a) Spectrum from $^{14}N^{14}N^{16}O$. (b) Spectrum from $^{14}N^{15}N^{16}O$. See text for discussion.

ions depends upon the molecular site. The widths of ^{15}N and ^{14}N signals are also quite different, with the former exhibiting a considerably narrower value. This conforms to the expectation[32,33] that the atom in the center (^{15}N) would develop, in the process of fragmentation, a relatively smaller kinetic energy than the atoms on the ends of the molecule.

Informative behavior is also exhibited by the observation of diatomic fragments such as N_2^+ and NO^+. The data on these ions for $^{14}N^{15}N^{16}O$ are shown in Fig. (2). In this case, the $^{15}N^{16}O^+$ yield appears to be slightly greater than the $^{14}N^{15}N^+$ yield. This result provides an interesting comparison with the yield ratio of these fragments formed by other known processes. For example, it is considerably different than that observed by photoionization[34,35] of N_2O at 30.4 nm and also appears to differ from the ratio observed in other multiphoton studies[32], conducted under somewhat different conditions of irradiation, at least for the $N_2O^{2+} \longrightarrow N^+ + NO^+$ channel. However, the observed ratio is rather close to that seen[36] when monochromatic soft x–rays (~ 405 eV) are used to excite N_2O. With a quantum energy of ~ 405 eV, the 2σ core electron on the central nitrogen atom is selectively excited to an empty 3π orbital, thereby initially concentrating the deposited energy at the molecular midpoint. A similar conclusion follows[36] when the $1s \longrightarrow 2p^*$ orbital is excited on the O atom at ~ 536 eV. Interestingly, it is also known that Auger spectra in N_2O are sensitive to the site of excitation[37]. Finally, the comparison of the results in Fig. (2) with those obtained from electron scattering[38] is at present uncertain, since the latter reveal an unexpected difference in the NO^+/N_2^+ ratio in relation to the findings observed with soft x–ray excitation[36]. Overall, the comparison of the multiphoton data with the results of other experiments on N_2O indicates that the experimental findings are consistent with a picture of the interaction involving a significant level of localization of the deposited energy in the NO region of the molecule.

The appreciable yield of NO^+ gives rise to a further interpretation[39]. The NO^+ system represents a triple bond, whereas, initially in the N_2O system, the considerably stronger bond occurs between the two nitrogens. This implies the movement of a significant electron density along the axis of the molecule in the mechanism leading to the formation of the NO^+ fragment. Given the large number of molecular state crossings that are expected to exist at the level of excitation pertinent to such a transition, it seems unlikely that this flow of charge would occur adiabatically[39] and channels leading to the production of excited states may be significant. The application of a sufficiently strong external field could be effective in generating such a motion.

III. GENERAL CONSIDERATION OF THE STRONG–FIELD MODE OF EXCITATION IN MOLECULES

Important conclusions for the achievement of x–ray amplification stem from the character of strong–field multiquantum coupling to molecules discussed above. In addition to providing specific data on N_2, N_2O, and their ions, the information obtained from the studies of ion production[6] and fluorescence leads to a general hypothesis concerning the basic character of molecular multiphoton interactions in the strong–field ($\varepsilon > e/a_0^2$) regime. Through comparison with electron and ion collisional processes, it will be shown that there is a large class of molecular excitations which are expected to be strongly excited only by strong–field coupling. Therefore, new states of excitation uniquely produced by the strong–field mode of interaction are generally anticipated for all molecular materials regardless of structure or composition. Furthermore, it will be seen that the genre of excited states produced by this form of interaction will naturally embody highly asymmetric charge distributions.

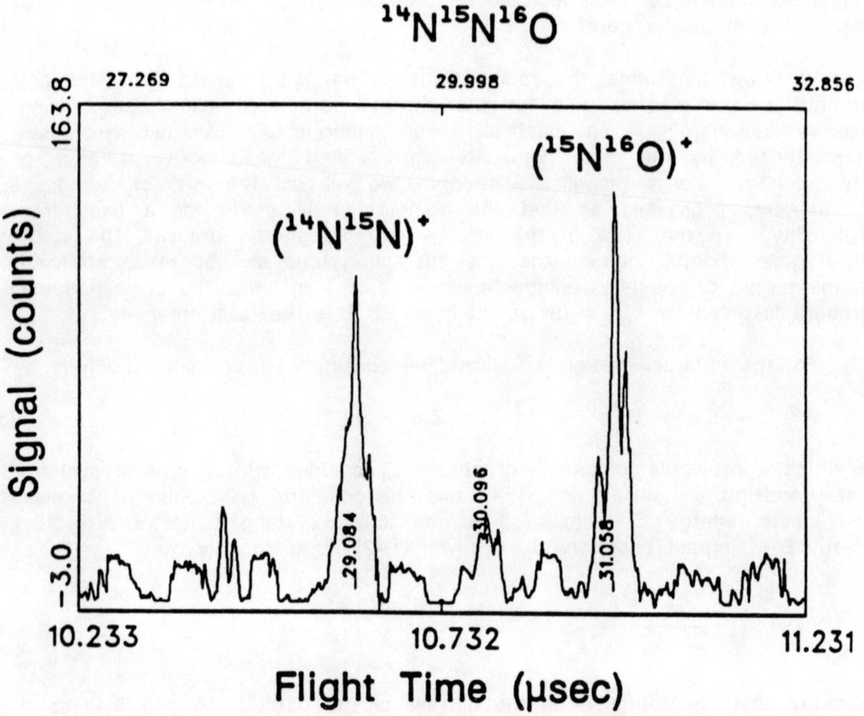

Fig. (2): Charge state spectrum of diatomic ions $(^{14}N^{15}N)^+$ and $(^{15}N^{16}O)^+$ observed from $^{14}N^{15}N^{16}O$. For discussion see text.

The comparison[40] of the excitation of molecules with high–intensity lasers and charged–particle collisions can be visualized by using a simple description which graphically represents the interaction in terms of variables describing the maximum electric field strength ε_m produced and the kinetic energy transfer E_t communicated to the system during the process of excitation. Although this comparison must be regarded as very qualitative, it forms a simple basis for viewing the molecular coupling.

We now determine the relationships between the energy transfer and the strength of the electric field for the charged–particle collisions and strong–field laser excitations[40]. To establish these relationships, the two processes are parameterized by the peak field strength ε_m and the effective duration of the interaction τ. For a given field strength, we will set the interaction times equal for the two processes so that the comparison is made for a fixed transition probability. In the case of the strong–field multiphoton process, the studies of ion fragmentation[9] indicate that the interaction time can be taken approximately as the period of the electromagnetic wave, $\tau \simeq \lambda/c$, while the corresponding field strength is given by $\varepsilon_m = (8\pi I/c)^{1/2}$ in which I is the laser intensity.

For the charged–particle collisions, we consider a process of the form

$$A^{Z_1+} + B \longrightarrow A^{Z_1+} + B^{Z_2+} + Z_2 e^- \tag{3}$$

in which a projectile ion A^{Z_1+} of charge Z_1 collides with a neutral system B at rest producing a recoiling ion B^{Z_2+} and the collateral generation of Z_2 electrons. The kinetic energy E_t transferred to the target system B, for a projectile with energy E at impact parameter b, is given[41,42] approximately by

$$E_t \simeq \frac{Z_1^2 Z_2^2 e^4 M_A}{2EM_B b^2}, \tag{4}$$

with M_A and M_B representing the masses of the particles A and B, respectively, and e denoting the electronic charge. With similar notation, the maximum collisional electric field ε_m experienced by system B can be written as

$$\varepsilon_m = \frac{Z_1 e}{b^2}. \tag{5}$$

This treatment does not determine how the kinetic energy transferred will be shared by the two atoms; one atom will presumably receive most of the energy. Given the approximate time–scale of the collisional interaction as

$$\tau \simeq \left(\frac{b^2 M_A}{2E}\right)^{1/2}, \tag{6}$$

the combination of Eqs. (4), (5) and (6) yields the desired relationship between E_t and ε_m, namely,

$$E_t = \frac{Z_2^2 e^2}{M_B} \tau^2 \varepsilon_m^2. \tag{7}$$

Parametrized in this way, the energy transfer E_t has no explicit dependence on the properties of the charged projectile, and hence, Eq. (7) holds generally for both ions and electrons. The final connection between the strong–field multiphoton interaction and the charged–particle collision is obtained by setting the time scale of the collision equal to λ/c which yields

$$E_t \simeq Z_2^2\left(\frac{e^2a_o^2}{M_Bc^2}\right)\left(\frac{\lambda}{a_o}\right)^2\varepsilon_m^2 \tag{8}$$

The validity of Eq. (8), for the simple collisional picture used, requires that

$$\frac{\Delta q}{M_Av_A} << 1, \tag{9}$$

where Δq is the momentum transfer occurring in the collision and v_A is the projectile velocity. Using Eqs. (5), (6), and (8), we find that

$$\frac{\Delta q}{M_Av_A} = \left(\frac{2Z_2^2}{Z_1}\right)^{1/2}\left(\frac{\varepsilon_m}{e/a_o^2}\right)^{3/2}\left(\frac{\lambda}{a_o}\right)^2\left(\frac{e^2/a_o}{M_Ac^2}\right). \tag{10}$$

As an example of typical values arising in collisional experiments, we set $2Z_2^2/Z_1 \sim 1$ and $M_A \sim 100$ amu. Using $\lambda = 248$ nm, Eq. (9) is then satisfied for $\varepsilon_m \lesssim 30\ e/a_o^2$. However, if Eq. (9) is not satisfied, the kinetic energy transfer will exceed that given by Eq. (8) so that the expression actually represents a lower bound for charged–particle collisions. These results can be summarized by plotting the energy transfer E_t against the electric field strength ε_m using Eq. (8). Figure (3) represents this relationship for the case of N_2^{2+} discussed above ($Z_2=2$, $M_B=28$ amu, $\lambda=248$ nm).

In order to establish an upper limit for the kinetic energy transfers associated with strong laser fields, we have examined certain physical processes which can impart kinetic energy to the molecular system during the interaction. These include (a) the direct absorption of photons in an N–quantum process with N sufficiently large to correspond to the removal of a K–electron, (b) the emission of Auger electrons, (c) the emission of K_α radiation, and (d) the influence of the driven motion of an ionized electron in a process equivalent to inverse bremsstrahlung in the field of the ion. The bound associated with the latter could be connected with experimental results[9] on ion production while the former three are readily estimated from the kinematics of two–body decay. A final consideration was the fact that any ion will generally experience a ponderomotive potential in the high intensity wave. However, this energy is fully returned[43–45] to the radiation field during the course of the pulse so that no net displacement of any molecular ions, regardless of charge or mass, will occur. The outcome of this analysis is the finding that, for a very broad range of conditions and materials for irradiation up to an intensity of ~ 10^{21} W/cm^2 (~ 170 e/a_o^2), the recoil energies would fall in the range $E_t \lesssim 60$ meV.

In addition to the energy transfer, there is another important difference between strong–field interactions and charged–particle collision, namely, the geometrical pattern of the electric field developed during the interaction. The electric field of linearly polarized radiation is a unidirectional field fixed in space. Since molecular rotation is negligible during a sufficiently short pulse (τ < 1ps), the field is also fixed in the molecular frame. One of the main conclusions of previous work on the high power irradiation of nitrogen and other molecules was that the interaction is strongly anisotropic[15,16,32] and that this type of electric field can drive electrons into states correlating to charge–asymmetric dissociation limits[9]. Charged particle collisions, on account of the varying geometrical configuration occurring during the interaction as the field rapidly rotates with respect to the molecular axis, are clearly less effective in producing this kind of unidirectional force. This fact may contribute to the reason why certain excited states are generated with substantial probability in strong–field experiments and apparently not in charged–particle collisions.

Ion collisions present strong forces ($\varepsilon_m >> e/a_0^2$) and result in large kinetic energy transfers while electron collisions generate relatively weak forces and correspondingly small kinetic energy transfers. In contrast, since the short–pulse strong–field excitation (1) generates a high field strength ($\varepsilon_m >> e/a_0^2$), (2) communicates a recoil energy which is very small in relation to molecular bond strengths, and (3) readily excites electronic motions along specific molecular directions, it represents a unique domain of interaction. These respective regions of interaction are conveniently displayed in the E_t versus ε_m plane shown in Fig. (3). In this picture, the strong–field multiphoton coupling ($\varepsilon_m \gtrsim e/a_0^2$) clearly represents a distinct regime of interaction that is unachievable with charged–particle collisions and permits strong electronic forces to be applied to a system whose molecular frame can remain intact. It is this situation that enables the strong–field coupling to efficiently produce new modes of excitation in molecular materials. The region corresponding to the experiments both on N_2 and N_2O, as shown in Fig. (3), clearly falls in this new zone. It is this feature of the interaction which can account for the N_2^+ and NO^+ fragment yields from N_2O and the otherwise anomalous appearance of the $2\sigma_g$ vacancies and the intense 55.8–nm emission from N_2^{2+} produced from N_2. Although N_2 and N_2O have served as a specific examples, such interactions producing strongly charge asymmetric states are expected to occur generally for all materials regardless of structure or composition.

VI. MOLECULAR DESIGN PRINCIPLE FOR X–RAY LASERS

The results described in Sections II and III above lead directly to the formulation of a new principle of molecular design for x–ray lasers. The specific findings that lead to this conclusion are summarized in Table I.

TABLE I
Features of Strong–Field Multiquantum
Interaction Which Lead to Molecular X–Ray Laser Design

Property of Interaction	References to Discussion
(1). Fields $\varepsilon >> e/a_0^2$ can be readily produced	(1). Ref. (2) and Ref. (3)
(2). Rapid energy transfer to molecule, $\Delta\tau \sim \lambda/c$, Δr small ~ 0.1 Å	(2). Ref. (9) and Ref. (10)
(3). Coupling exhibits strong anisotropy, mainly parallel to molecular axis	(3). Ref. (15), Ref. (16), and Ref. (32)
(4). Substantial e^- displacements Δx driven for $\varepsilon >> e/a_0^2$; $\Delta x \gtrsim r_0$	(4). Ref. (9) and Ref. (10)
(5). Inner molecular shells can be directly excited [e.g. $N_2(2\sigma_g)$].	(5). Ref. (10)
(6). Site selective molecular energy deposition (N_2O/N_2).	(6). Ref. (29) and Section II.C above.

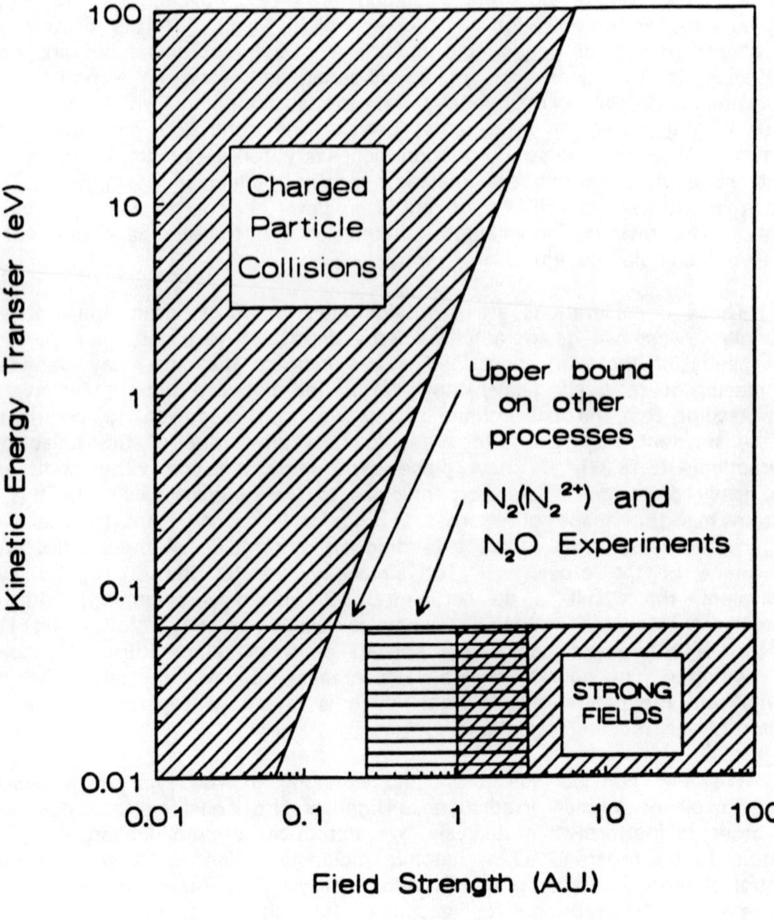

Fig. (3): Display of kinetic energy transfer versus electric field strength for various processes. The region labelled "charged–particle collisions" corresponds to the example of N_2^{2+} described in the text. The upper bound (~ 60 meV) associated with the kinetic energy transfer occurring in strong–field multiphoton processes is shown, and the strong–field zone is defined by $\varepsilon_m \geqslant e/a_0^2$. Also indicated is the regime in which the new N_2^{2+} excitations were experimentally produced in N_2 and the NO^+ and N_2^+ fragment distributions shown in Fig. (2) were generated from N_2O.

The properties of the strong-field coupling compiled in Table I are fundamental to the production of coherent radiation of x-ray wavelengths primarily because they enable the combination of (1) very highly excited forms of matter with (2) the characteristics of dense cold materials. Such circumstances are exceptionally conducive to the generation of amplification at x-ray wavelengths. It is the achievement of this conventionally paradoxical condition, which can be made to occur in the molecular frame, that permits the controlled generation of the very high rate of excited state production necessary for x-ray amplification. Moreover, since molecular systems are utilized, specific molecular designs can be created whose properties selectively govern the flow of energy into particular excited states. The resulting wavelength of emission will thereby be dependent upon the detailed molecular design.

These considerations are now illustrated within the framework of a particular example. Since we desire a high energy deposition localized on a particular atom, the results of the work described above indicate that this may be accomplished by placing a relatively heavy atom on a line between two lighter systems. We now assume that the axis defined in this way coincides with the polarization vector of the incident radiation. In order to take advantage of the anisotropy of the interaction[9,15,16,32], we now place two atoms, one on either side adjacent to the heavy atom on a line perpendicular to the original axis. In this form, the system has the shape of a cross. Since it is inconvenient to orient molecules, we now add two more atoms extending on either side of the central atom out of the plane of the cross. If, for simplicity, we let the six lighter systems be equivalent, the result is an octahedral molecule (O_h symmetry) with the heavy atom in the center. Specific examples would be SF_6, MoF_6, and UF_6, all of which are convenient gas phase species at room temperature. A molecule with O_h symmetry, of course, will always present an axis involving the two lighter atoms with the central heavy atom which is favorably aligned with the polarization vector of the incident radiation.

We now consider, as a specific example, the behavior of a molecule (SF_6) as a result of intense irradiation in light of the considerations described above. In order to perform this analysis, we introduce a simplification that enables the system to be regarded as a diatomic molecule. Hence, we will assume that the central S atom interacts primarily with a single F atom. A slight breaking of the O_h symmetry occurring, for example, through vibrational excitation could be envisioned as the physical basis for this dynamical choice. Although we anticipate that the remaining F atoms will influence the molecular behavior, we expect that many of the important features of the dynamics will be revealed by the simpler diatomic picture.

Consider the SF diatom in the charge state SF^{17+}. Due to the ability to produce site-selective energy deposition (point (6) in Table I), we will assume that the ionization is distributed as $S^{12+} - F^{5+}$. For ground state atoms, this gives the configurations $S^{12+}(1s^2 2s^2)$ and $F^{5+}(1s^2 2s^2)$, both of which give $^1S_0$ terms and, consequently, a single $^1\Sigma^+$ molecular state. Furthermore, since the energy transfer producing the ionization occurs very rapidly (point (2) in Table I), we take the initial internuclear separation of the SF^{17+} system as equivalent to the S-F internuclear distance occurring in the neutral parent SF_6 molecule. This distance is $r_0 \simeq 1.56$ Å. The point (A) representing this situation can now be placed on the molecular energy level diagram shown in Fig. (4) along with the corresponding $^1\Sigma^+$ coulomb state. From this initial configuration, point (A), the system will begin to evolve along the $^1\Sigma^+$ curve. Furthermore, since essentially no momentum was communicated to atoms in the molecule during the strong-field multiquantum

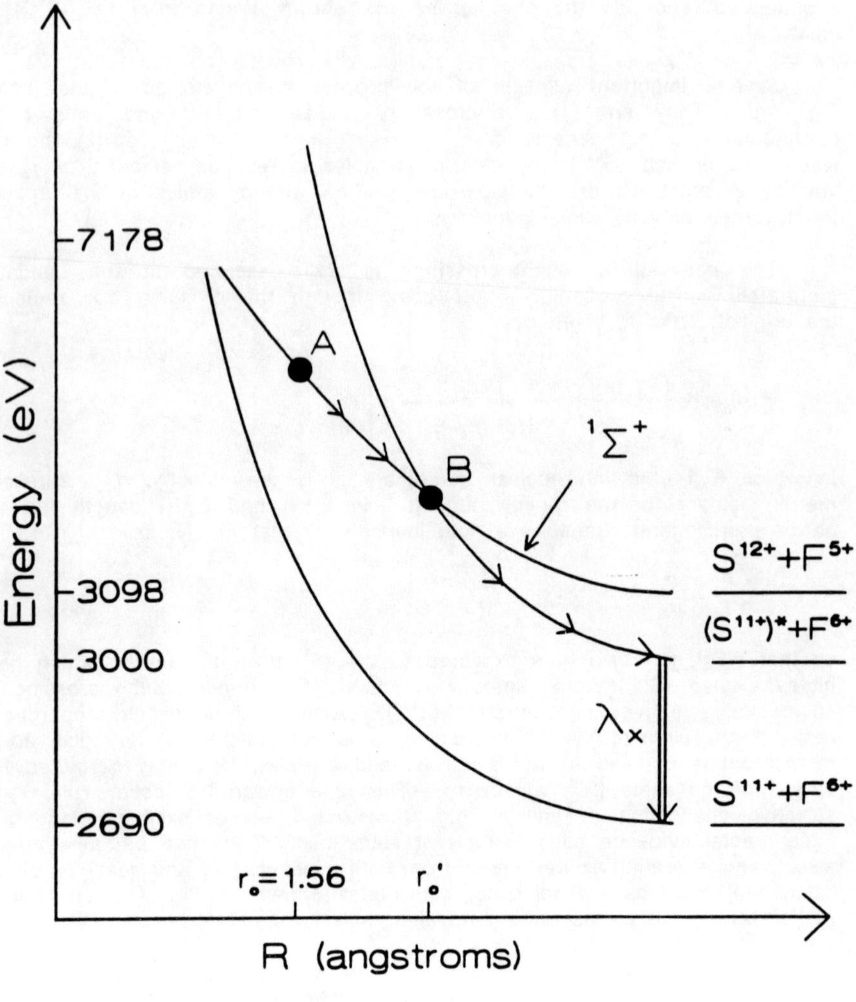

Fig. (4): Partial potential energy diagram of $S^{q_1+} + F^{q_2+}$ ($q_1 + q_2 = 17$) system. Note that the electron transfer crossing at r_0' is such that $r_0' > r_0$. For S^{11+}, $\lambda_X \simeq 40$ Å (~300eV). The ordinate is not drawn to a linear scale.

excitation, the dissociative motion along the $^1\Sigma^+$ curve begins with a vanishing velocity ($v \simeq 0$). As the diatomic system dissociates, curve crossings will be encountered, such as the one labeled (B), shown arising from the $(S^{11+})^* + F^{6+}$ curve.

Several important aspects of the process are illustrated in the diagram of Fig. (4). They are (1) a crossing to the ground state $S^{11+} + F^{6+}$ configuration is not accessible; (2) the crossing leading to excited $(S^{11+})^*$ by electron transfer is reached, since $r_o' > r_o$, (3) the velocity at point (B) can be very low, and (4) the dynamics of the crossing will be governed only by radial couplings.

The behavior at such crossings is well described in the Landau–Zener picture[46]. The probability P of going though the crossing and remaining on the original curve is given by

$$P = \exp\left[-\left(\frac{\hbar}{2\pi v}\right)\left(\frac{|V_{12}|^2}{d/dR(E_1-E_2)}\right)\right] \tag{11}$$

in which R is the internuclear coordinate, v is the velocity, V_{12} represents the matrix element for the radial coupling, and $E_1(R)$ and $E_2(R)$ denote the energies of the participating states. We note immediately that

$$\lim_{v\to 0} P = 0, \tag{12}$$

so that efficient crossing is expected to the $(S^{11+})^* + F^{6+}$ curve, which generates highly excited $(S^{11+})^*$ ions, since v is small. Furthermore, such crossings confer substantial selectivity on the state $(S^{11+})^*$ produced through the properties of the radial matrix element V_{12}. For example, since the interaction is radial, no angular momentum is coupled to the electrons and crossings are only permitted to other curves of the same ($^1\Sigma^+$) symmetry. This is a powerful selection rule[47] for the selective channeling of energy into a restricted set of excited quantum states. Experimental evidence has shown that such mechanisms can be quite effective in selectively generating excited states from both molecular[48] and quasi-molecular/collisional[49] systems. Finally, for the states shown in Fig. (4) representing the $(S^{11+})^*$ levels, the corresponding wavelength is $\lambda_X \simeq 40$ Å (~ 300 eV).

The most important aspect of Fig. (4) is the fact that the nature of the particle interactions is such that no single step enables a direct transition from the initially excited configuration to the ground state of the atomic system of interest (S^{11+}). Therefore, the dynamics of the excited material force the energy to flow in an orderly pathway that leads inevitably to the excited states of the system. For the free ion, the only available downward channel involves radiation. In summary, a dominant energy pathway in the mechanism of molecular relaxation of the excited system is established which contains a segment that can be traversed only by the emission of radiation. When this general circumstance is present, stimulated emission from the resulting highly nonequilibrium distribution of states can be readily generated. This situation is directly analogous to that occurring in excimer media[50].

V. STRONG–FIELD PROPAGATION

A fundamentally new regime of electromagnetic propagation is expected to arise in plasmas for subpicosecond radiation at a sufficiently high intensity[11,51–

55]. Among the important new phenomena that may occur is a mode of channeled propagation. When a very intense pulse of radiation enters a cold medium, multiphoton ionization produces a local plasma, and the following qualitative behavior involving a relativistic charge–displacement mechanism develops. The conversion of electrons from bound to free states by the ionization induces a strong reduction in the refractive response of the atomic material. The resulting plasma exhibits a further decrease in index from the free–electron component collaterally produced. If the ambient electron density n_o is such that the plasma frequency $\omega_p = \sqrt{4\pi n_o e^2/m}$ is below the frequency of the radiation ω, an interesting mode of channeled propagation could develop. The presence of both relativistic motion[56] and charge–displacement can lead to self focusing. The influence of the charge–displacement can be seen in the following way. For a sufficiently short pulse (~ 100 fs), the ions remain spatially fixed while the relatively mobile electrons are expelled by the ponderomotive potential from the central high–intensity region of the beam. A state of equilibrium can then be established between the ponderomotive and the electrostatic force densities owing to the net charge displacement. Since the electrons, which embody a negative contribution to the index, are expelled, an on–axis region of relatively high refractive index is formed which can support channeling of the laser beam. Preliminary analysis concerning the channeling indicates (1) that the Compton intensity (~ 4.5 x 10^{19} W/cm^2 at λ = 248–nm) establishes the scale for the prominence of this phenomenon and (2) that the frequency scaling strongly favors the use of the shortest wavelength available at high power[52]. A channel radius of several wavelengths also emerges from this analysis. Moreover, recent dynamical calculations[56] support the conclusion that such modes of propagation can be formed under stable conditions. The ability to stably propagate high intensity radiation in this manner would be a fundamental advance in the control and concentration of power in materials, a key question relevant to the production of amplification in the x–ray range. It furthermore facilitates the spatial control necessary to produce a high brightness output.

VI. SUMMARY AND CONCLUSIONS

In terms of the basic question of the controlled deposition of energy at high specific powers, the answer that emerges from this analysis involves three separate components. They are (1) a new ultraviolet pulsed power technology, (2) an energy deposition mechanism based on highly nonlinear coupling, and (3) a condition for channeled propagation. It can be seen that these three elements, which appear capable of producing conditions comparable to, or possibly exceeding those of a thermonuclear environment, fit together in a remarkably congenial way. Figure (5) illustrates these relationships. A principal finding is that the radiative conditions needed for the strong multiphoton coupling governing the energy transfer rate[5] are essentially identical to those required for the channeled propagation and that the laser technology can readily produce the regime of irradiation necessary. The compatibility of these three factors is a key feature of this method for the attainment of very high energy density states of matter and stimulated emission in the x–ray range.

VII. ACKNOWLEDGEMENTS

The technical assistance of P. Noel and J. Wright and fruitful discussions with J. C. Solem are acknowledged. We also appreciatively acknowledge the assistance of R. Jensen, R. M. Moriarty, and A. Awasthi in the preparation of the isotopic N_2O. Support for this research was provided by the U. S. Air Office of Scientific Research, the U. S. Office of Naval Research, and the Strategic Defense Initiative Organization.

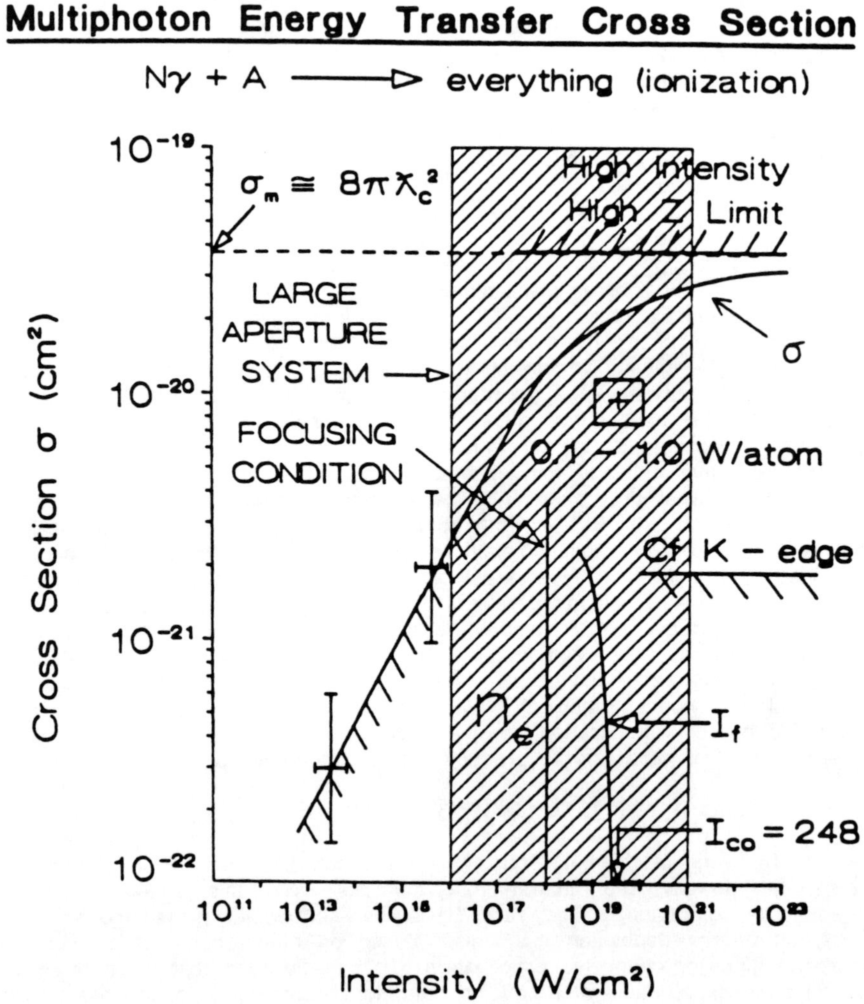

Fig. (5): Illustration of how the energy transfer cross section σ, the channeling condition, and the pulse power technology all fit together. The significant fact is that the conditions needed for the strong multiphoton coupling governing the energy transfer ($I \simeq I_{CO} \simeq 10^{19}$ W/cm^2) are <u>identical</u> to those required for the channeled propagation ($I_f \simeq I_{CO} \simeq 10^{19}$ W/cm^2), and that the developing large aperture ultraviolet laser technology (shaded zone) moves us into the desired range. For further discussion of the cross section σ, the channeled propagation, and the pulse power technology, see References [5], [52], and [2], respectively.

VIII. REFERENCES

[1] A. V. Vinogradov and I. I. Sobel'man, "The Problem of Laser Radiation Sources in the Far Ultraviolet and X-Ray Region," Sov. Phys.-JETP 36, 1115 (1973).

[2] T. S. Luk, A. McPherson, G. Gibson, K. Boyer, and C. K. Rhodes, "Ultrahigh Intensity KrF* Laser System," Opt. Lett. 14, 1113 (1989).

[3] A. J. Taylor, C. R. Tallman, J. P. Roberts, C. S. Lester, T. R. Gosnell, P. H. Y. Lee, and G. A. Kyrala, "High-Intensity Subpicosecond XeCl Laser System," Opt. Lett. 15, 39 (1990).

[4] C. K. Rhodes, "Multiphoton Ionization of Atoms," Science 229, 1345 (1985).

[5] K. Boyer, H. Jara, T. S. Luk, I. A. McIntyre, A. McPherson, R. Rosman, C. K. Rhodes, "Limiting Cross Sections for Multiphoton Coupling," Revue Phys. Appl. 22, 1793 (1987).

[6] C. K. Rhodes, "Ordered Many-Electron Motions in Atoms and X-Ray Lasers," in Giant Resonances in Atoms, Molecules, and Solids, ed. J.-P. Connerade, J. M. Esteva, and R. C. Karnatak (Plenum Press, New York, 1987) p. 533.

[7] K. Boyer and C. K. Rhodes, "Atomic Inner-Shell Excitation Induced by Coherent Motion of Outer-Shell Electrons," Phys. Rev. Lett. 54, 1490 (1985).

[8] C. K. Rhodes, "Physical Processes at High Field Strengths," Physica Scripta T17, 193 (1987).

[9] K. Boyer, T. S. Luk, J. C. Solem, and C. K. Rhodes, "Kinetic Energy Distributions of Ionic Fragments Produced by Subpicosecond Multiphoton Ionization of N_2," Phys. Rev. A 39, 1186 (1989).

[10] G. Gibson, T. S. Luk, A. McPherson, K. Boyer, and C. K. Rhodes, "Observation of a New Inner-Orbital Molecular Transition at 55.8 nm in N_2^{2+} Produced by Multiphoton Coupling", Phys. Rev. A 40, 2378 (1989).

[11] "Charge-Displacement Self-Channeling as a Method for Energy Concentration," K. Boyer, T. S. Luk, J. C. Solem and C. K. Rhodes, OSA Proceedings on Short Wavelength Coherent Radiation: Generation and Application, R. W. Falcone and J. Kirz, editors, (Optical Society of America, Washington, D. C., 1988) p. 233.

[12] G. J. Pert "Model Calculations of XUV Gain in Rapidly Expanding Cylindrical Plasmas II," J. Phys. B 12, 2067 (1979).

[13] D. L. Matthews, P. L. Hagelstein, M. D. Rosen, M. J. Eckart, N. M. Ceglio, A. U. Hazi, H. Medecki, B. J. MacGowan, J. E. Trebes, B. L. Whitten, E. M. Campbell, C. W. Hatcher, A. M. Hawryluk, R. L. Kauffman, L. D. Pleasance, G. Rambach, J. H. Scofield, G. Stone, and T. A. Weaver, "Demonstration of a Soft X-Ray Amplifier," Phys. Rev. Lett. 54, 110 (1985).

[14] R. A Smith, V. Barrow, J. Edwards, G. Kiehn, and O. Willi, "High Density Plasmas for Recombination X-Ray Lasers," Appl. Phys. B 50, 187 (1990).

[15] K. Codling, L. J. Frasinski, and P. A. Hatherly, "On the Field Ionization of Diatomic Molecules by Intense Laser Fields," J. Phys. B 22, L321 (1989).

[16] L. J. Frasinski, K. Codling, and P. A. Hatherly, "Multiphoton Multiple Ionization of N_2 Probed by Covariance Mapping," preprint 1988.

[17] W. Eberhardt, E. W. Plummer, I.–W. Lyo, R. Carr and W. K. Ford, "Auger–Electron–Ion Coincidence Studies of Soft X–Ray–Induced Fragmentation of N_2," Phys. Rev. Lett. 58, 207 (1987).

[18] H. Ågren, "On the Interpretation of Molecular Valence Auger Spectra," J. Chem. Phys. 75, 1267 (1981).

[19] A. Lofthus and P. H. Krupenie, "The Spectrum of Molecular Nitrogen," J. Phys. Chem. Ref. Data 6, 113 (1977).

[20] R. W. Wetmore and R. K. Boyd, "Theoretical Investigation of the Dication of Molecular Nitrogen," J. Phys. Chem. 90, 5540 (1986).

[21] C.–M. Liegener, "Calculations on the Auger spectrum of N_2," J. Phys. B 16, 4281 (1983).

[22] G. Herzberg, private communication.

[23] R. L. Kelly, "Atomic and Ionic Spectrum Lines below 2000 Angstroms: Hydrogen through Krypton," Part I, J. Phys. Chem. Ref. Data 16, Suppl. No. 1 (1987).

[24] R. S. Mulliken and W. C. Ermler, Diatomic Molecules, Results of ab initio Calculations (Academic Press, New York, 1977) p. 127.

[25] A. C. Wahl, Science 151, 961 (1966).

[26] W. Eberhardt, T. K. Sham, R. Carr, S. Krummacher, M. Strongin, S. L. Weng, and D. Wesner, "Site–Specific Fragmentation of Small Molecules Following Soft X–ray Excitation," Phys. Rev. Lett. 50, 1038 (1983).

[27] K. Müller–Dethlefs, M. Sander, L. A. Chewter, and E. W. Schlag, "Site–Specific Excitation in Molecules at Very High Energies: Changes in Ionization Patterns of CF_3CH_3," J. Phys. Chem. 88, 6098 (1984).

[28] P. Morin, G. G. B. de Souza, I. Nenner, and P. Lablanquie, "Selective Resonant Photoionization Processes Near the Si 2p Edge of Tetramethyl–silane," Phys. Rev. Lett. 56, 131 (1986).

[29] A. McPherson, T. S. Luk, G. Gibson, K. Boyer, and C. K. Rhodes, "Ion Production and Molecular Excitation Occurring in Multiphoton Processes," Nucl. Instrum. Meth. B 43, 468 (1989).

[30] L. Friedman and J. Bigeleisen, "Oxygen and Nitrogen Isotope Effects in the Decomposition of Ammonium Nitrate," J. Chem. Phys. 18, 1325 (1950).

[31] T. S. Luk, and C. K. Rhodes, "Multiphoton Dissociative Ionization of Molecular Deuterium," Phys. Rev. A 38, 6180 (1988).

[32] L. J. Frasinski, K. Codling, and P. A. Hatherly, "Covariance Mapping: A Correlation Method Applied to Multiphoton Multiple Ionization," Science 246, 1029 (1989).

[33] T. Matsuo, T. Tonuma, M. Kase, T. Kambara, H. Kumagai, and H. Tawara, "Production of Multiply Charged Ions from CO and CO_2 Molecules in Energetic Heavy-Ion Impact," Chem. Phys. 121, 93 (1988).

[34] D. M. Curtis and J. H. D. Elands, "Coincidence Studies of Doubly Charged Ions Formed by 30.4 nm Photoionization," Int. J. Mass. Spectrom. Ion Processes 63, 241 (1985).

[35] S. D. Price, J. H. D. Elands, P. G. Fournier, J. Fournier, and P. Millie, "Electronic States and Decay Mechanisms of the N_2O^{2+} Dication," J. Chem. Phys. 88, 1511 (1988).

[36] J. Murakami, M. C. Nelson, S. L. Anderson, and D. M. Hanson, "Fragmentation of Nitrous Oxide by Monochromatic Soft X-Rays," J. Chem. Phys. 85, 5755 (1986).

[37] F. P. Larkins, W. Eberhardt, In-Whan Lyo, R. Murphy, and W. E. Plummer, "The Core Hole Decay of N_2O Following Core to Bound State Excitation," J. Chem. Phys. 88, 2948 (1988).

[38] A. P. Hitchcock, C. E. Brion, and M. J. Van der Wiel, "Ionic Fragmentation of Inner Shell Excited States of CO_2 and N_2O," Chem. Phys. Lett. 66, 213 (1979).

[39] A. D. Bandrauk, private communication.

[40] K Boyer, G. Gibson, H. Jara, T. S. Luk, I. A. McIntyre, A. McPherson, R. Rosman, J. C. Solem, C. K. Rhodes, "Corresponding Aspects of Strong-Field Multiquantum Processes and Ion-Atom Collisions," IEEE Trans. Plasma Science 16, 541 (1988).

[41] K.-H. Schartner, "Highly Charged Recoil Ions," in Fundamental Processes in Energetic Atomic Collisions, edited by H. O. Lutz, J. S. Briggs, and H. Kleinpoppen, Series B, Vol. 103 (Plenum Press, New York, 1983) p. 637.

[42] A. S. Schlachter, W. Groh, A. Müller, H. F. Beyer, R. Mann, and R. E. Olson, "Production of Highly Charged Rare-Gas Recoil Ions by 1.4 MeV/amu U^{44+}," Phys. Rev. A 26, 1373 (1982).

[43] T. S. Luk, T. Graber, H. Jara, U. Johann, K. Boyer, and C. K. Rhodes "Subpicosecond Ultraviolet Multiphoton Electron Spectroscopy of Rare Gases," J. Opt. Soc. B 4, 847 (1987).

[44] T. W. B. Kibble, "Refraction of Electron Beams by Intense Electromagnetic Waves," Phys. Rev. Lett. 16, 1054 (1966).

[45] A. Szöke, "Interpretation of Electron Spectra Obtained from Multiphoton Ionization of Atoms in Strong Fields," J. Phys. B 18, L427 (1985).

[46] P. H. Mokler and F. Folkmann, "X-Ray Production in Heavy Ion-Atom Collisions" in Structure and Collisions of Ions and Atoms, edited by I. A. Sellin (Springer-Verlag, Berlin, 1978) p. 201.

[47] U. Fano and W. Lichten, "Interpretation of Ar^{+}–Ar Collisions at 50 keV, Phys. Rev. Lett. 14, 627 (1965).

[48] G. Dujardin, L'Hellner, B. J. Olsson, M. J. Besnard–Ramage and A. Dadouch, "Negative–Fragment–Ion Formation by Photon Excitation of Molecules in the Vincinity of the Core–Ionization and Direct–Double Ionization Thresholds," Phys. Rev. Lett. 62, 745 (1989).

[49] T. J. M. Zouros, D. H. Lee, and P. Richard, "Projectile 1s → 2s Excitation Due to Electron–Electron Interaction in Collisions of O^{5+} and F^{6+} Ions with H_2 and He Targets," Phys. Rev. Lett. 62, 2261 (1989).

[50] Excimer Lasers, Second Edition, edited by C. K. Rhodes (Springer–Verlag, Berlin, 1984).

[51] P. Sprangle, Cha–Mei Tang, and E. Esarey, "Relativistic Self–Focusing of Short–Pulse Radiation Beams in Plasmas," IEEE Trans. Plasma Science 15, 145 (1987).

[52] J. C. Solem, T. S. Luk, K. Boyer, and C. K. Rhodes, "Prospects for X–Ray Amplification with Charge–Displacement Self–Channeling," IEEE J. Quantum Electron. QE–25, 2423 (1989).

[53] P. Sprangle, E. Esarey, and A. Ting, "Nonlinear Interaction of Intense Laser Pulses in Plasmas," Phys. Rev. A 41, 4463 (1990).

[54] P. Sprangle, E. Esarey, and A. Ting, "Nonlinear Theory of Intense Laser–Plasma Interaction," Phys. Rev. Lett. 64, 2011 (1990).

[55] T. Kurki–Suonio, P. J. Morrison, and T. Tajima, "Self–Focusing of an Optical Beam in a Plasma," Phys. Rev. A 40, 3230 (1989).

[56] A. B. Borisov, A. V. Borovskiy, V. V. Korobkin, A. M. Prokhorov, C. K. Rhodes, and O. B. Shiryaev, "Stability of Relativistic Self–Focusing of Intense–Subpicosecond Pulses in Plasmas," Phys. Rev. Lett., submitted.

PART I

X-RAY AND VUV LASERS
THEORY AND EXPERIMENT

PHOTO-PUMPED X-RAY LASERS THEORY AND EXPERIMENTS

C. Bergmann, B.N. Chichkov,* E.E. Fill, and D.G. Goodwin**
Max-Planck-Institut für Quantenoptik, D-8046 Garching, Germany

ABSTRACT

We report on recent theoretical and experimental work on photo-pumped X-ray lasers at our institute. First, numerical results comparing photoionization pumping and photoresonant pumping are given. Then the perspectives of photoresonant pumping are discussed with special emphasis on the idea of eliminating the line coincidence problem by using the same kind of ion for pumping and lasing. Results of experimental implementations of photo-pumping are presented.

1. INTRODUCTION

Despite the success of electron collisional and recombination pumping of X-ray lasers there is good reason to investigate other excitation methods as well, since the present achievements are still short of meeting many of the goals of soft X-ray laser research. For example there is still no X-ray laser pumped by a medium-size laboratory installation, or a laser with a high enough gain length product to saturate the transition.

We believe that photo-pumping, i.e. excitation of the medium with appropriate electromagnetic radiation has a high potential in this direction. The essential advantage of photo-pumping schemes compared to electron collisional and recombination schemes is the possibility of independently optimizing the parameters of the excitation source and the active medium, offering a high degree of variation of experimental conditions. Furthermore photo-pumping seems to be the only conceivable way to ex-

---

*A.v.Humboldt fellow 1988/90. Permanent address: P.N.Lebedev Physical Institute, Moscow

**Present address: California Institute of Technology, Pasadena, Ca 91125, USA cite the medium on a time scale which is transient with respect to the decay times of the ionic levels.

There are at least five different kinds of photo-pumping, notably photoresonant pumping, photoionization pumping, inner shell photoionization, multi-photon pumping and photon-assisted schemes. In this paper we will deal with only the first two of these since they are most suitable for an experimental realization at the facilities of our institute.

Recently our emphasis has been placed on photoresonant pumping since in addition to the above mentioned possibility of independent parameter optimization this pumping method offers the following advantages:

1. A favourable scaling of the pumping rate W_p with ionic charge Z,

$$W_p = N_1 \vartheta_{1u} A_{u1} g_u / g_1 \sim Z^4, \tag{1}$$

where N_1 is the number density of ions in the ground level, ϑ_{1u} is the number of photons per mode of the transition to the upper laser level, A_{u1} is the Einstien-A coefficient of the transition and the g_i are the degeneracy factors of the corresponding levels. Note that the electron collisional pumping rate scales approximately as Z^{-3}.

2. A low electron density of the active medium leads to a small susceptibility to refraction effects. In addition, since collisions play almost no role level populations in the active medium may often be determined by analytic modelling.

It is surprising that despite of a large number of works (see for example [1-6]) neither photoresonant pumping nor ionization pumping gain in the soft X-ray region has yet been demonstrated, the shortest laser wavelength obtained so far being 216.3 nm [7]. In view of this deficiency it seemed to be necessary to improve our understanding of the basic physics of photo-pumping, to work on solving the line coincidence problem and to perform experimental investigations on favourable geometries for coupling the pumping photons into the active medium.

NUMERICAL SIMULATIONS

In order to gain more insight into the process of photo-pumping a collisional radiative code was developed which treats hydrogenic ions under the irradiation by an X-ray pump. Details of the code may be found in [8]. Special features of the code are a relatively detailed determination of the effect of He-like ions and the calculation of the gain taking full consideration of the fine structure splitting of hydrogenic levels. Hydrodynamics and the effects of radiation transfer are not taken into account, in order to concentrate on the dynamics of the various ionic levels.

The radiation of the pump is characterized by its brightness temperature and by a filter function which takes care of the spectral composition of the pumping radiation. In case of line pumping an ideal filter function would be a flat hat centered around the line to be pumped. In case of ionization pumping the ideal filter function is a unit step function at the ionization frequency of the hydrogenic ions.

Results of gain calculations using idealized spectra of this kind are illustrated in fig. 1. The figure shows three-dimensional plots of the gain coefficient vs. total

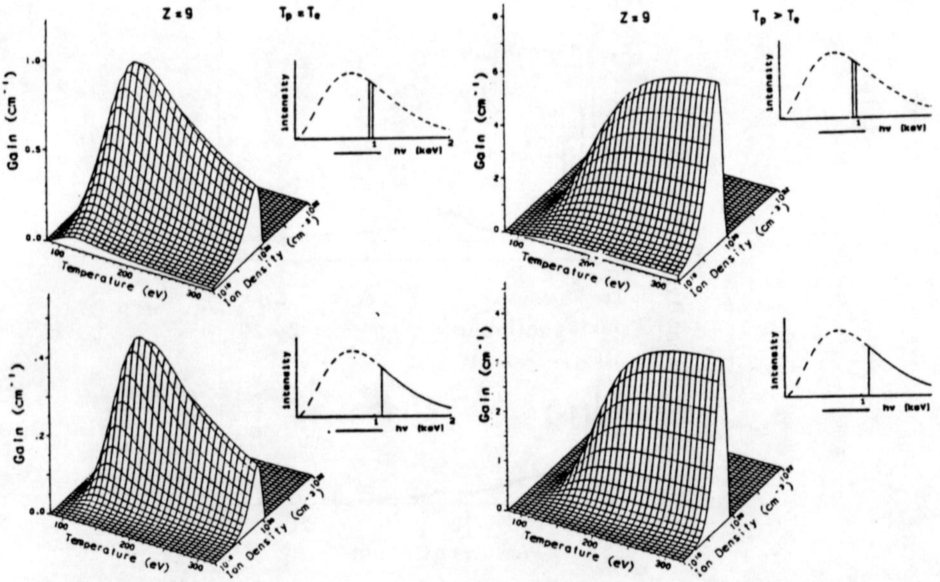

Fig. 1: Three-dimensional plot of 3-2 gain coefficient in hydrogenic fluorine. Excitation is by line pumping (top) and ionization pumping (bottom) with the corresponding spectra schematically shown in the insert. In the left part of the figure the brightness temperature T_p of the pump is assumed to be equal to the electron temperature T_e of the active medium. In the right part $T_p > T_e$ and the temperature scale applies to T_p. $T_e = 81$ eV.

ion density and brightness temperature T_p of the pump. The upper and lower parts of the figure compare line- and ionization pumping. Two distinctly different assumptions are made: In the left part of fig. 1 the electron temperature T_e of the active medium is assumed to be equal to the brightness temperature of the pump, whereas in the right part the electron temperature of the active medium is fixed.

The main features of the plots can be understood as follows: The gain coefficient increases with increasing density (quite similar in all cases) until a limiting density is reached at which electron collisions drive the laser levels to LTE. The temperature dependence is quite different for the cases with variable and with fixed electron temperature: If the electron temperature equals the brightness temperature of the pump there is a maximum temperature above which collisional ionization of the hydrogenic ions reduces the gain since less and less active ions are available. If the electron temperature is fixed there is no such maximum, the gain coefficient is typically a factor of 6 higher (for $T_e = Z^2 eV$)and saturates at high brightness temperatures.

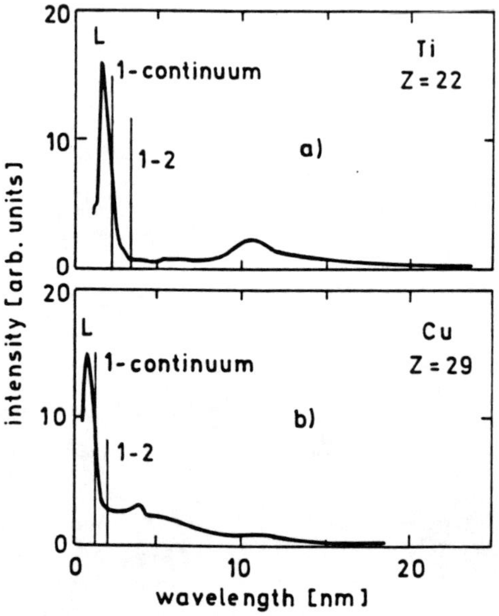

Fig. 2: Low-resolution spectra of Ti and Cu laser plasmas. Laser pulse duration 3 ns, intensity 3×10^{13} W/cm^2. The vertical bars correspond to the ionization threshold and to the 1-2 transition of hydrogenic carbon and fluorine, respectively.

The situation demonstrated in fig. 1 applies to an idealized pump spectrum which contains only the useful pump radiation. Any real X-ray 'flashlamp' however, will emit in other wavelength regions as well. Radiation which populates the lower laser level or depopulates the upper one may be called 'spoil-radiation'. Examples are given in fig. 2 which shows low-resolution spectra of titanium and copper laser plasmas, corrected for grating efficiency and film sensitivity.[9] The wavelengths corresponding to the ionization threshold and to the 'spoil-radiation' (L_α) of suitable active ions (hydrogenic carbon and fluorine respectively) are marked in the figure.

In order to quantify the amount of 'spoil-radiation' which can be tolerated computer simulations were also made with various kinds of more realistic spectra. An example is shown in fig. 3, in which the useful pump radiation is assumed to sit on a background intensity of fraction f of the maximum pump intensity. The plots show the maximum gain coefficient versus f for both, line and ionization pumping and indicate the high sensitivity of the gain to any amount of background radiation.

As mentioned in the introduction photo-pumping offers the possibility of transient

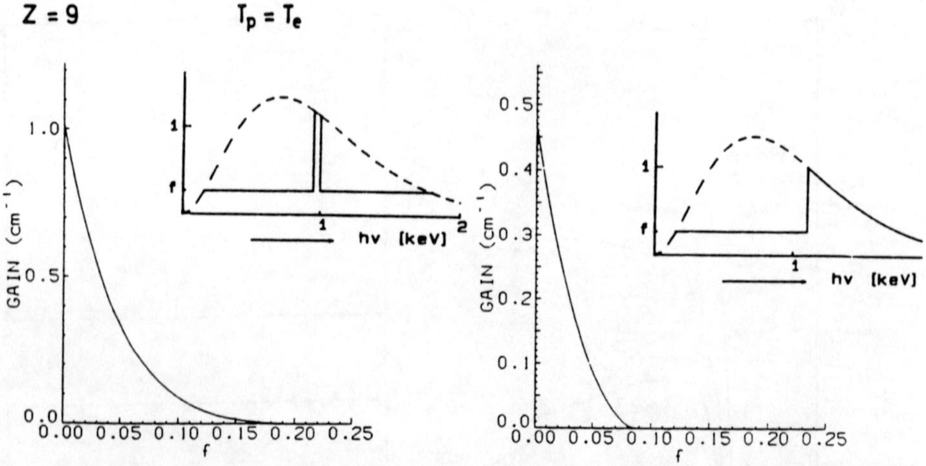

Fig. 3: Gain coefficient at optimum ion density for a pump spectrum with pump radiation on a uniform background with fractional intensity f. The figure applies for line- and ionization-pumping. The temperatures (optimized for each case) are 153.9 ev and 162 eV respectively.

excitation of an X-ray laser , in a sense that the pumping time of the upper level is shorter than its decay time. In the example of a hydrogenic 3-2 laser the excitation time must be shorter than the lifetime of the level $n = 3$, which (neglecting collisions) is given by [10]

$$\tau_3 = (A_{31} + A_{32})^{-1} \simeq 10^{-8} Z^{-4} s. \qquad (2)$$

For hydrogenic fluorine (Z=9) one has $\tau_3 \simeq 1.5\,ps$.

It is clear that in this case only photoresonant pumping would be applicable since ionization pumping requires the population of the upper level by three-body recombination, the time scale of which is orders of magnitude larger.

In computer simulations with ultrashort pump pulses we found that gain on the 3-2 transition in hydrogenic ions was generated even if the pump spectrum was black-body i. e. non selective. The mechanism of this process is illustrated in fig. 4, which shows the evolution of the ionic levels n=1,2 and 3 of hydrogenic fluorine if a black-body pump pulse of 500 fs duration was applied. The electron temperature of the medium before pumping was 81 eV and therefore the ionic state of the fluorine ions was mainly helium-like. As shown in the figure the pump pulse not only ionizes

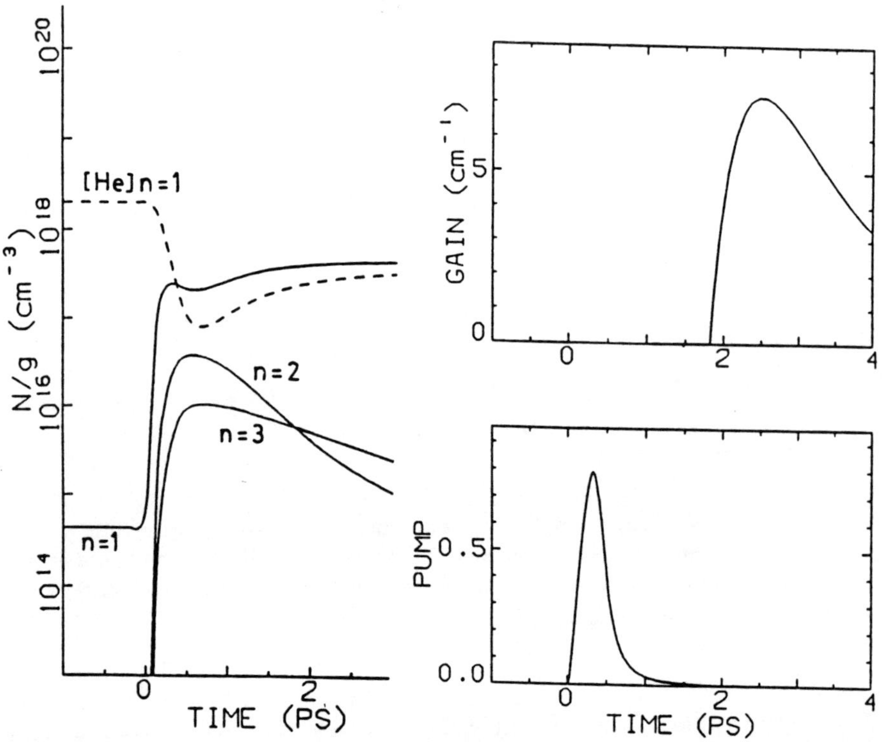

Fig. 4: The left part shows the temporal evolution of the hydrogenic levels 1,2 and 3 (reduced by their statistical weights) after application of a black-body pump pulse of 500 fs duration. Plasma and pump parameters are Z=9, T_p =500 eV, T_e = 81 eV initial ion density 10^{18} cm^{-3}. The right part of the figure shows the pump pulse and gain coefficient.

the helium-like ions to the hydrogenic state but also pumps the levels n=2 and n=3, n=2 being excited much stronger because of its higher Einstein-A-coefficient. After a time given by the decay time of the lower level, however, the population of n=2 goes below that of n=3 and a transient inversion is generated. The time evolution of the pump pulse and gain are also shown in fig. 4. Figure 5 shows a three-dimensional plot of the maximum gain coefficient versus ion density and brightness temperature of the pump. As can be seen quite substantial gains are obtained if ways can be found to achieve brightness temperatures of a few 100 eV. It should be mentioned that a black-body pump spectrum as assumed in these calculations is somehow a worst case assumption: If the spectrum would have a dip at the frequency of the 1-2 transition

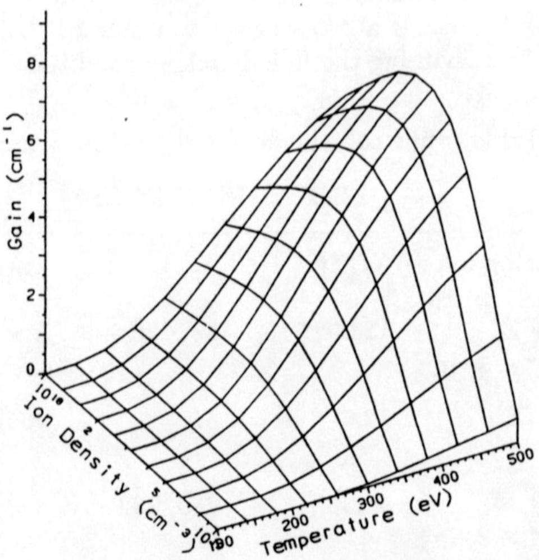

Fig. 5: Three-dimensional plot of maximum transient gain coefficient vs. pump temperature and ion density. Laser transition is 3-2 in hydrogenic fluorine. $T_e = 81$ eV. A black-body pump spectrum is assumed.

the gain would appear earlier and would be higher than shown in fig. 5.

A more detailed account of the numerical simulations has been given in [10]. The results obtained may be summarized as follows:

- As a general rule to obtain high gain the brightness temperature of the pump radiation must be higher than the electron temperature of the active medium.

- Line pumping leads to a higher gain than ionization pumping and the gain apears under a wider range of parameters.

- The 'spoil-radiation' problem is a serious one.

- Significant transient gain can be generated even with a non-selective (black-body) pump spectrum if the pump pulse duration is shorter than the lifetime of the upper laser level.

SPECIFIC PHOTO-PUMPING SYSTEMS

One of the main tasks in designing a photo-pumping experiment is to find an ion pair suitable for the X-ray flashlamp and the active medium. Examples relevant to ionization pumping have already been given in fig. 2. In table I some useful line coincidences and ionization pump pairs are listed. In order to render the X-ray flashlamp 'hotter' than the active medium it is desirable that the pumping ion be of a higher ionization stage than the one of the active medium.

In the case of line pumping the wavelength mismatch should be less than a Doppler width which yields a $\delta\lambda/\lambda$ of a few times 10^{-4}. Both Stark and opacity broadening help in alleviating the line matching condition.

Table I: line coincidences and ionization pump pairs

LINE COINCIDENCES

Pump ion	Lasing ion	$\lambda_p\,[A]$	$\lambda_l\,[A]$	$\Delta\lambda/\lambda$	Ref.
Na X $1s^2 - 1s2p$	Ne IX $1s^2 - 1s4p$	11.0027	58, 230	2×10^{-4}	[11]
Mn XXII $2p^2 - 2p3d$	F IX $1 - 3$	12.6	81	$< 10^4$	[12]
Al XI $2s - 3p$	Mg IX $2s^2 - 2s4p$	48.3	250, 230	5×10^{-5}	[13]
Al XIII $1 - 2$	Er XLI $3d - 4f$	7.2	54.3 - 64.8	$< 10^{-3}$	[14]
K XIX $1 - 2P_{1/2}$	Cl XVII $1 - 4P_{1/2}$	3.35	16.76, 64.72	3×10^{-4}	[11,15]

IONIZATION PUMP PAIRS

Ti → C VI
Ti → Al XI
Cu→ F IX
Al XIII L_α → Na XI
(and similar cases where $10.2\,Z_s^2 \simeq 13.6\,Z_a^2$)

Quite a number of accidential line coincidences seems to exist and many of them are probably not known yet. Nevertheless it would be desirable to find ways to systematically solve the line coincidence problem. A possible solution was put forward recently by proposing to use the same kind of ion for pumping and lasing.[16,17] It was shown that such 'automatic' coincidences could successfully be used for ions of the configuration $2s^2 2p^k$ $(k \leq 6)$.

The case of neon-like ions $(k = 6)$ is shown in fig. 7. Note that the upper laser level $3p$ cannot be photopumped directly but may be populated by radiative decay from the level $4d$ which can be photo-pumped quite efficiently.

A rate equation analysis shows [16] that an inversion between $3p$ and $3s$ is generated despite the fact that strong 'spoil-radiation' corresponding to the resonant line $3s - 2p$ is present in the pump radiation. The reason for this is the large ratio between the Einstein-A coefficients of the transitions $4d - 3p$ and $3p - 3s$ in neon-like

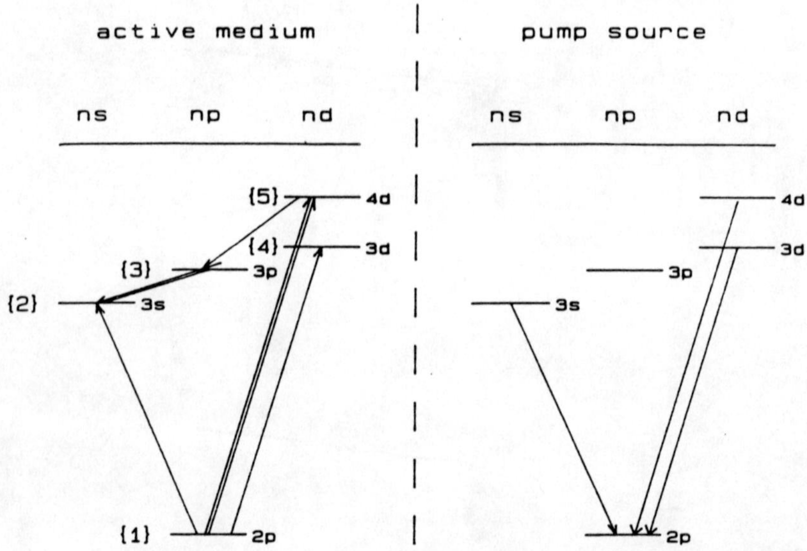

Fig. 7: Simplified level scheme of neon-like ions. The laser transition is 3p-3s. Arrows indicate photo-pumping paths in the active medium and the main pumping transitions in the source ion.

ions. The analysis yields a condition for inversion which may be written as

$$\vartheta_{12} < \vartheta_{15} \frac{g_5}{g_3} \frac{A_{53}}{A_{32}}, \tag{3}$$

where ϑ_{ik} and A_{ki} are the number of photons per mode and the Einstein-A coefficient of the transition $i - k$ $(i < k)$ and the indices apply to the level number as indicated in fig. 7.

The condition of eq. (3) shows that a sizeable amount of 'spoil-radiation' can be tolerated since A_{53}/A_{32} is of the order of 100. Note that the degeneracy ratio g_5/g_3 is also > 1 but its contribution is not so significant.

EXPERIMENTAL IMPLEMENTATIONS OF PHOTO-PUMPING

Experiments have been performed to investigate favourable geometries for the realization of photo-pumping schemes. The question of an optimum pump geometry arises due to the conflicting requirements of separating the pump plasma and active medium (in order to avoid heat conduction by the electrons) but still providing an optimum radiative coupling of the two plasmas.

The experiments have been carried out using the high-power iodine laser ASTERIX IV. [18] This laser has a fundamental wavelength of 1.315 μm and provides pulses with a duration of 350 ps. Special features of the laser are an excellent beam

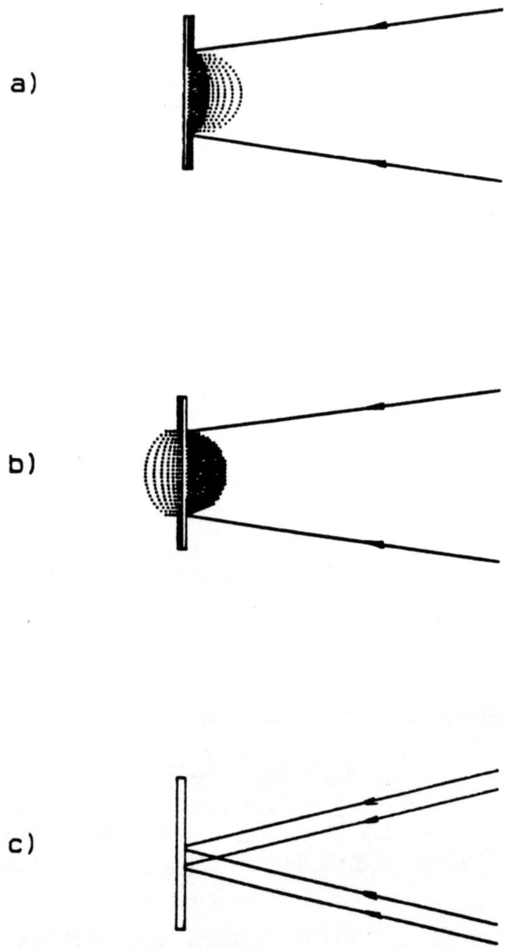

Fig. 8: Three different target geometries for the realisation of a photo-pumped X-ray laser: a) Thin layer target; b) X-ray converter; c) Two parallel line foci.

quality (about 2.5 times diffraction limited) and a high shot rate of 1 shot per 20 minutes.

Three different pumping geometries were investigated (see fig. 8): The first one (fig. 8a) implies the evaporation of a thin layer of the active medium onto the X-ray flashlamp substrate. After burning through the thin layer the laser irradiates the substrate thus generating the strong pump radiation. It was anticipated that by choosing appropriate layer thicknesses both plasmas could be generated with parameters favourable for photo-pumping.

The second geometry (fig. 8b) involves the evaporation of the active medium on the rear side of a thin foil. This foil acts as an X-ray flashlamp and at the same time separates the active medium from the pump plasma. Such an arrangement may be called an X-ray converter. The active medium plasma could be generated by a second laser beam or by the pumping X-rays themselves.

In the third geometry (fig. 8c) plasma and active medium are generated by two parallel line foci separated by a distance of a few 100 μm. This geometry provides the highest degree of variation of the parameters of the two plasmas but the radiative coupling is lower than in the other two examples.

1. Thin Layer Targets

Photoionization experiments using layered targets were performed with the pump laser focussed to a line focus 10 mm long and 100 μm wide. The pump pairs investigated were Ti-C, Ti-Al and Cu-F. Layer thicknesses of 300 A, 600 A and 1000 A of the lower-Z material were used at laser intensities up to 10^{13} W/cm^2. Time integrated and time resolved spectra were taken in an axial direction. The homogeneity of irradiation was checked using a time integrating spatially resolving spectrometer transverse to the line focus.

A typical time resolved spectrum is shown in fig. 9. In this figure the left side shows a pure carbon spectrum from a 2 μm thick polyethylene foil. The spectrum on the right side is obtained from a titanium substrate covered with a 600 A layer of carbon. The 3-2 transition in hydrogenic carbon is quite prominent in both spectra and may be used to verify a pumping effect.

Unfortunately in none of these experiments could a pumping effect be demonstrated unambiguously. For example in case of the streaked spectrum shown in fig. 9 the analysis of the data reveals that the emission of the carbon layer on titanium is slighty weaker than the one of pure carbon and that it also decays faster. These findings can be explained by the cooling effect of the high-Z titanium ions.

It is interesting to compare the time sequence of the carbon and titanium emissions: Contrary to expectations it turns out that both emissions begin virtually at the same time, indicating turbulent mixing of the two components.

As a general result of these experiments it seems that the relatively simple layered targets do not provide a useful geometry for photo-pumped X-ray lasers.

2. X-ray Converter

Experiments related to a geometry in which the active medium is on the other side of the target were carried out at our laboratory by Celliers and Eidmann. [19] They used the second harmonic of a Nd-glas laser at 10^{12} and 10^{13} W/cm^2 (pulse duration 3 ns) to irradiate gold foils with thicknesses ranging from 500 to 2000 A. A symmetric set of diagnostics including a soft X-ray spectrometer and X-ray photodiodes was used to monitor the front and the rear emission of the target.

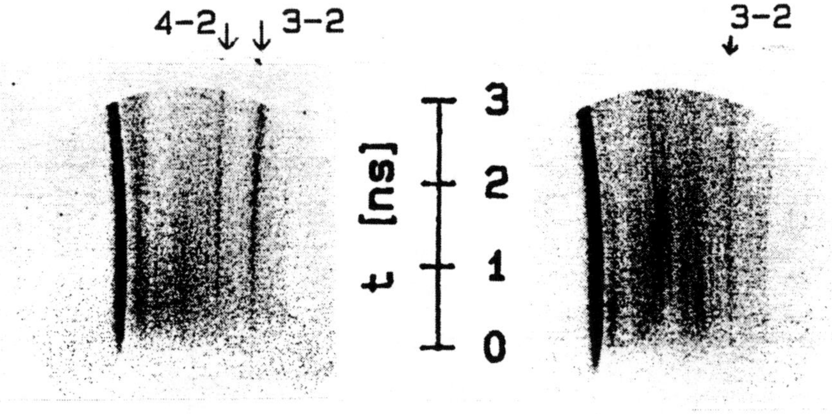

Fig. 9: Streaked soft X-ray spectrum of polyethylene (left) and 600 A carbon layer on titanium (right). The dark line at the left of each spectrum is the zeroth order. The arrows mark the 3-2 and 4-2 transitions in hydrogenic carbon at wavelengths of 182 and 135 A. As evident from the streaks both lines are weaker and decay faster in case of layer target.

These investigations showed that for the thinnest foils (thickness 500 A) the ratio of the front to rear X-ray emission was almost unity while the transmitted laser intensity was reduced by a factor of five. Using thicker foils the front to rear ratio increased but the transmitted laser intensity was dramatically reduced.

From these experimental results it appears that there is a tradeoff between X-ray emission and blocking of the laser light. In a photo-pumping experiment conditions would have to be found to provide adequate values for both.

3. Geometry with Parallel Line Foci

The third geometry investigated involves the irradiation of a plasma from a second plasma on the same side of the target but displaced laterally. To this purpose a glass wedge was inserted into part of the beam in front of the focussing lens. The wedge angle was so chosen as to provide a distance between the line foci of 500 μm. The emission of the line foci was monitored in a transverse and in an axial direction. By moving the wedge in and out of the beam the ratio of the intensities generating the two line foci could be varied. A slit of 100 μm width was placed close to the target between the line focus and the axial spectrometer in order to transmit only the emission of the weaker line focus.

In the experiment reported the target material was fluorine for both plasmas. A 2000 A layer of LiF was evaporated onto an aluminum substrate. The intensity for generating the pump plasma was 2×10^{13} W/cm^2. One third of this intensity provided the active medium plasma. Axial spectra of the active medium with and without pumping are shown in fig. 10. It is obvious that the presence of the pump

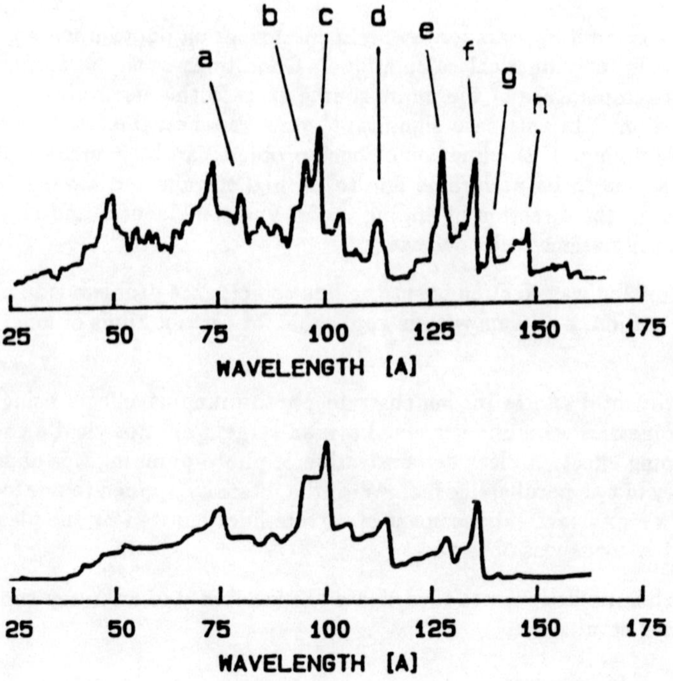

Fig. 10: Spectrum of 2000 A LiF on Al with (top) and without pumping from adiacient line focus. The lines correspond to the following transitions: (a) F IX 3-2; (b) F VII 4d-2p; (c) F VIII 3d-2p; (d) F VII 3p-2s; (e) F VII 3d-2p; (f) F VII 3s-2p; (g) F VI 2s3d-2s2p; (h) F VI 2p3d-2p^2.

leads to a considerable enhancement of many lines of the pumped plasma. For example the 3-2 line of hydrogenic fluorine at 81 A appears only in the presence of the pump. The relatively large distance of 500 μm between the two plasmas gives confidence that the effect is not due to heat conduction from one electron cloud to the other or any other kind of coupling (such as by particle emission, magnetic fields etc.) Moreover a close inspection of the pumping effect on the various lines reveals that the enhancement is strongest for the lines with the highest Einstein-A coefficient, characteristic for photo-pumping.

Further experiments of this kind, varying parameters such as the distance between the two line foci, the intensity ratio, target shape, etc. are under way to find optimum conditions for successful line pumping experiments.

CONCLUSION

In conclusion it appears that experimental work on photo-pumping is encouraged by the results of numerical calculations. Conditions must be found in which the brightness temperature of the pump source exceeds the electron temperature of the active medium. In this case significant gains can be achieved by both ionization and line pumping. If the line coincidence problem can be overcome photo-resonant pumping seems to be preferable due to its higher gain and lower electron density. More work in the direction of finding useful line coincidences and of verifying them experimentally seems to be necessary.

A promising way to circumvent the line coincidence problem is to pump with the same kind of ion, a scheme with is applicable for certain kinds of ions, e.g. neon-like ones.

Experimental efforts to demonstrate photo-pumping effects using two kinds of target geometries were undertaken. Layered targets did not yield a conclusive proof of a pumping effect. A clear demonstration of photo-pumping was obtained by using a geometry of two parallel line foci. With the intensity applied to one focus a factor of 3 higher, a significant enhancement of certain lines emitted by the plasma generated by the other focus was observed.

Experiments designed to investigate the feasibility of an X-ray converter yielded encouraging results.

Acknowledgement: This work was supported, in part, by the Commision of the European Communities in the framework of the Association Euratom/IPP. One of the authors (B.Ch.) acknowledges support by the A.v.Humboldt foundation, Bonn, W.Germany.

REFERENCES

[1] V.A. Bhagavatula, IEEE J. Quantum Electron. QE-16, (1980) 603.

[2] R.C. Elton and R.H. Dixon, Physical Review A28, (1983) 1886.

[3] R.E. Corbett et al., Rutherford Laboratory Report No. RAL 85-047, (1985) p. A4.7 (unpublished).

[4] S.J. Stephanakis et al., IEEE Trans. Plasma Science 16, (1988) 472.

[5] P. Monier, C. Chenais-Popovics, J.P. Geindre, and J.C. Gauthier, Physical Review A38, (1988) 2508.

[6] C.A. Back, R.W. Lee, and C. Chenais-Popovics, Phys. Rev. Lett. 63, (1989) 1471.

[7] N. Qi and M. Krishnan, Phys. Rev. Lett. 59, (1987) 2051.

[8] D.G. Goodwin and E.E. Fill, J. Appl. Phys. 64, (1988) 1005.

[9] K. Eidmann and T. Kishimoto, Appl. Phys. Lett. 49, (1986) 377.

[10] D.G. Goodwin and E.E. Fill, Appl. Phys. B50, (1990) 177.

[11] A.V. Vinogradov, I.I. Sobelman, and E.A. Yukov, Sov. J. Quant. Electron. 5, (1975) 59.

[12] P.G. Burkhalter et al., J. Opt. Soc. Am. B2, (1985) 1894.

[13] M. Krishnan and J. Trebes, Appl. Phys. Lett. 45, (1984) 189.

[14] J. Nilsen, Phys. Rev. A40, (1989) 5440.

[15] J.D. Garcia and J.E. Mack, J. Opt. Soc. Am. 55 (1965) 654.

[16] B.N. Chichkov and E.E. Fill, Optics Commun. 74 (1989) 202.

[17] B.N. Chichkov and E.E. Fill, Phys. Rev. in print.

[18] H. Baumhacker et al., CLEO Conf. Proc., Baltimore (1989), paper MG4.

[19] P. Celliers and K. Eidmann, Phys. Rev. A41 (1990) 3270.

X-RAY LASER STUDIES AT ILE OSAKA UNIVERSITY

H. Daido, Y. Kato, H. Azuma, K. Yamakawa, H. Shiraga, H. Takabe, E. Miura, S. Nakai, and C. Yamanaka*
Institute of Laser Engineering, Osaka University
**Institute for Laser Technology,*
2-6 Yamada-oka, Suita, Osaka, 565, Japan

S.A. Ramsden*, G.J. Pert†, M.H. Key‡, and S.J. Rose
**University of Hull, Hull HU67RX England, UK*
†University of York, York, YO15DD England, UK
‡Rutherford Appleton Laboratory, Chilton Didcot, OX11OQX England, UK

In Gekko XII x-ray laser experiment, a stripe and a foil targets were irradiated with a 351 nm, 130 ps laser pulse with the maximum energy of 150 J line focused to a 7 mm length and 50 μm width, which we call "long pulse experiment," giving time-integrated gain of 1.2 cm^{-1} on sodium Balmer alpha line at 5.42 nm and 1.9 cm^{-1} on magnesium Balmer alpha line at 4.55 nm. Since the pulse width in this experiment was long compared with characteristic plasma expansion time of 30 ps for sodium, we have implemented a revised experiment where a foil target was irradiated with a 28 ps pulse using Gekko MII laser. The measured time-integrated gain of 4 cm^{-1} was 3 times higher than the value obtained in longer pulse experiment. This result convinced us the concept of short pulse experiment for explosive mode of recombination Balmer alpha laser.

I. INTRODUCTION

Recently, a significant amount of gain and intense amplified spontaneous emissions at soft x-ray wavelengths have been successfully demonstrated with several approaches[1]. In this paper, we present x-ray laser studies at ILE(Institute of Laser Engineering), Osaka University. Several studies to attain x-ray laser gain have been performed using solid state and gas lasers. In our Institute, large scale twelve beam glass laser Gekko XII[2] is working. In 1988, Japan-UK joint experiment was done using Gekko XII. Main topics was extension of recombination pumped Balmer alpha lasers toward water window spectral region[3].

Medium size laser fusion system Gekko MII[2] is also used for x-ray laser study. This system provides ultra-short intense laser pulses. The recombination pumped Balmer alpha lasers under short laser pulse irradiation condition was also performed. These experiments were also the collaboration work of Japan-UK joint experiment[4].

We also have a medium size CO_2 laser system[5] and a compact repetitive glass laser system[6]. These systems are also used for x-ray laser studies[7] and laser plasma x-ray source development[8]. In this paper, we concentrate on the recombination pumped Balmer alpha lasers produced by Gekko XII laser and Gekko MII laser. The other two topics are presented in other papers of this Proceedings.

The motivation of this work is that how far we can extend the recombination laser with the Balmer alpha transition to shorter wavelengths. Approach to this question is to irradiate a target with a high intensity, short pulse and short wavelength laser.

First of all, we show atomic number Z scaling of recombination Balmer alpha lasers. Figure 1 shows the energy level diagram of the hydrogen-like ion. Population inversion is generated between principal quantum number of 3 and 2 due to the three body recombination pumping of level three and fast radiative decay of level two populations. The laser wavelength is proportional to Z^{-2}. The electron temperature T_e is proportional to Z^2 and electron density n_e is proportional to Z^7 which is determined by the thermal limit. Required irradiation intensity I is proportional to Z^4 which is determined by the relation between I and T_e, i.e. I is proportional to T_e^2. The characteristic expansion time is proportional to Z^{-4}, which is based on many computer runs over various Z materials[10]. Therefore, higher Z material gives shorter wavelength lasers. The required irradiation intensity and wavelength are higher and shorter. Furthermore, the time scale should be short enough to maintain population inversion.

II. LONG LASER PULSE EXPERIMENT

Figure 2 shows the schematic diagram of the experimental arrangement of Gekko XII system. We usually used one arm of the system and sometimes, used both two beams. Three spectrographs were installed in the vacuum chamber whose pressure was below 10^{-5} Torr. The laser beam came from the final amplifier and was converted to 0.35μm light via two KDP crystals and was focused with line focusing optics.

During Japan-UK joint experiment, two types of targets i.e., stripe and foil targets were mainly irradiated by one arm of the Gekko XII laser. Sometimes, fiber targets were irradiated. We briefly describe the scenario of the laser action. During ionization phase, the key issue is to produce large number of fully ionized ions efficiently by laser irradiation. The heating stops, the highly ionized plasma goes into cooling phase, (recombination phase) caused by rapid expansion, radiation and heat conduction to the cold dense matter. Note that motional Doppler shift due to rapid expansion prevents the pumping of the lower lasing level. Then finally, the population inversion occurs via three-body recombinations.

Before gain determination, we describe how to calibrate the relative sensitivity of the two spectrometers. Figure 3 shows the relative sensitivity of the spectrometers as a function of soft x-ray wavelength.

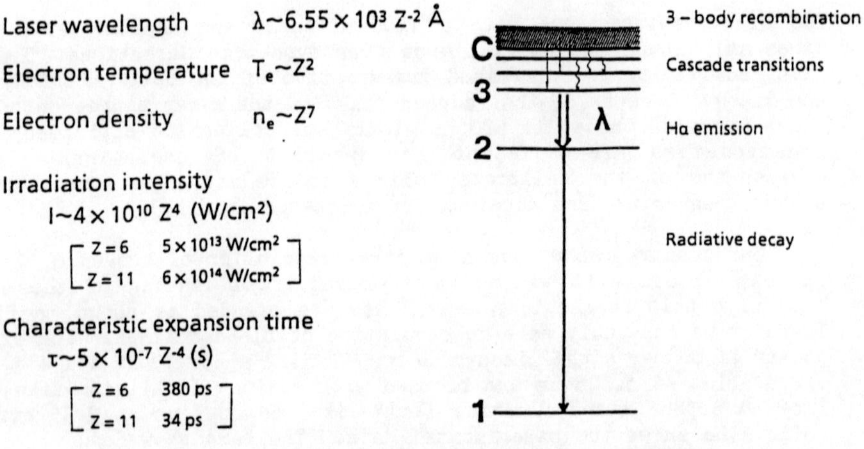

Fig. 1 Energy level diagram of the hydrogen-like ion.

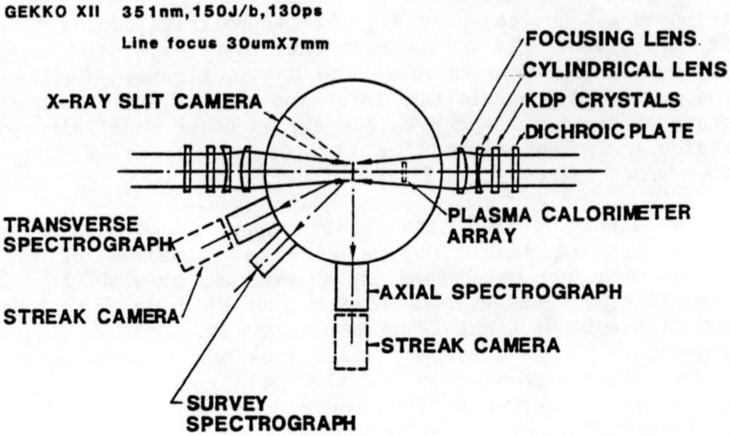

Fig. 2 Schematic diagram of the laser beams and the diagnostic equipments.

Two spectrometers, one with a 1200 lines/mm and another with a 2400 lines/mm, viewed the small plasma from symmetric directions. The relative sensitivity is determined irrespective of the optical thickness of the lines. Because of the reduced image of the x-ray source onto a slit which was wide opened to 500 μm width, the collection efficiency of the spectrometers are constant irrespective of the target length. Wavelengths of the calibration lines and Balmer alpha transitions of sodium, magnesium, and aluminum are indicated in Fig.3.

Our spectrographs have a spectral resolution of around 0.015 nm in spite of broad slit width due to demagnified imaging of the source. Even with this resolution, many lines are blended as shown in Fig. 4. Therefor we carefully make deconvolution of blended lines. For example, sodium XI Balmer alpha lines of J = 5/2 - 3/2 at 5.419 nm and J = 3/2-1/2 doublet at 5.405 nm are blended with sodium IX (Lithium-like) 4p-2s line at 5.386 nm and fluorine IX (H-like) 5d-2p line at 5.35 nm. The solid line shows the experimental data. The bars show deconvolved peak intensity of each line. Dotted line is the convolved spectral profile. These procedure is applied to evaluate accurate intensities of emission lines.

Densitometer traces of the axial and transverse spectra of an NaF stripe target are shown in Fig. 5. Approximate scales linear in intensity are given. The enhancement factors of axial and transverse intensity ratios are also shown in the bottom figure. Sodium Balmer alpha line shows the population inversion and other resonance lines show strong absorption. For MgF_2 target, we observe definite population inversion at Balmer alpha line at 4.55 nm which is very close to the water window as shown in Fig.6.

We also performed time resolved gain measurement in which the two x-ray streaked spectrometers were used instead of films. The two spectrographs are calibrated in the same way as with the film detection. Time history of the gain coefficient as shown in Fig. 7 was determined from time dependent axial and transverse Balmer alpha intensities. The gain starts 100 ps after the laser peak and becomes maximum at 150-200 ps and decays gradually as the Balmer alpha emission intensity decreases. The error bars in these data arise from granular noises of the image intensifier. The maximum gain of 3.2 ±1.0 cm^{-1} was observed in this data.

Time integrated gain coefficients of sodium XII Balmer alpha line of J = 5/2 - 3/2 transition are summarized in Table 1. When the target is irradiated with 70 J laser energy, the gain of around 1 cm^{-1} is obtained. The target geometry does not strongly affect the gain. Increasing the laser energy may slightly reduce the gain. As for $Na_2B_4O_7$ target, although dilution effect which is expected to reduce Lyman alpha trapping will cause higher gain coefficient, the experimental value shows opposite result, which is qualitatively consistent with the recent theoretical calculations[11].

III. SHORT LASER PULSE EXPERIMENT

In Gekko MII which is the module of the Gekko XII laser, ultrashort high intensity laser pulses can be delivered to study x-ray lasers under

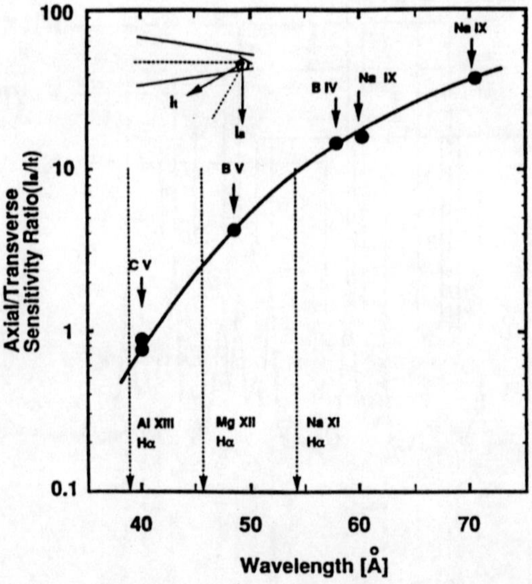

Fig. 3 Relative sensitivity of the axial and transverse spectrometers determined by irradiating a narrow stripe target coated with $Na_2B_4O_7$ in an arrangement shown in the upper part of the figure. Wavelengths of the calibration lines and the Balmer alpha transition of sodium, magnesium, and aluminum are indicated.

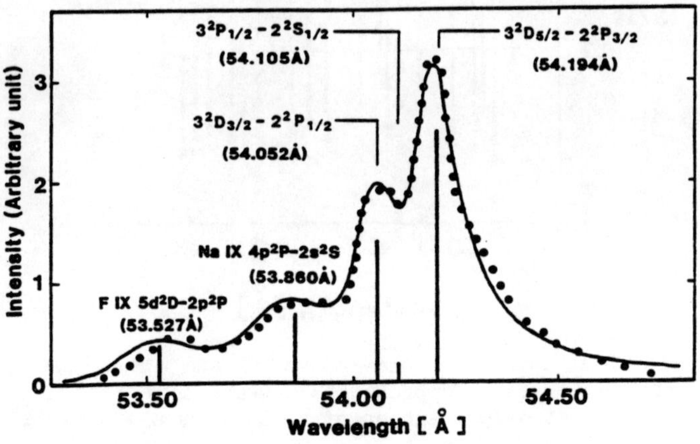

Fig. 4 An example of the deconvolution of blended lines at sodium XI Balmer alpha lines. The solid and the doted lines denote the experimental data and the convoluted spectrum, respectively. The bars show the peak intensity of the each individual line.

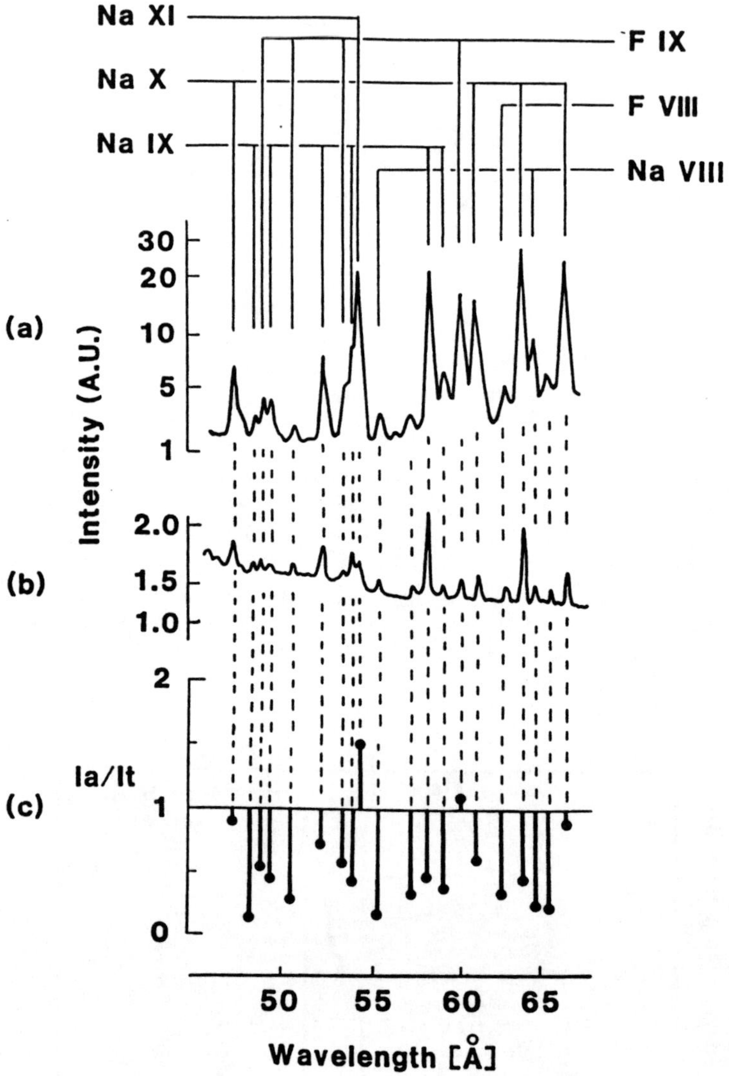

Fig. 5 Densitometer traces of the (a) axial and (b) transverse spectra of an NaF stripe target. Approximate scales linear in intensity are given. The I_a/I_t ratio for each spectral line is shown in (c).

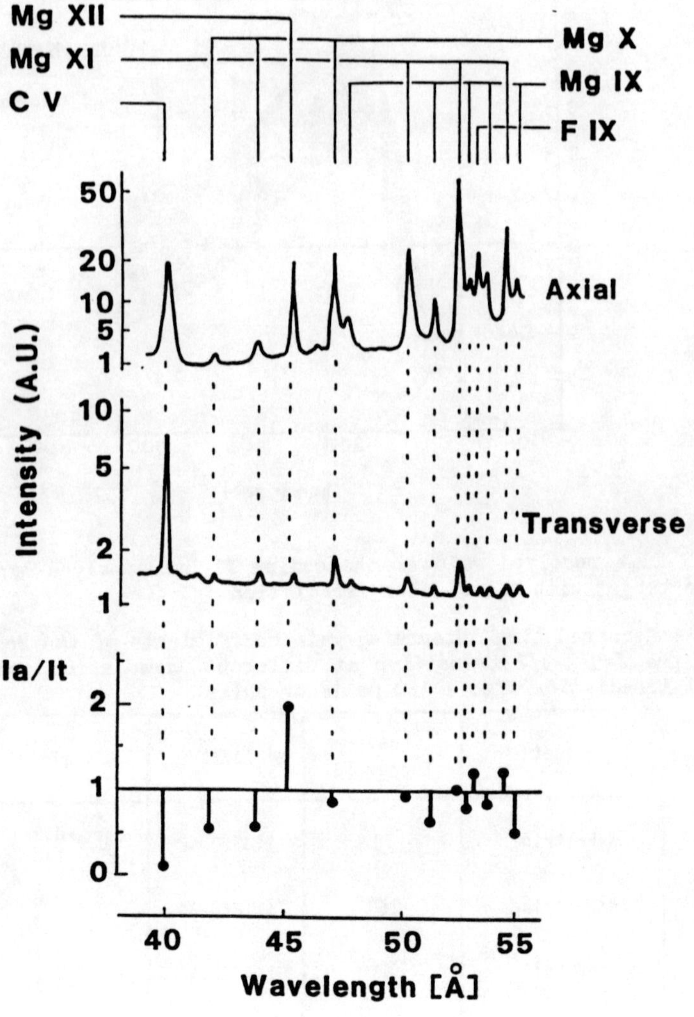

Fig. 6 Densitometer traces of the (a) axial and (b) transverse spectra of an MgF_2 stripe target. The I_a/I_t ratio for each spectral line is shown in (c).

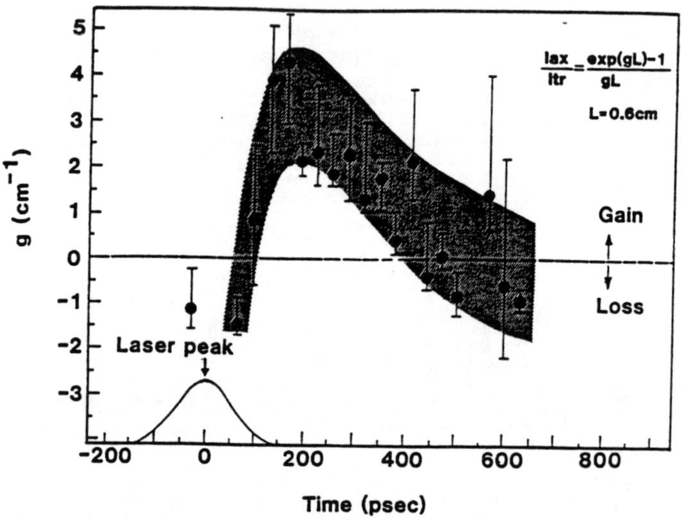

Fig. 7 Time resolved gain of the sodium XI Balmer alpha line for a NaF stripe target with 148 J laser irradiation.

Table 1 Measured time-integrated gain coefficients of the sodium XI Balmer alpha J=5/2-3/2 transition at different irradiation conditions for target irradiation with a 130 ps laser pulse.

Target	Laser Energy [J]	Ia/It	g [cm^{-1}]
NaF Stripe	73	1.33±0.40	$0.9^{+0.8}_{-1.2}$
NaF Stripe	153	1.12±0.34	$0.4^{+0.8}_{-1.3}$
NaF Foil	63	1.47±0.44	$1.2^{+0.8}_{-1.1}$
$Na_2B_4O_7$ Stripe	69	0.74±0.22	$-1.1^{+0.9}_{-1.4}$

short pulse excitation scheme. Between the mode-locked oscillator and the rod type preamplifiers, the pulse compression system was installed. The picosecond laser pulse can be generated in this section, then it goes through the amplifier chain. The laser output energy was 50 J in 3.5 ps at fundamental wavelength of 1.05 μm. KDP crystal was installed just behind the amplifier chain to produce a 30 J in 28 ps at 0.53 μm wavelength pulse. The principle of the pulse compression is described as follows[12]. The laser pulse enters into the polarization preserving single mode optical fiber, the self-phase modulation occurs due to the optical Kerr effect leading to the frequency shift. In the leading edge of the Gaussian pulse, we can observe down shift of the frequency while in the trailing edge, the up shift of the frequency. The chirped pulse enters into the grating pair, blue shifted component of the light arrives earlier at the exit of the grating pair than red one according to the dispersion of the gratings. Then the width of such a chirped pulse is compressed significantly by the grating pair. In addition, a spectral window which eliminates the nonlinear chirp component provides a short pulse with a reduced pedestal[13] which produces the preformed plasma before a main laser pulse arrives[13]. The streaked records of the laser pulses clearly show the efficient suppression of the pedestal with the spectral windowing. The laser delivers such a pulse for recombination pumped x-ray laser studies.

Figure 8 shows the schematic diagram of the experimental setup. One beam of Gekko MII laser irradiates the x-ray laser target with line focus optics. The laser is 526 nm wavelength with energy of 25 J in a 28 ps pulse. The intensity is 6×10^{14} W/cm^2 on the target. We have two spectrographs of axial and transverse one which have been cross-calibrated. We obtain the gain coefficient g by,

$$I_a/I_t = [\exp(gl)-1]/gl \quad , (1)$$

where I_a, I_t and l are the observed intensities of axial and transverse directions and the gain length, respectively.

Sodium Balmer alpha line is also blended with Li-Like 4p-2s line as shown in Fig. 9. We carefully deconvolved each line emission in the same way that was described before. The enhancement factors of each individual transition are shown in the bottom of Fig.9. Typical densitometer traces of the NaF target over 4-9 nm is shown in Fig. 10. Only the sodium Balmer alpha line indicate the positive gain.

Figure 11 shows the dependence of gain in sodium XI Balmer alpha line on laser energy. The gain is observed reproducibly when the laser energy exceeds 20 J, and becomes strongly negative when the irradiation energy is reduced to half.

Table 2 shows the comparison of experimental conditions of long pulse and short pulse irradiation cases. Observed gain coefficient of around 4 cm^{-1} in the short pulse experiment is about three times higher than that of long pulse experiment for a NaF target, while for a MgF_2 target no gain was observed in short pulse experiment due to insufficient ionization.

Computer simulation of the short pulse experiment was made using the code HYBD3 including two dimensional hybrid model to calculate the

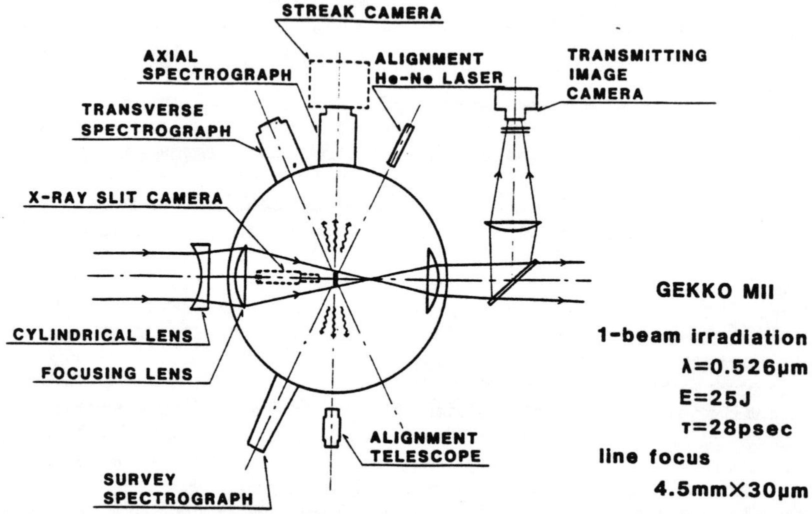

Fig. 8 Schematic diagram of the Gekko MII laser beam and the measurement system for the x-ray laser study.

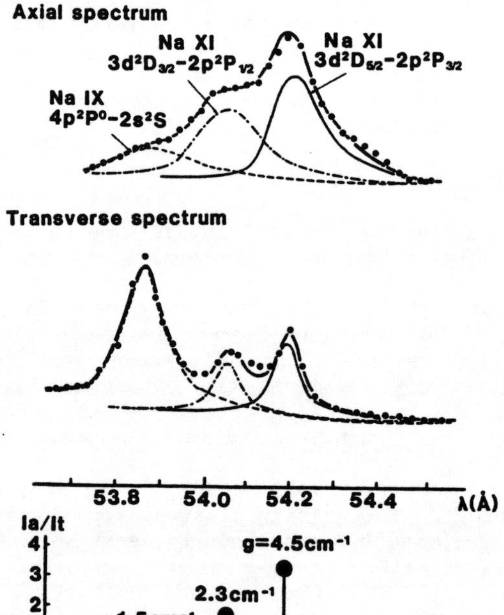

Fig. 9 Axial and transverse spectra near sodium XI Balmer alpha line. Thick solid curves are the experimental data. Thin solid, doted line and dashed lines are deconvolved line shapes for each component with the convoluted spectra shown by doted lines. The I_a/I_t ratio for each component is given in (c).

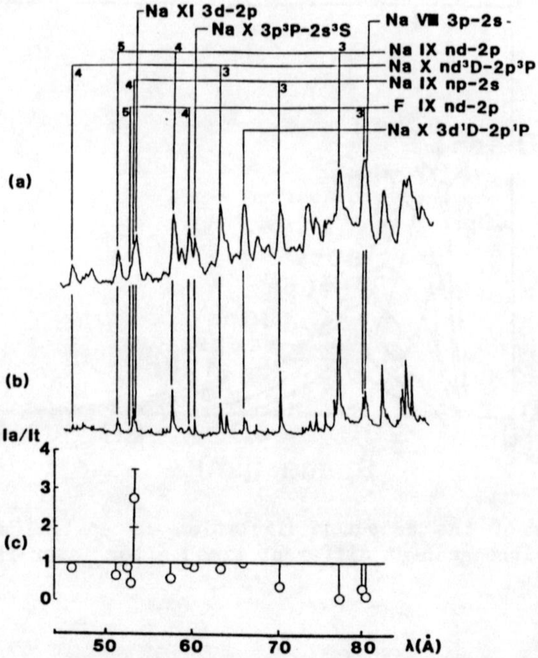

Fig. 10 Axial (a) and transverse (b) spectra of NaF target. The axial-to transverse intensity ratios for the spectral lines are shown in (c). Only the sodium Balmer alpha line indicates the positive gain.

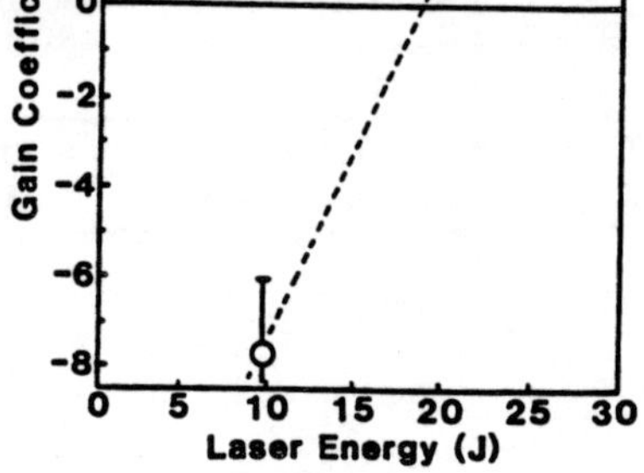

Fig. 11 Dependence of the time integrated gain coefficient of sodium XI Balmer alpha J=5/2-3/2 on laser energy. A typical error in the gain value is shown by a vertical bars.

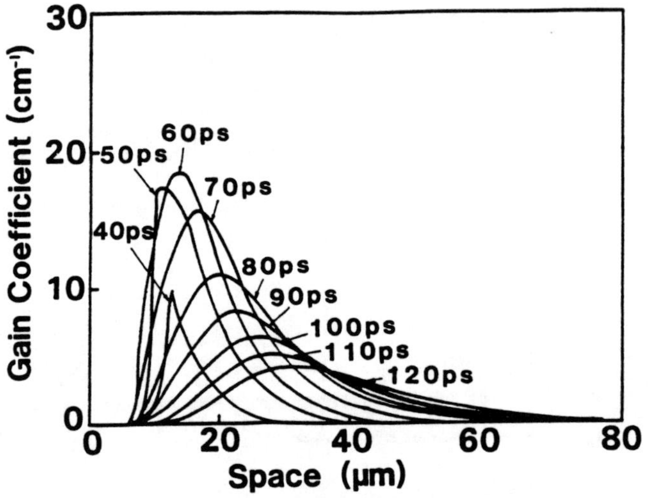

Fig. 12 Result of the computer simulation on spatial distribution of sodium Balmer alpha gain at different times after laser irradiation.

Table 2 The laser and the target parameters of the x-ray laser experiments.

		Phase I (long pulse)	Phase II (short pulse)
Laser		Gekko XII	Gekko MII
Wavelength		351 nm	526 nm
Pulse width		130 ps	28 ps
Laser energy (max)		152 J	25 J
Line focus		7 mm × 30 μm	4.5 mm × 30 μm
Max intensity		5×10^{14} W/cm^2	6×10^{14} W/cm^2
Target		Stripe NaF, MgF_2, Al	Foil NaF, Mg, Al, LiF
Obserbed gain	Na XI Hα	1.2±0.7/cm(62.5J)	3.9±0.9/cm(23J)
	Mg XII Hα	1.9±0.8/cm(153J)	-3.1/cm(23.7J)

hydrodynamics. Laser and target parameters are listed below.

Laser Energy: 23 J, Wavelength: 526 nm
Pulse Width: 28 ps, Size of the Line Focus: 4.5 mm x 30 μm
Target: 1 μm thick NaF,

Figure 12 shows the spatial distribution of the gain coefficient of sodium XI Balmer alpha line at different times, where t=0 corresponds to the peak of the laser pulse. The gain increases sharply from 40 ps and becomes maximum at 60 ps taking a gain value of $g=18\ cm^{-1}$. The maximum gain is produced at 15 μm from the initial target surface. Space and time averaged gain is calculated to be 5 cm^{-1}, which is in close agreement with the experimental result.

IV. SUMMARY

A gain coefficient of 4 cm^{-1} was obtained for sodium XI Balmer alpha at 5.42 nm by irradiation with a 28 ps laser pulse. The gain is three times higher than the value obtained with 130 ps pulse irradiation. The simulation shows peak gain local in time and space of 18 cm^{-1}. The space and time averaged value is 5 cm^{-1} which agrees with the experimental result.

REFERENCES

[1] M. H. Key, J. Mod. Opt. 35,(1988)575, and references therein.
[2] C. Yamanaka, Y. Kato, Y. Izawa, K. Yoshida, T. Yamanaka, T. Sasaki, M. Nakatsuka, T. Mochizuki, J. Kuroda and S. Nakai, IEEE J. Quantum Electron. QE-17,(1981)1639.
[3] Y. Kato, E. Miura, T. Tachi, H. Shiraga, H. Nishimura, H. Daido, M. Yamanaka, T. Jitsuno, M. Takagi, P. R. Herman, H. Takabe, S. Nakai, C. Yamanaka, M. H. Key, G. J. Tallents, S. J. Rose, and P. T. Rumsby, Appl. Phys. B50,(1990)247; P. R. Herman, T. Tachi, K. Shihoyama, H. Shiraga, and Y. Kato, IEEE Trans. Plasma Sci. 16,(1988)520.
[4] H. Azuma, K. Yamakawa, S. A. Ramsden, Y. Kato, T. Tachi, M. Nishio, H. Shiraga, G. J. Pert, S. J. Rose and S. Nakai, to be published in Opt. Lett.
[5] C. Yamanaka, S. Nakai, M. Matoba, H. Fujita, Y. Kawamura, H. Daido, M. Inoue, F. Fukumaru, and K. Terai, IEEE J. Quantum Electron. QE-17,(1981)1678.
[6] B. T. Kim, Y. Oishibashi, T. Kanabe, M. Nakatsuka, and S. Nakai, IEEE Photo. Technol. Lett. 2,(1990)27.
[7] E. Miura, H. Daido, Y. Kitagawa, K. Sawai, Y. Kato, K. Nishihara, and S. Nakai, Appl. Phys. Lett. 55,(1989)223.
[8] H. Daido, K. Nishihara, E. Miura, and S. Nakai, J. Opt. Soc. Am. B7,(1990)266.
[9] G. M. Zeng, M. Takahashi, H. Daido, T. Kanabe, H. Aritome, M. Nakatsuka, and S. Nakai, J. Appl. Phys. 67,(1990)3597.
[10] G. J. Pert, unpublished.
[11] G. J. Pert and S. J. Rose, Appl. Phys. B50,(1990)307.
[12] P. Maine, D. Strickland, P. Bado, M. Pessot, and G. Mourou, IEEE J. Quantum Electron. 24,(1988)398.
[13] K. Yamakawa, Y. Kato et al., to be published.

LITHIUM-LIKE RECOMBINATION SCHEME FOR X-UV LASERS

P. Jaeglé, G. Abdelatif, A. Carillon, P. Dhez, B. Gauthé, G. Jamelot, A. Klisnick, and J. P. Raucourt
LSAI, CNRS and Université Paris-Sud, Bât. 350, 91405, Orsay Cedex and LULI, CNRS and Ecole Polytechnique, 91128, Palaiseau Cedex, France

Abstract: Recombination in lithiumlike ions in a rapidly cooling plasma is an attractive way towards X-ray lasers because, at a given lasing wavelength, it requires less pumping energy than most other pumping schemes. In this paper we describe the status of experiments relative to the Li-like scheme, and we discuss the problem of increasing the gain-length product necessary to achieve efficient X-ray lasers.

I - INTRODUCTION

The first proposal by Gudzenko and Shelepin of recombination pumping[1] considered the relaxation of an hydrogen plasma produced by a gas discharge. For lasing at short wavelengths, the recombination scheme in hydrogenlike ions has been extensively studied since 1974 with C^{5+} ions in laser-produced plasmas[23] . The first evidence of the exponential increasing of a aluminium lithium-like line has been presented in 1981 at the same conference[4]. In the case of Li-like ions, the plasma temperature must reach the point where the ionization is such as the plasma contains mainly helium-like ions. So, during expansion and cooling the free electrons recombine with ions leading to Li-like ions in highly excited states. This is the condition for the production of population inversions. The ionization energy of Li-like ions being much less than for H-like ions, the lithium-like scheme requires less laser energy than most other schemes.

Several elements from magnesium to chlorine have proved to give amplification for lines lying between 80 Å and 200 Å. Aluminium has been studied much more extensively than any other element and the results obtained with Li-like aluminium allow to give a good account of what has been achieved till now regarding the lithium-like scheme and of the problems which remain to be solved.

Let us recall the diagram of levels of Li-like ions. Figure 1 shows transitions of interest for soft X-ray amplification. The wavelengths are given for aluminium. In a plasma of intermediate density (10^{17} to 10^{21} cm^{-3}) when the pumping laser turns off, the temperature fastly falls down and three-body recombination populates strongly the upper energy levels of the recombined Li-like ion. Recombination is followed by a radiative-collisional cascade to intermediate levels n = 4,5. At the same time, owing to the low electron temperature, electron-ion collisions from the ground state cannot balance the fast radiative decay of the lower excited

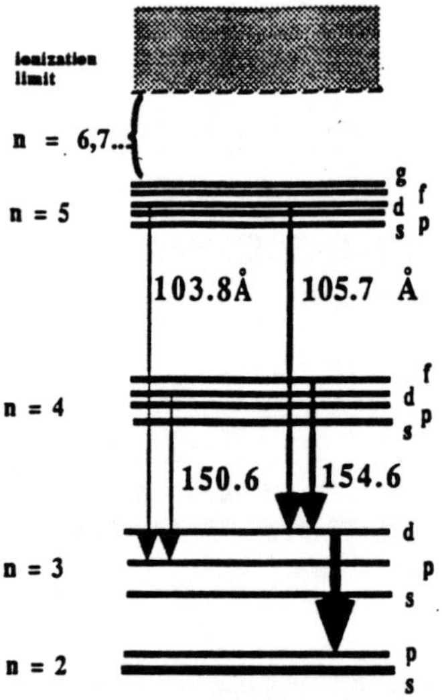

Figure 1: Simplified energy-level diagram of lithiumlike ions. The wavelengths correspond to aluminium.

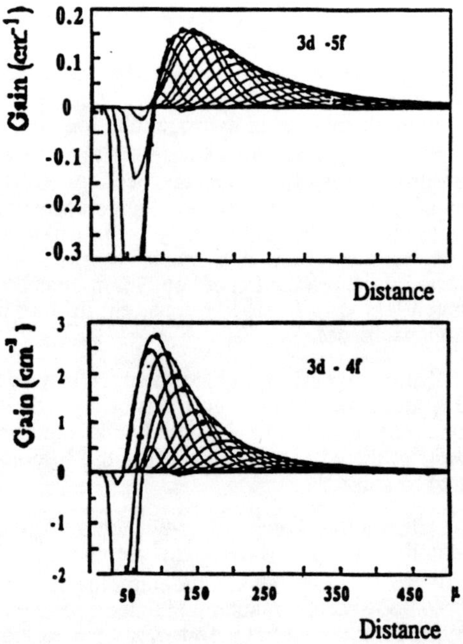

Figure 2: Gain calculated at various times as a function of the target distance for the 105.7 Å (above) and the 154.7 Å lines of Li-like aluminium. Two successive curves are separated by 50 ps.

levels (n = 3). These processes result in transient population inversions, especially between levels 3d on the one hand and 4f and 5f on the second hand.

Figure 2 gives examples of population inversions calculated in a recently improved plasma model[5]. Electron temperature and density are deduced from a one-dimensional lagrangian code. This last takes account of laser-radiation absorption in underdense plasma corona by inverse Bremsstrahlung, collisional ionization, radiative and collisional recombination, and thermal conduction [6]. Each curve on the figure displays the gain versus the distance to the target at given time. Time interval between the curves is 50 ps. The gain maximum occurs at 400 ps for the 4f line (470 ps for the 5f line) after the maximum of the laser pulse. It is noticeable the gain calculated at 154 Å is one order of magnitude larger than at 105 Å. This is a constant result of calculations which is not systematically verified in experiments. Plasma temperature at the gain maximum is of 45 - 50 ev. The value of gain is very sensitive to this parameter. Temperature reduction by means of an adjustable parameter results in a strong enhancement of the gain. Thus one can expect that increasing of plasma cooling, either by optimizing target shape or with using radiation loss from heavy element, will enhance amplification.

II - AMPLIFICATION IN SHORT PLASMA COLUMNS

Several experiments performed in different laboratories with different techniques have shown that aluminium plasma recombination actually provides gain at the wavelengths of the 5f or 4f lines. In this section we summarize a first set of experiments differing each other in many aspects but having in common the magnitude of the plasma column length used for measuring the gain. In these experiments the plasma length was 1 centimetre or less. We do so because - as we will see in the next section - new problems seem to be adressed from recent experiments when one attempts to increase the plasma length up to several centimetres. These experiments are described in section III.

The first measurements of amplification by Li-like ions have been performed by our group in the first eighties. The neodymium-glass laser (1.06 m) of GILM at Palaiseau was used with a pulse duration comprised between 0.6 and 25 ns. The laser beam was focused on the target surface by combining either spherical and cylindrical lenses or two crossed cylindrical lenses. The focal-spot size was 0.1 mm x10 mm, later extended up to 20 mm with the use of two superimposed laser beams. The target used for producing the plasma was a plane sheet of spectroscopically pure metal. Emission from the plasma axis was analyzed using an X-UV spectrometer based on a 1 m radius concave grating of 3600 grooves per mm used at 6° angle of grazing incidence. Time integrated and time resolved spectra could be recorded owing to the use of either a scintillator coupled with an optical multichannel analyser (OMA) or an X-UV streak camera as detectors [7].

An extensive study of time-integrated emission of aluminium plasma has been made for laser irradiance comprised between 0.6×10^{11} and $3. \times 10^{12}$ W/cm^2 [8].A wavelength band wide of about 10 Å round the position of 3d - 5f transition was recorded by OMA. Strong absorption lines were observed, corresponding to transitions of lower charge ions. Regarding the 3d - 5f transition at 105.7 Å and, to a lower extent, the 3p - 5d one at 103.8 Å they give absorbing lines at very low irradiance, which turn progressively into negative peaks in the plasma absorption background when the irradiance rises above a threshold value of about $3. \times 10^{11}$ W/cm^2. This behaviour is characteristic of the lithiumlike structure and not only of aluminium. Analogous negative peaks were observed in the absorption spectra of magnesium at 127.85 Å and sulphur at 65.2 Å the wavelengths of the 3d - 5f transition[9],.

Amplification time history has been studied in time-resolved experiments [10]. The X-UV streak camera provided a time resolution about 100 ps. It was connected either with photographic plate or with the OMA. A typical result corresponding to a laser irradiance of $3. 10^{12}$ W/cm2 and to a pulse duration of 2 ns is presented in figure 3. Time-resolved emission of two aluminium plasma columns, at the wavelength of 3d - 5f transition, is shown on the upper part of the figure. The column lengths are respectively 7 and 3 mm. Let us notice that the time of maximum intensity is delayed with the column lengthening. The lower part of the figure shows the time evolution of gain, deduced at each time from the above intensities. The plasma is at first optically thick, and then it goes continuously from absorption to amplification; gain is decreasing again at the end of plasma recombination, when density has fallen. The maximum of gain is near 2 cm^{-1}. It occurs about 5 ns after the top of the laser pulse.

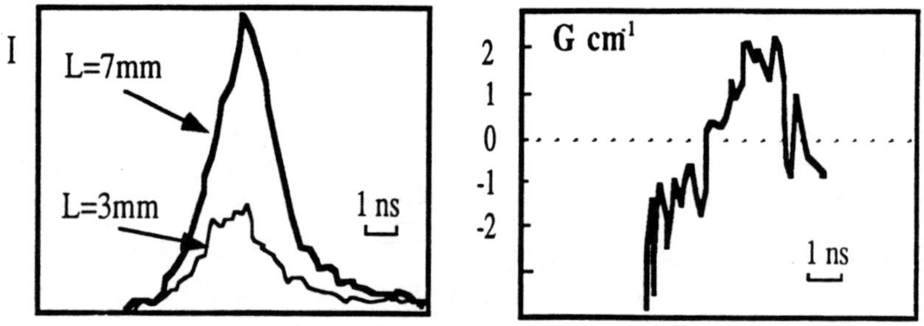

Figure 3: Time-resolved gain measurement at 105.7 A in Li-like aluminium. On the left: time variations of radiation emitted by plasma columns of lengths 3 mm and 7 mm. On the right: gain deduced from these intensities. Target: 0.2 μ of Al on a graphite plate. Laser wavelength 1.06 μ, pulse duration 2.0 ns, irradiance 3. 10^{12} W/cm^2.

The transition 3d - 4f is theoretically a better candidate than the 3d - 5f one for amplification: its radiative decay probability is about 3 times larger, and populations deduced from simulations are almost the same for both levels 4f and 5f. Evidence of amplification due to transition 3d - 4f has been found in time-resolved experiments but, in our experiments the gain value is not found significantly higher than for 3d-5f.

Since a few years, a number of X-ray laser experiments using the Li-like scheme have been carried out in various laboratories [11 12 13 14]. Experimental parameters are very different from one experiment to another one, what does not permit close comparisons of gain values, but all experiments have shown evidence of gain involving one or both 3d - 4(5)f transitions.

The recombination scheme is studied at Princeton with particular characteristics of plasma column production including the possibility of magnetic confinement[15]. Heavy impurities may be introduced to radiatively increase the cooling. Owing to the use of a CO^2 pump laser, the plasma heating arises in a zone of electron density 10^{19} cm^{-3}, instead of 10^{21} cm^{-3} with a Nd-glass laser. Then, population inversions take place in a region of lower density they have a longer duration (owing to the magnetic confinement) than with other devices. The amplification in Li-like magnesium, aluminium and silicon has been studied in this confined plasma column [16]. Time-integrated plasma emission was recorded along both axial and transverse directions, the axial spectrometer being placed at two different transverse positions corresponding to gain and non-gain regions. GxL values about 1.5 at 187.2 Å in Mg X, at 129 Å in Si XII and 3 at 154.7 Å in Al XI have been reported. The 3d --5f transitions were not observed. That might be related to the low initial density of the plasma.

Amplification for 3d - 5f transitions has been observed in other experiments realized with frequency doubling and tripling of Nd-glass radiation. The Osaka group chose to study, in the same experiment, amplifications from H-like carbon and oxygen and Li-like aluminium by using targets made of 600 A-thick foils of formvar coated with 200 A of aluminium[17]. A 100 J-150 ps-laser beam at λ = 0.53 m was focussed to a linear spot of 0.035 mm width and 15 mm length, which gives an irradiance of 1.3x10^{14} W/cm^2. Time integrated spectra covering the wavelength region of 100-300 Å exhibited amplification for the 3d - 5f and 3p - 5d transitions of Li-like aluminium, with gain coefficients of 1.5 cm^{-1} and 1 cm^{-1} respectively. Surprisingly, amplifications occurred only when the laser irradiated the formvar surface opposite to the aluminium, and not when the aluminium was directly irradiated. No amplification was detected at the wavelength of the 3d - 4f line.

In a similar experiment performed at Rochester by Moreno et al , amplifications were detected for 3 --4 as well as for 3 - 5 transitions [18]. The wavelength of the pump laser was shorter and the laser pulse has a longer duration (650 ps) than in the former experiment, but the irradiance on the target surface was analogous. Targets consisted of 5000 Å layer of aluminium and a 100 Å layer of gold on a 10 μ thick mylar backing. They were irradiated on both sides. Gold was included in targets in order to enhance the plasma cooling, but it did not have significant effect on aluminium-line intensities. Amplifications were deduced from time-integrated spectra with high-spectral resolution. Figure 4 shows a large increase of the amplifying lines from 3 mm- to 6 mm-long plasma columns. Gain values of 4.5 cm^{-1} and 3.5 cm^{-1} were found

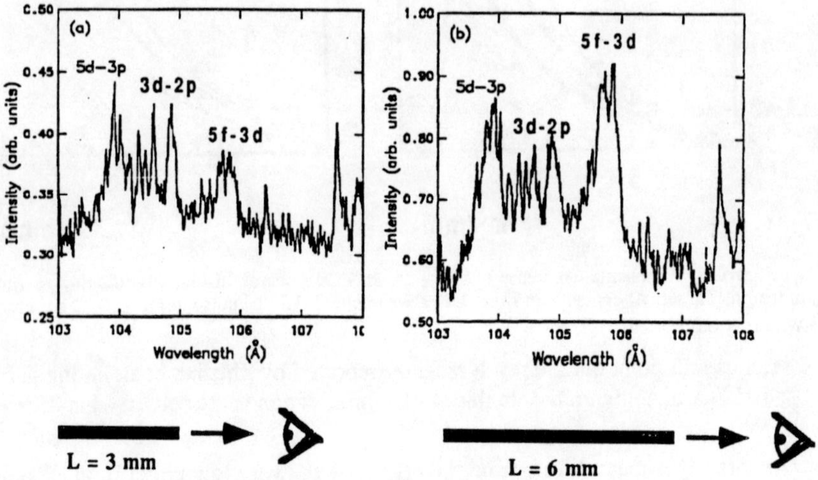

Figure 4: Time integrated spectra showing the 5f - 3d, 5d - 3p lines of Li-like aluminium to increase with the plasma column length. Gain coefficient: 3.5 cm^{-1}.Target: Al 0.5 μ on Au 0.01 μ and 10 μ mylar foil. Laser wavelength 0.35 μ, pulse duration 0.6 ns, irradiance 8. 10^{13} W/cm^2 (from ref. 18)

respectively for the 3d - 4f and 3d - 5f transitions. They are the largest experimental values measured up to now. However, the plasma-column length was short and the GxL values, 2.7 and 2.1, are of the same order as in other experiments.

Pumping by recombination in H-like systems, based on short-laser pulses and fibre targets is under investigation since years at Rutherford Appleton Laboratory [19] A collaborative work has been undertaken between the British group and our own in order to test the Li-like scheme in the same experimental conditions [20]. Targets were 7 μ diameter and 7 to 12 mm-long carbon fibres coated by 3500 Å of aluminium. They were irradiated by two or four frequency-doubled beams of the VULCAN laser, which delivered 15 J per beam in 120 ps. The size of the focussing spot for each beam was 0.03 mm x 7 mm, which corresponds to an irradiance of 6. 10^{12} W/cm^2. The X-UV emission of the plasma was detected by two time-resolving spectrometers, one along column axis and the other one in the transverse direction. The transverse spectra exhibited intense resonance lines 2s - 3p and 2p - 3d, but 3d - 5f and 3p - 5d are barely discernible and their transverse intensities could not be measured. The exponential increase of 3d - 4f and 3d - 5f line intensities, given by the on-axis spectrometer at the time of peak gain for each transition, is shown in figure 5. It shows gain coefficients of 2.6 cm^{-1} and 1.6 cm^{-1}, respectively for the 3d - 4f and 3d - 5f transitions. The corresponding GxL values are 3.1 and 1.9.

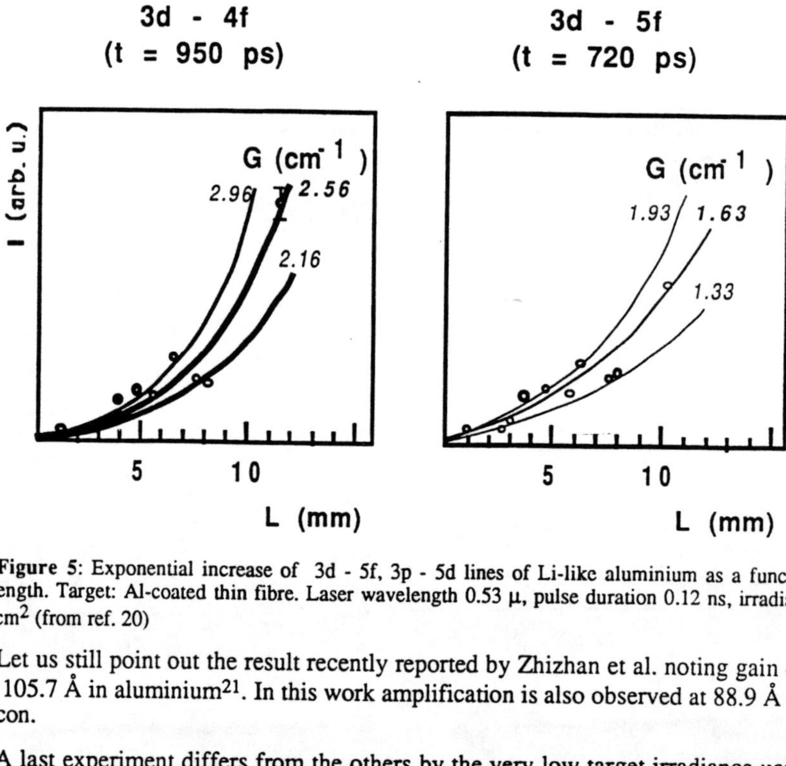

Figure 5: Exponential increase of 3d - 5f, 3p - 5d lines of Li-like aluminium as a function of plasma length. Target: Al-coated thin fibre. Laser wavelength 0.53 μ, pulse duration 0.12 ns, irradiance 5. 10^{13} W/cm^2 (from ref. 20)

Let us still point out the result recently reported by Zhizhan et al. noting gain of 3.1 cm^{-1} at 105.7 Å in aluminium[21]. In this work amplification is also observed at 88.9 Å in Li-like silicon.

A last experiment differs from the others by the very low target irradiance used for producing the plasma (2. 10^{11} W/cm^2) .from a Nd-glass laser (1.06 μ) with a long pulse (5 ns) [22]. A large increase of line intensities against the plasma column length is reported in this work. For the 3d - 4f and 3d - 5f transitions, this increase fits gain coefficients of 4.5 cm^{-1} and 3.4 cm^{-1} what corresponds to GxL products as high as 5.4 and 4.1. This result has given rise to some surprise However, at the Conference Laser's 81 we reported the first evidence of non-linear intensity increasing at $\lambda = 105.7$ Å in aluminium, obtained for an irradiance of 3-5x10^{11} W/cm^2 .and a 20 ns-laser pulse. A gain coefficient between 0.5 $cm^{-1}$and 2 cm^{-1} was deduced from line intensity measurements, this large uncertainty coming from the detector calibration. [23] In a further work (ref. 7 above) we studied the threshold of population inversion emergence in aluminium in using 3.5 ns-pulses[24].. We found it to be near 2x10^{11}W/cm^2, that is the value of Hara's experiment. But, near this threshold, the gain was very weak, (i.e. gain coefficient ≤ 0.01 cm^{-1}). On the other hand, computational simulations never revealed us a possible large gain in similar conditions. Thus although our previous works are in agreement with the production of gain in the condition of Hara's experiment, the large values reported in this work remain to be explained. It seems us unlikely that a 2-D effect in plasma expansion will suffice to clear completely the point.

As a conclusion, disregarding now some differences between experiments - like the fact that gain is either for only one or for both 4f- and 5f- lines - one can deduce from the above results that, at laser irradiance from 10^{12} W/cm^2 to 10^{14} cm^2, plasma lines of length around 1 cm provide gain coefficients between 1 cm^{-1} and 4 cm^{-1}.

III - AMPLIFICATION IN LONG PLASMA COLUMNS

All the results summarized in the previous section make up a first stage of the research about the lithium-like scheme. It is obvious that they are not sufficient to carry out real X-ray lasers. The goal to be reached is to put the gain-length product at a value of 10 or 15, when it is only about 2 or 3 in the previous experiments. This requires new experiments in which the length of the plasma line is significantly increased, either directly or by using multipass devices made with multilayers mirrors.

We recently carried out preliminary experiments with a 6 cm-long plasma column [25]. Five of the six beams of the Nd-glass laser of LULI at Palaiseau were used to produce the plasma. They were located in a plane perpendicular to the column axis on one and the same side of the target. After focusing by cylindrical lenses of focal length 450 mm, the five beams were superposed on a rectangular spot of 0.2 mm x 64 mm. The target was 60 mm wide in order to avoid cold plasma at the ends of the column.The energy of each beam was about 100 J with a pulse duration of 2 ns. So the irradiance was of 2.x10^{12} W/cm^2 on the target surface.

The grazing incidence spectrometer used the Wadsworth circle (with radius equal to the fourth part of the radius of curvature of the concave grating) instead of the classical Rowland circle (half radius of curvature)[26]. This configuration has been chosen because it privileges the analysis of radiation having a strong directivity with respect to isotropic emission. An optical multichannel analyser was used as a detector. An analyser of beam divergence could be inserted between the plasma column and the spectrometer. It enabled to

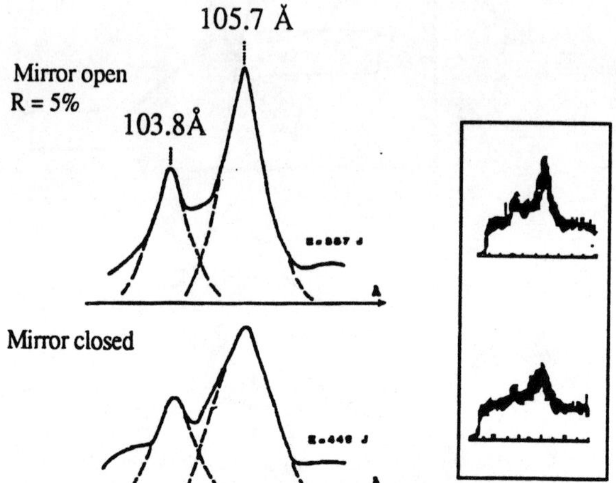

Figure 6: Effect of double-pass in a 6 cm-long plasma column on the intensities of 3p - 5d and 3d - 5f lines of Li-like aluminium. (Top) Mirror open (reflectivity 5 %). (Bottom) Mirror closed. Effective gain-length product: 4. Target: 0.1 μ of Al on a massive polymer slab. Laser wavelength 1.06 μ, pulse duration 2. ns, irradiance 2. 10^{12} W/cm^2 (from ref. 25)

record, in a single shot, several spectra emitted along the column axis and along oblique directions. A multilayers tungsten/carbon mirror, with a 5 % synchrotron-radiation-controlled reflectivity [27] in a narrow band around 105.7 Å, was set perpendicularly to the axis, at the end opposite to the detector. So 3d - 5f amplification was determined by comparison of single- and double-pass intensities.

To reduce the radiative trapping, the target was a 1000 Å layer of aluminium coating the surface of a massive plate made either of a polymer free of aluminium or of graphite. Experiments have shown amplification effectively to occur, the increasing of intensity from single- to double-pass case being of 1.7 (figure 6). This corresponds to a GxL value of 3, i.e. the gain coefficient was not larger than 0.5 cm^{-1}. As regards the "effective" GxL value, which takes the double-pass into account, it was of 4.

Although this value is large enough to conceive an X-ray laser using for instance a six-pass cavity, it is of importance to understand why the gain-coefficient comes to such small values in the case of long plasma columns, a fact that we have observed so far in all attempts we made to increase the plasma length. In the present case, radiation trapping must likely be ruled out because, with thin layer targets, the 2p-3d emission near the target surface cannot be strong. Moreover, the refraction of the amplified beam is negligible because the plasma density in the active region, as well as the density gradient in the direction perpendicular to the axis, are very small having regard to the value of the wavelength of the amplified line. Therefore we must turn to new considerations regarding the production of the plasma column.

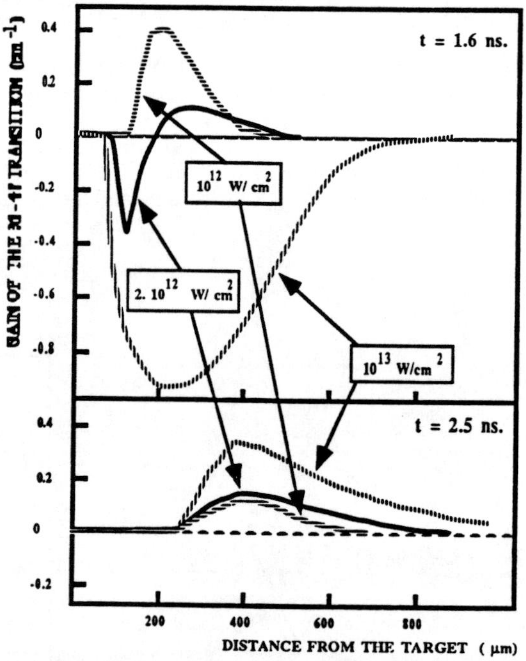

Figure 7: Calculation of gain versus target distance for the 154.7 Å line of Li-like aluminium. Laser wavelength: 1.06 µ, pulse duration: 0.6 ns, irradiance 1., 2 and $10.x10^{12}$ W/cm^2. Curves are plotted at fixed time, i.e. 2. ns (upper curves) or 3.1 ns (lower curves) after laser pulse maximum.

We must keep in mind for instance that, in all gain measurement procedures, one assumes the plasma to be homogeneous in temperature and density along the column axis. Now the distribution of energy may be far to be constant from a point to another through the laser beam section and one is founded to wonder about the effects of inhomogeneities on the gain coefficient. Gain calculations in aluminium illustrated by figure 7 are intended to give a preliminary and yet suggestive response to this question. The figure displays the gain for the 3d - 4f transition as a function of target distance, in a plasma produced by a 600 ps laser pulse, for three different irradiances ($2x10^{12}$, $4.x10^{12}$, and $2x10^{13}$ W/cm^2). Results are given at two fixed times, i.e. 2 ns and 3.1 ns after laser pulse maximum. Let us consider now a plasma column produced by an inhomogeneous beam such as regions of low, middle and high irradiance are distributed along the focal spot. It appears from figure 7 the low irradiance region to give a large amplification at 2 ns, even when the high irradiance region, being still strongly absorbing, reduces or destroys the gain. This region becomes amplifying later and at larger distance from the target. From figure 7 we see this to occur at a time where the gain of the low irradiance region is already well reduced. The middle irradiance region has an intermediate behaviour. It is clear that the plasma column will provide much less amplification than if it was an homogeneous one. The point which is emphasized by this calculation is laser-beam inhomogenities to disynchronize the gain production along the plasma column and consequently to reduce the total gain. It has been often assumed that plasma inhomogeneities are quickly smoothed by electron-ion collisions. Nevertheless very few 2D-simulations are presently available to support this statement especially during plasma recombination.

Now it has been shown laser-beam inhomogeneity to actually produce long-scale plasma irregularities of such a type that they could result in a reduction of gain when plasma length increases [28]. In the case the LULI laser a complete mapping of the plasma is still missing but an image of a small part of the column has also revealed the presence of strong irregularities. This is shown in figure 8. Therefore the long-scale plasma inhomogeneity increasing is a good candidate for being the cause of the gain coefficient decreasing which is observed when the plasma column is lengthened above 1 or 2 cm..

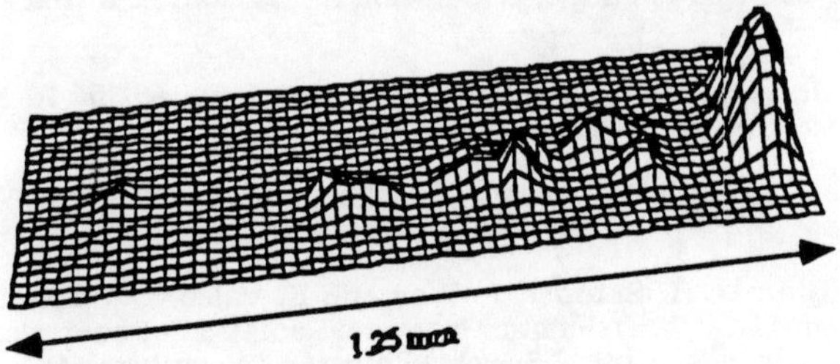

Figure 8: Pinhole image for 500 ev radiation of a small part (1.25 mm) of the plasma line produced in an experiment performed at LULI. Strong irregularities are visible.

IV - CONCLUSION

The recombination scheme in lithiumlike ions has proved to give gain at the wavelengths of 4f and 5f lines. It is an attractive way to an X-ray laser. Gain coefficients are very sensitive to details of experimental techniques which cannot be included in theoretical simulations. It is very likely that making the plasma cooling faster, either by adding heavy elements or by a suitable target design, will increase the gain. However large gain-length products are not yet achieved although attempts have been made with long plasma lines. Target parameters and target composition must be optimalized. A reasonable uniformity (20%-30%) of the plasma along its axis turns out to be a major requirement of future experiments.

This work has been supported by DRET under contract 88-236.

References:

[1]L.I. Gudzenko, L.A. Shelepin, Sov. Physics - Doklady 10 (1965) 147

[2] F.E. Irons and N.J. Peacock, J. Phys. B: Atom Molec Phys 7 (1974) 2084

[3] G.J. Pert, S.A. Ramsden, Optics Comm. 11 (1974) 270

[4] G. Jamelot ,P. Jaeglé, A. Carillon, A. Bideau, C. Möller, ·Proc. of the International Conference on Lasers'81 , C.B. Collins ed., STS Press 1982 p. 178

[5]A. Klisnick, A, Sureau, H. Gennou, C. Moller, J. Virmont, Applied Physics B, **50** (1990), 153

[6] J.C. Gauthier, J.P. Geindre, N. Grandjouan, and J. Virmont, J. Phys. D **16** (1983) 321

[7] A. Carillon, G. Jamelot, A. Klisnick, G. Tévanian, P. Jaeglé, and P. Dhez, Proceedings of the SPIE, Vol. 688, N.M. Ceglio and P Dhez, editors, (1987) p. 12

[8] G. Jamelot, A. Klisnick, A. Carillon, H. Guennou, A. Sureau, and P. Jaeglé, J. Phys. B: Atom. Molec. Physics, **18** (1985) 4647

[9] P. Jaeglé, A. Carillon, A. Klisnick, G. Jamelot, H. Guennou, and A. Sureau, Europhys. Lett. **1** (1986) 555 and P. Jaeglé, G. Jamelot, A. Carillon, A. Klisnick, J. de Physique **47** (1987) C6-31

[10] P. Jaeglé, A. Carillon, A. Klisnick, G. Jamelot, H. Guennou, and A. Sureau, Europhys. Lett. **1** (1986) 555 and P. Jaeglé, G. Jamelot, A. Carillon, A. Klisnick, J. de Physique **47** (1987) C6-31

[11] P. Jaeglé, in Physics of Highly-Ionized Ions, ed. by R. Marrus, NATO ASI Series, Series B: Physics vol. 201 p. 209, Plenum Press1988

[12]S. Suckewer, C.H. Skinner, H. Milchberg, C. Keane, and D. Voorhes, Phys. Rev. Lett. **55** (1985) 1753

[13]D. Kim , C.H. Skinner, A. Wouters, E. Valeo, D. Voorhees, and S Suckewer, OSA Proceedings on Short Wavelength Coherent Radiation: Generation and Applications, R.W. Falcone and J. Kirz eds. (Optical Society of America, Washington DC, 1988), Vol.2, p. 116

[14]Y. Kato, P.R. Herman, T. Tachi, K. Shihoyama, K. Kamei, and H. Shiraga, Springer Proceedings in Physics, Springer-Verlag, Vol. 30, C. Yamanaka editor, 1988, p. 57

[15] S. Suckewer, H. Fishman, J. Appl. Phys. 51 (1980) 1922

[16] see ref. 11

[17] Y. Kato, P.R. Herman, T. Tachi, K. Shihoyama, K. Kamei, and H. Shiraga, Springer Proceedings in Physics, Springer-Verlag, Vol. 30, C. Yamanaka editor, 1988, p. 57

[18] J.C. Moreno, H.R. Griem, S. Goldsmith, and J. Knauer, Phys. Rev. A **39** (1989) 6033

[19] C. Chenais-Popovics, R. Corbett, C.J. Hooker, M.H. Key, G.P. Kiehn, C.L.S. Lewis, G.J. Pert, C. Regan, S.J. Rose, S. Sadaat, R. Smith, T. Tomie, and O. Willi, Phys. Rev. Lett. **59** (1987) 2161

[20] C.L.S. Lewis, R. Corbett, D. O'Neill, C. Regan, S. Sadaat, C. Chenais-Popovics, T. Tomie, J. Edwards, G.P. Kiehn, R. Smith, O. Willi, A. Carillon, H. Guennou, P. Jaeglé, G. Jamelot, A. Klisnick, A. Sureau, M. Grande, C. Hooker, M.H. Key, S.J. Rose, I.N. Ross, P.T. Rumsbv, G.J. Pert, and S.A. Ramsden, Plasma Phys. and Cont. Fus. **30** (1988) 35 and A.

Carillon, J. Edwards, P. Jaeglé, G. Jamelot, M.H. Key, G.P. Kiehn, A. Klisnick, C.L.S. Lewis, D. O'Neill, G.J. Pert, S.A. Ramsden, C. Regan, S.J. Rose, and T.Tomie, J. Phys. B: Atom Molec Physics, **23** (1990) 147

21 Zhizhan Xu, Zhenquan Zhang, Pinzhong Fan, Shisheng Chen, Lihuang Lin, Peixiang Lu, Xiangping Feng, Xiaofeng Wang, Jinzhi Zhou, Aidi Qian, A.P.S Conference 1989, Abstracts **34**, (1989) 2096

22 T. Hara, K. Ando, N. Kusakabe, H. Yashiro, and Y. Aoyagi, Jap. J. of Appl. Physics **28** (1989) L 1010

23 G. Jamelot, P. Jaeglé, A. Carillon, A. Bideau, C. Möller, H. Guennou, and A. Sureau, Proceedings of the International Conference on Laser'81, Carl B. Collins, editor, STS Press, p. 178, 1982

24 see ref. 7.

25 P. Jaeglé, A. Carillon, P. Dhez, B. Gauthé, F. Gadi, G. Jamelot, and A. Klisnick, Europhysics Letters, **7** (1988) 337

26 A. Carillon, P. Dhez, B. Gauthé, P. Jaeglé, G. Jamelot, A. Klisnick, J. C. Lagron, X-Ray Instrumentation in Medecine and Biology, Plasma Physics, Astrophysics, ans Synchrotron Radiation, Proceedings of the SPIE, Vol. 1140, R. Benattar Editor (1989) , p.271

27 M. Kühne , P. Müller, Proceedings of the SPIE, Vol. 1140, R. Benattar Editor (1989) , p.220

28 J.C. Kieffer, M. Chaker, M. Hébert, H. Pépin D.M. Villeneuve, J.E. Bernard, H.A. Baldis, 19th Anomalous Absorption Conference, Fort Lewis College, Durango, Collorado, June 19-23, 1989.

NEW RESONANTLY PHOTO-PUMPED X-RAY LASER SCHEMES

Joseph Nilsen
Lawrence Livermore National Laboratory, Livermore, California 94550

Resonant photo-pumping is one of the methods originally proposed for demonstrating X-ray lasers and still is being actively pursued. However, the shortest wavelength at which significant gain has been measured using resonant photo-pumping is at 2163 Å in Be-like C. This paper presents several new laboratory X-ray laser schemes which show great promise based on resonant photo-pumping of Ne-like and Ni-like ions.

The first scheme lases on several 3p → 3s transitions near 300 Å in neon-like copper(Z = 29) and is resonantly photo-pumped by the $1s2p\ ^3P_1 \rightarrow 1s^2\ ^1S_0$ emission line in helium-like magnesium(Z = 12). The second scheme uses two 3d → 2p emission lines in fluorine-like vanadium(Z = 23) to resonantly photo-pump two 2p → 4d lines in neon-like titanium(Z = 22) with the subsequent lasing on several 3p → 3s transitions in Ne-like Ti. Calculations show that five of the Ti laser lines have gains greater than 5 cm^{-1} for reasonable values of the pump lines, with the shortest wavelength of the five being at 326.5 Å . Experiments are currently being planned to test this scheme. Both Ne-like schemes have the tremendous advantage over other resonantly photo-pumped schemes in that they have higher gain per ion and are less susceptible to radiation trapping effects.

For Ni-like systems, a similar scheme lases on several 4d → 4p transitions ranging from 54 - 89 Å in nickel-like erbium(Z = 68) and is resonantly photo-pumped by the $2p_{1/2} \rightarrow 1s_{1/2}$ emission line, Ly-α, in hydrogen-like aluminum(Z = 13). The Al-Er scheme has the tremendous advantage in that it enhances the gain of laser lines which would be expected to lase from collisional excitations and recombination processes, therefore it should be much easier to observe the photo-pumping process as it will increase an already finite exponential gain. A second scheme lases on several $3p_{3/2} \rightarrow 3s_{1/2}$ transitions near 33 Å in nickel-like dysprosium and is photo-pumped by a $2p_{3/2} \rightarrow 1s_{1/2}$ emission line in hydrogen-like silicon. This laser achieves very short wavelengths by lasing between inner shell transitions. Finally, a new class of laboratory X-ray lasers which lase on several 4d → 4p transitions ranging from 30 - 73 Å in nickel-like ions(Z = 75 - 90) and which are resonantly photo-pumped by Ly-α radiation from hydrogen-like ions(Z = 18 - 23) is discussed. At high Z, there is a fortuitous resonance between the 3d → 6f transition in the Ni-like ions and the corresponding Ly-α lines in the H-like ion with (Z - 21)/3. This scheme is unique in that it consists of a family of resonant pairs. Calculations of the kinetics and gain of these five laser schemes will be presented.

1. INTRODUCTION

Over the last several years there has been tremendous progress in the development of X-ray lasers.[1] However, only two pumping mechanisms have been used to create gain and produce laser output, these have been collisional excitation, in the case of the Ne-like[2-4] and Ni-like[5,6] lasers, and recombination, in the case of the H-like[7,8] and Li-like[9,10] lasers. Other mechanisms such as photoionization[11-15] and resonant photo-pumping[16-31] have been proposed as possible candidates to build an X-ray laser. However, the resonant photo-pumping mechanism has not yet been demonstrated in the soft X-ray region. The shortest wavelength at which significant gain has been measured using resonant photo-pumping is at 2163 Å in Be-like C[32]. This paper presents several new laboratory X-ray laser schemes which show great promise based on resonant photo-pumping of Ne-like and Ni-like ions.

2. NEON-LIKE SCHEMES

The first neon-like scheme[30] lases on several 3p → 3s transitions near 300 Å in neon-like copper(Z = 29) and is resonantly photo-pumped by the $1s2p\ ^3P_1 \rightarrow 1s^2\ ^1S_0$ emission line in helium-like magnesium(Z = 12). This scheme will be analyzed in detail to demonstrate the techniques used to calculate gain in the other neon-like and nickel-like schemes. Experiments are currently being planned to test this scheme.

The second neon-like scheme[31] uses two 3d → 2p emission lines in fluorine-like vanadium(Z = 23) to resonantly photo-pump two 2p → 4d lines in neon-like titanium(Z = 22) with the subsequent lasing on several 3p → 3s transitions in Ne-like Ti. Calculations show that five of the laser lines have gains greater than 5 cm^{-1} for reasonable values of the pump lines, with the shortest wavelength of the five being at 326.5 Å . Experiments are currently being done to test this scheme. Both Ne-like schemes have the tremendous advantage over other resonantly photo-pumped schemes in that they have higher gain per ion and are less susceptible to radiation trapping effects.

2.1. Basic Laser Scheme for Mg-pumped Cu Laser

Figure 1 shows the basic lasing scheme[30] which consists of the He-like Mg intercombination line, $1s2p\ ^3P_1 \rightarrow 1s^2\ ^1S_0$, resonantly photo-pumping an electron in the ground state of the Ne-like Cu ion to the $\overline{2p_{3/2}}\,4d_{5/2}(J = 1)$ level. The bar over the 2p state indicates a hole or vacancy in the closed L shell[$1s^2 2s^2 2p^6$] or neon-like core. The $\overline{2p_{3/2}}\,4d_{5/2}\ (J = 1)$ level decays preferentially to the $\overline{2p_{3/2}}\,3p_{3/2}\ (J = 0)$ and $\overline{2p_{3/2}}\,3p_{3/2}\ (J = 1)$ levels, which are the upper laser states. These upper laser levels lase to the $\overline{2p_{3/2}}\,3s_{1/2}\ (J = 1)$ level, which is the lower laser level. The lower laser level decays back to the Ne-like ground state. Other levels are also fed by the decay of the 4d level and result in additional laser lines. All the laser transitions, along with their wavelengths, are given in Table 1. Figure 2 shows the relevant kinetic rates for the laser line at 235 Å , assuming electron and ion temperatures of 150 eV for the Cu plasma, a Cu ion density of 1.0×10^{18} ions/cc, and a pump strength of 0.0004 photons per mode for the Mg line.

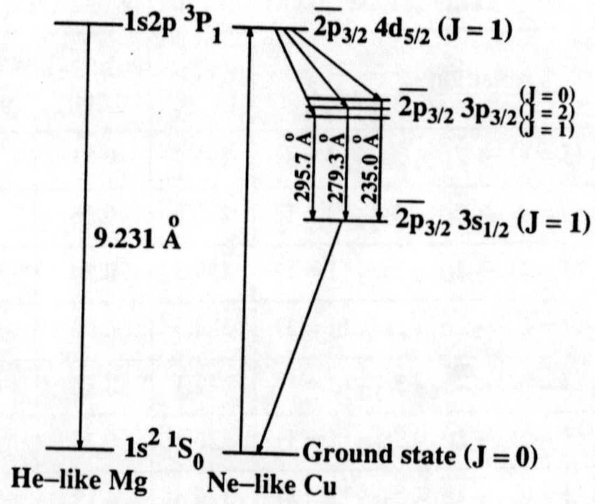

Fig. 1. Mg-pumped Cu X-ray laser.

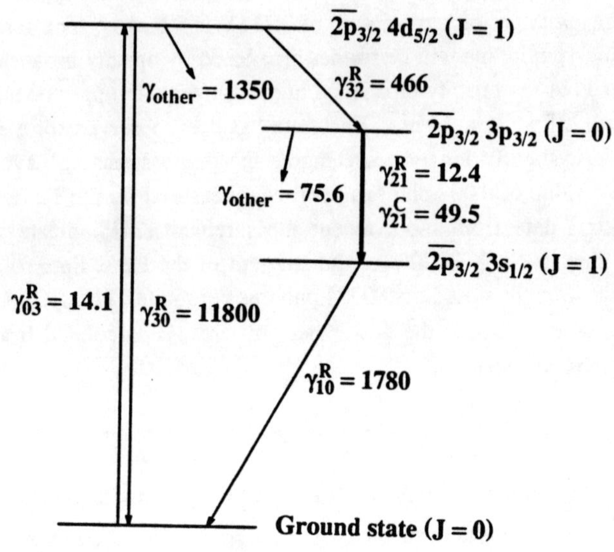

Fig. 2. Kinetic rates($nsec^{-1}$) relevant for the 235 Å Mg-pumped Cu lasing transition with $n_\lambda = 0.0004$.

Table 1. Laser transitions for Ne-like Cu				
Transition	$\lambda(\mathring{A})$	Gain(cm^{-1}) for n_λ =		
		0.0001	0.0004	0.0016
$\overline{2p}_{3/2}\,3p_{3/2}\,(J=1) \rightarrow \overline{2p}_{3/2}\,3s_{1/2}\,(J=1)$	295.7	0.81	2.92	8.37
$\overline{2p}_{3/2}\,3p_{3/2}\,(J=0) \rightarrow \overline{2p}_{3/2}\,3s_{1/2}\,(J=1)$	235.0	0.56	2.00	5.73
$\overline{2p}_{3/2}\,3p_{3/2}\,(J=2) \rightarrow \overline{2p}_{3/2}\,3s_{1/2}\,(J=1)$	279.3	0.24	0.83	2.37
$\overline{2p}_{1/2}\,3p_{3/2}\,(J=2) \rightarrow \overline{2p}_{1/2}\,3s_{1/2}\,(J=1)$	284.7	0.03	0.07	0.20
$\overline{2p}_{1/2}\,3p_{1/2}\,(J=0) \rightarrow \overline{2p}_{1/2}\,3s_{1/2}\,(J=1)$	221.1	0.16	0.54	1.53
$\overline{2p}_{3/2}\,3p_{1/2}\,(J=2) \rightarrow \overline{2p}_{3/2}\,3s_{1/2}\,(J=1)$	330.4	0.35	1.24	3.54
$\overline{2p}_{3/2}\,3p_{3/2}\,(J=0) \rightarrow \overline{2p}_{1/2}\,3s_{1/2}\,(J=1)$	376.8	0.19	0.67	1.92

The He-like Mg $1s2p\ ^3P_1 \rightarrow 1s^2\ ^1S_0$ pump line is measured to be at 9.2310 Å [33-35] while the ground state to the $\overline{2p}_{3/2}\,4d_{5/2}$ (J = 1) transition in Ne-like Cu is measured to be at 9.233 Å [33,36,37]. The FWHM(full width half maximum) Doppler width for the Mg line in a plasma with an ion temperature of 300 eV is 2.5 mÅ. The actual linewidth for an optically thick pump line will be further broadened by opacity broadening. For the Na X He-α line at 11 Å an opacity broadened linewidth, with an upper bound of 35 mÅ, has been measured[22] while a 14 mÅ linewidth has been observed for a Al XIII Ly-β line at 6 Å [26]. At slightly longer wavelengths the pump strengths have been measured[20,25], with a value of 0.002 photons per mode measured for the 11 Å He-α line in Na X. The spectral data from more recent measurements[24] indicate that the He-intercombination line in Na X is 20% of the strength of the He-α line, so a reasonable estimate of the line strength would be 0.0004 photons per mode. The spectral intensity I_λ of the pump line, where $I_\lambda \Delta\lambda$ is the power per unit area, is calculated from the modal density n_λ, in photons per mode, by

$$I_\lambda = 8\pi c \frac{hc}{\lambda^5} n_\lambda = \frac{1.497 \times 10^{18}}{\lambda^5} n_\lambda \frac{\text{Watts}}{\text{cm}^2\ \text{m}\mathring{A}} \tag{1}$$

where λ is the photon wavelength in Å, c is the speed of light, and h is Planck's constant.

The ionization balance issue can easily be answered as He-like low Z plasmas, such as Na and Al, and Ne-like high Z plasmas have been made routinely in the laboratory both in laser produced plasmas and in Z-pinches. In spectroscopic studies of Na, He-like Na[21] lines have been generated in plasmas with the electron temperature ranging from 220 - 520 eV and the electron density 0.2 - 5.8 x 10^{20}/cc. Laser produced Ne-like Cu plasmas have been produced and even made to lase[4] while Z-pinches have produced Ne-like Cu plasmas[13,14] in an effort to create a laser. Another advantage of the Mg-pumped Cu scheme is that the strong Mg Ly-α and He-α lines have sufficient energy to

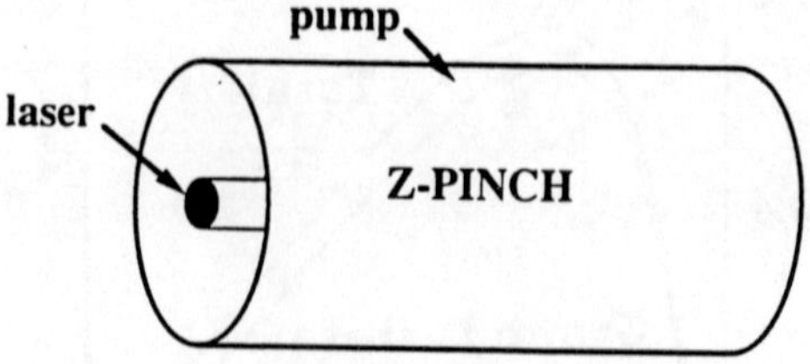

Fig. 3. Laser geometry using a Z-pinch.

photoionize Na-like Cu to Ne-like Cu but they do not have enough energy to photoionize the ground state of Ne-like Cu to F-like Cu. This facilitates making a large population of Ne-like Cu without the fear of overionizing.

To demonstrate an actual laser, one simple experimental arrangement would be to have two adjacent foils of Mg and Cu illuminated with optical lasers of the appropriate intensity and duration to enable the Mg and Cu to reach the proper ionization stages and densities. A more efficient geometry would be the soda straw geometry[13,14] used for the Ne-like Cu in the Z-pinch experiments where a Cu wire is placed in the center of a cylinder and the Mg pump would be deposited on the outside, as shown in Fig. 3. This would put the Cu in the center of the optically thick Mg pump material and allow the pump field to build up to a larger value than would be obtained using surface emission from a foil as the pump.

2.2. Laser Gain Calculation for Mg-pumped Cu

To estimate what gain might be achieved in Ne-like Cu by resonant photo-pumping, an atomic model of Cu was constructed using the atomic physics package YODA[38]. YODA was used to calculate energy levels, oscillator strengths, and collision strengths for the n=2, 3, and 4 Ne-like Cu states. These were combined with simple hydrogenic data for the Rydberg states and the other nearby sequences. Using this atomic model as input, several steady state XRASER calculations[39] were done assuming there was a modest He-like Mg line of 0.0004 photons per mode available to pump the Ne-like Cu ground state to $\overline{2p}_{3/2}\,4d_{5/2}$ (J = 1) transition. The calculations were done with a Cu ion density of 1.0×10^{18} ions/cc and an electron density determined by the ionization of the Cu. The ion temperature was kept equal to the electron temperature. Figure 4 shows the fraction of the Cu ions in all the Ne-like states and the fraction of the Cu ions in the Ne-like ground state versus electron temperature $\theta = kT$. The total Ne-like fraction peaks at 56% at an electron temperature of 200 eV. Across the entire temperature range shown, 86% - 95% of the total Ne-like population is in the ground state. Figure 5 shows the gain for the two strongest laser lines versus electron temperature. The gain has a peak value of 3.47 and 2.35 cm^{-1} at an electron temperature of 200 eV. Under these conditions the Cu resonance

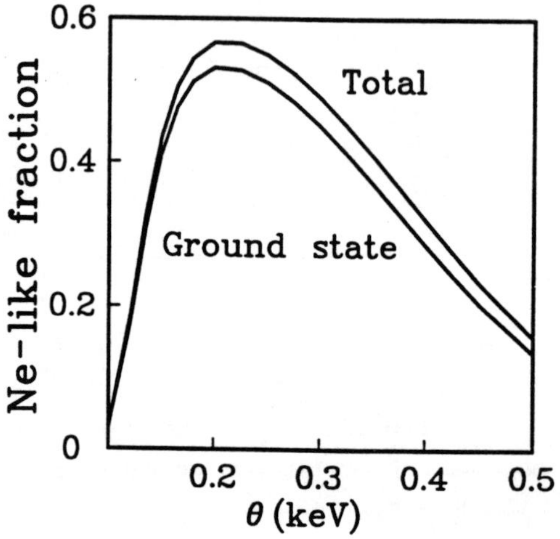

Fig. 4. Fraction of the Cu ions in all the Ne-like states and the fraction of the Cu ions in the Ne-like ground state versus electron temperature θ.

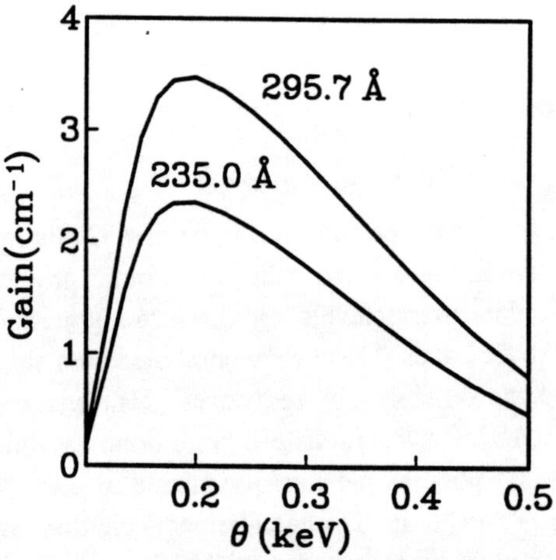

Fig. 5. Gain for the two strongest laser lines versus electron temperature θ with $n_\lambda = 0.0004$.

line being pumped has an absorption coefficient of 115 cm^{-1}, indicating that a 100 μ thick laser would be approximately one mean free path thick to the pump radiation. The Ne-like Cu $\overline{2p}_{3/2}\,3s_{1/2}$ (J = 1) to ground state transition which is responsible for radiation trapping effects which can lower the gain in the 295 Å, 235 Å, and 279 Å laser lines, has an even lower absorption coefficient of 49 cm^{-1}, which means it will be optically thin for a 100 μ thick laser.

In the remainder of this section, electron and ion temperatures of 150 eV are used for the Cu plasma. This is slightly on the cold side of the temperature which yields the optimum gain for the Ne-like Cu laser lines. At this temperature, 44% of the Cu population is in the Ne-like sequence, with 41% or 4.12×10^{17} ions/cc in the Ne-like ground state. Calculations were done at this temperature for several values of the pump line n_λ = 0.0001, 0.0004, and 0.0016. In Table 1 are shown the gains for the seven Ne-like Cu laser lines at these three values of the Mg pump line. The two strongest lines have gains which vary from 0.6 - 8.4 cm^{-1}, depending on the strength of the pump line. In addition, the presence of the other laser lines in the experimental spectrum, some with calculated gains as high as 3.5 cm^{-1}, would be further confirmation of the photo-pumping mechanism. The line at 284 Å, which is observed to lase in laser driven Ne-like Cu plasmas, is relatively uneffected by this pumping mechanism and would not be expected to lase at these low densities and temperatures.

To a good approximation, the gain and populations of the excited states of a particular laser line can be estimated analytically using a four-level model. For example, consider the 235 Å laser line as being part of a four level system, with level 0, the Ne-like ground state; level 1, the $\overline{2p}_{3/2}\,3s_{1/2}$(J=1) lower laser state; level 2, the $\overline{2p}_{3/2}\,3p_{3/2}$(J=0) upper laser state; and level 3, the $\overline{2p}_{3/2}\,4d_{5/2}$(J=1) state. If the only flux feeding the 4d state is photoexcitation from the ground state, then, in steady state, the population of the 4d state, N_3, can be estimated by knowing the population of the Ne-like ground state, N_0, and the strength of the He-like pump line, n_λ, using

$$N_3 = n_\lambda \frac{\gamma_{30}^R}{\gamma_3} \frac{g_3}{g_0} N_0 \qquad (2)$$

where g_i is the degeneracy(2J+1) of level i, γ_{ij}^R is the radiative rate from level i to j, and γ_i is the total destruction rate of level i including all processes such as radiative and collisional processes. Using the rates shown in Fig. 2, $n_\lambda = 4.0 \times 10^{-4}$, and $N_0 = 4.12 \times 10^{17}$/cc, Eq.(2) yields $N_3 = 4.28 \times 10^{14}$/cc, which is very close to the value of 4.31×10^{14}/cc calculated by XRASER.

If the upper laser state is fed only by decay from the 4d state, then its population N_2, can be estimated as

$$N_2 = \frac{\gamma_{32}}{\gamma_2} N_3 \ . \qquad (3)$$

The total rate from level 3 to 2, γ_{32}, is the sum of the collisional rate γ_{32}^C and the radiative rate γ_{32}^{R}. The collision rate is neglected since it is a hundred times smaller than the radiative rate. Inserting the results from Eq.(2) into Eq.(3) yields $N_2 = 1.45 \times 10^{15}$/cc, which is very close to the value of 1.51×10^{15}/cc calculated by XRASER, indicating that this decay route contributes almost all of the flux into the 3p state.

If the lower laser state is fed only by decay from the upper laser state, then its population N_1, can be estimated as

$$N_1 = \frac{\gamma_{21}}{\gamma_1} N_2 \; . \tag{4}$$

The total rate from level 2 to 1, γ_{21}, is the sum of the collisional rate γ_{21}^{C} and the radiative rate γ_{21}^{R}, which is also the spontaneous emission rate. The total destruction rate of level 1, γ_1, is dominated by the radiative decay rate from level 1 to 0, γ_{10}^{R}. Inserting the results from Eq.(3) into Eq.(4) yields $N_1 = 5.04 \times 10^{13}$/cc, which is 23% of the value of 2.18×10^{14}/cc calculated by XRASER. This indicates that other processes contribute a significant amount to the population of the 3s level, which is not surprising since the $\overline{2p}_{3/2}\,3p_{3/2}$ (J) levels with J = 1 and 2, as well as other levels, decay to this 3s state.

The small signal gain is then calculated as

$$g = \left[N_2 - \frac{g_2}{g_1} N_1\right] \frac{\lambda^2}{2\pi} \frac{\gamma_{21}^{R}\,\lambda^2}{2\pi c \Delta\lambda_L} \sqrt{\pi}\, a\, H(a,x) \; . \tag{5}$$

where H(a ,x) is the Voigt function[40]. The Voigt parameter a and detuning x are given by

$$a = \frac{\Delta\lambda_L}{2\,\Delta\lambda_D} \qquad x = \frac{\lambda - \lambda_0}{\Delta\lambda_D} \tag{6}$$

where λ_0 is the line center wavelength, $\Delta\lambda_L$ is the full-width half maximum Lorentzian linewidth, and $\Delta\lambda_D$ is the 1/e Gaussian half width. Using a = 0.182, x = 0, $\Delta\lambda_L = 6.05$ mÅ , H(a,x) = 0.824, and the results from Eqs.(3) and (4) in Eq.(5) yield a gain of 2.01 cm^{-1}, which agrees with the XRASER calculated gain of 2.00 cm^{-1} within 1%. This demonstrates how knowing only a few quantities such as the Ne-like ground state population, the strength of the pump line, the ion temperature to calculate the Doppler width, and a handful of kinetic rates one can accurately estimate the small signal gain.

2.3. V-pumped Ti Laser

In addition to the Ly-α and He-α lines there are strong L-shell lines which can be used as pump lines. In the case of the V-pumped Ti laser, the basic lasing scheme consists of the F-like V 3d → 2p lines resonantly photo-pumping an electron in the ground state of the Ne-like Ti ion to the $\overline{2p}_{3/2}\,4d_{5/2}$(J = 1) and $\overline{2p}_{1/2}\,4d_{3/2}$(J = 1) levels, as shown in Fig. 6. The kinetics are very similar to Ne-like Cu except that both of the strong 4d lines in Ne-like Ti may be pumped which leads to the possibility of many lasing transitions as shown in Table 2. These calculations were done with an electron density of 1.85×10^{18}

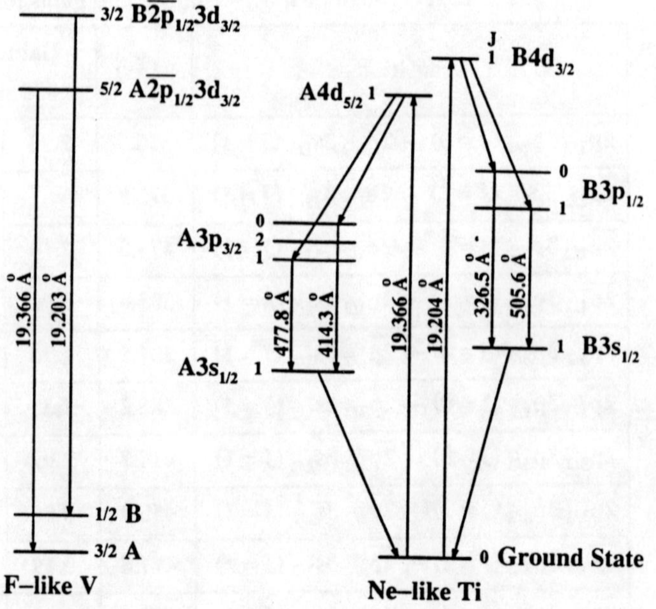

Fig. 6. V-pumped Ti X-ray laser. The notation A represents a $2p_{3/2}$ hole in the neon-like core while B represents a $2p_{1/2}$ hole.

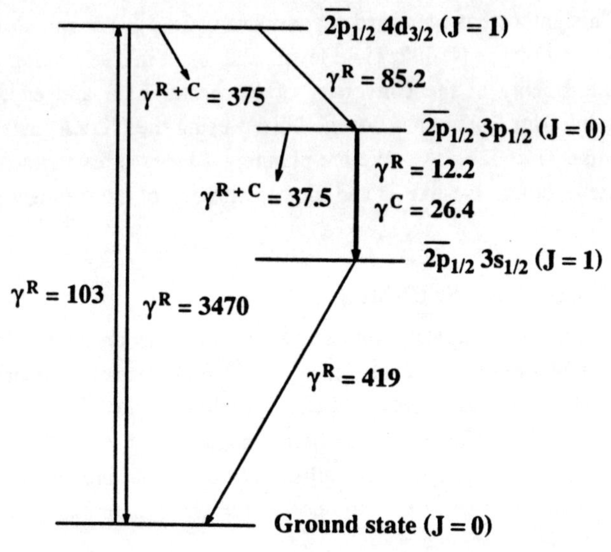

Fig. 7. Kinetic rates($nsec^{-1}$) relevant for the 326.5 Å V-pumped Ti lasing transition with $n_\lambda = 0.01$ for both F-like V lines.

Table 2. Laser transitions, wavelengths, and gains for Ne-like Ti				
Transition	λ(Å)	Gain(cm^{-1}) for n_λ =		
		0.005	0.010	0.020
$\overline{2p}_{1/2}\, 3p_{1/2}\, (J=0) \rightarrow \overline{2p}_{1/2}\, 3s_{1/2}\, (J=1)$	326.5	2.90	6.81	9.91
$\overline{2p}_{1/2}\, 3p_{3/2}\, (J=2) \rightarrow \overline{2p}_{1/2}\, 3s_{1/2}\, (J=1)$	471.9		0.56	2.18
$\overline{2p}_{1/2}\, 3p_{3/2}\, (J=1) \rightarrow \overline{2p}_{1/2}\, 3s_{1/2}\, (J=1)$	473.0	2.10	5.45	8.78
$\overline{2p}_{1/2}\, 3p_{1/2}\, (J=1) \rightarrow \overline{2p}_{1/2}\, 3s_{1/2}\, (J=1)$	505.6	2.90	7.44	11.88
$\overline{2p}_{3/2}\, 3p_{3/2}\, (J=0) \rightarrow \overline{2p}_{1/2}\, 3s_{1/2}\, (J=1)$	507.5	1.26	2.92	4.13
$\overline{2p}_{1/2}\, 3p_{1/2}\, (J=0) \rightarrow \overline{2p}_{3/2}\, 3s_{1/2}\, (J=1)$	285.2	1.35	3.14	4.48
$\overline{2p}_{3/2}\, 3p_{3/2}\, (J=0) \rightarrow \overline{2p}_{3/2}\, 3s_{1/2}\, (J=1)$	414.3	2.99	6.83	9.43
$\overline{2p}_{3/2}\, 3p_{3/2}\, (J=2) \rightarrow \overline{2p}_{3/2}\, 3s_{1/2}\, (J=1)$	458.9	0.56	2.13	4.33
$\overline{2p}_{3/2}\, 3p_{3/2}\, (J=1) \rightarrow \overline{2p}_{3/2}\, 3s_{1/2}\, (J=1)$	477.8	3.19	8.38	13.09
$\overline{2p}_{3/2}\, 3p_{1/2}\, (J=2) \rightarrow \overline{2p}_{3/2}\, 3s_{1/2}\, (J=1)$	507.4	1.30	4.15	7.75

electrons/cc and an electron temperature of 20 eV for the Ti plasma. For the line at 326.5 Å, Fig. 7 shows the kinetics for the lasing transition.

This author has identified several strong 3 → 2 lines in C-like and N-like Ti which may also pump the Ne-like Ti 4d levels and result in a self-pumped Ti laser. Experiments by Tom Boehly at the University of Rochester LLE and experiments at LLNL done jointly by Tom Boehly and Joseph Nilsen using the NOVA laser have seen strong lasing in Ne-like Ti at 326.5 Å. We are planning a series of experiments to understand the lasing mechanism in Ne-like Ti and the contribution of the resonant photo-pumping mechanisms.

3. NICKEL-LIKE SCHEMES

For Ni-like systems, a similar scheme[28] lases on several 4d → 4p transitions ranging from 54 - 89 Å in nickel-like erbium(Z = 68) and is resonantly photo-pumped by the $2p_{1/2} \rightarrow 1s_{1/2}$ emission line, Ly-α, in hydrogen-like aluminum(Z = 13). The Al-Er scheme has the tremendous advantage in that it enhances the gain of laser lines which would be expected to lase from collisional excitations and recombination processes, therefore it should be much easier to observe the photo-pumping process as it will increase an already finite exponential gain.

A second Ni-like scheme[27] lases on several $3p_{3/2} \rightarrow 3s_{1/2}$ transitions near 33 Å in nickel-like dysprosium and is photo-pumped by a $2p_{3/2} \rightarrow 1s_{1/2}$ emission line in hydrogen-like silicon. This laser achieves very short wavelengths by lasing between inner

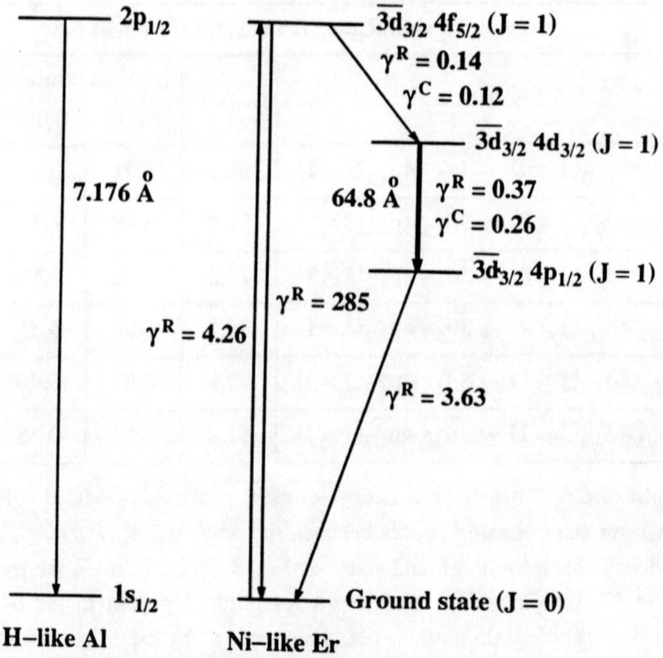

Fig. 8. Kinetic rates(psec^{-1}) relevant for the 64.8 Å Al-pumped Er lasing transition with $n_\lambda = 0.005$.

shell transitions. The resonance for this scheme has only been calculated theoretically, experimental verification still needs to be done. Experimental extrapolation of the Ni-like data for other Z's suggests this resonance may not be adequate.

Finally, a new class of laboratory X-ray lasers which lase on several 4d → 4p transitions ranging from 30 - 73 Å in nickel-like ions(Z = 75 - 90) and which are resonantly photo-pumped by Ly-α radiation from hydrogen-like ions(Z = 18 - 23) is discussed. At high Z, there is a fortuitous resonance between the 3d → 6f transition in the Ni-like ion and the corresponding Ly-α lines in the H-like ion with (Z - 21)/3. This scheme is unique in that it consists of a family of resonant pairs and offers the most promise for a resonantly photo-pumped laser. We will briefly mention the first two schemes and concentrate on the last scheme.

3.1. Al-pumped Er Scheme

The basic resonant photo-pumping scheme, using the 64.8 Å laser line as an example, is shown in Fig. 8. The process consists of the H-like Al $2p_{1/2} \rightarrow 1s_{1/2}$ emission line resonantly photo-pumping an electron in the ground state to the $\overline{3d}_{3/2}\,4f_{5/2}(J = 1)$ level of Ni-like Er. This transition in Ni-like Er is the strongest transition with an oscillator

Table 3. Laser transitions for Ni-like Er

Transition	λ(Å)	Gain(cm^{-1}) for n_λ =			
		0.000	0.002	0.005	0.010
$\overline{3d}_{3/2}\,4d_{3/2}(J=0) \rightarrow \overline{3d}_{3/2}\,4p_{1/2}(J=1)$	54.3	3.04	3.57	4.32	5.44
$\overline{3d}_{3/2}\,4d_{3/2}(J=0) \rightarrow \overline{3d}_{5/2}\,4p_{3/2}(J=1)$	59.9	3.45	4.15	5.12	6.58
$\overline{3d}_{3/2}\,4d_{3/2}(J=0) \rightarrow \overline{3d}_{3/2}\,4p_{3/2}(J=1)$	75.3	0.71	0.86	1.07	1.39
$\overline{3d}_{5/2}\,4d_{5/2}(J=2) \rightarrow \overline{3d}_{5/2}\,4p_{3/2}(J=1)$	86.0	6.50	6.59	6.72	6.91
$\overline{3d}_{5/2}\,4d_{5/2}(J=1) \rightarrow \overline{3d}_{5/2}\,4p_{3/2}(J=1)$	89.6	4.62	5.40	6.49	8.12
$\overline{3d}_{3/2}\,4d_{3/2}(J=1) \rightarrow \overline{3d}_{3/2}\,4p_{1/2}(J=1)$	64.8	*	0.75	3.41	7.38

strength of 6.6. This 4f level decays directly to the $\overline{3d}_{3/2}\,4d_{3/2}(J=1)$ level, which is the upper laser state. Lasing occurs between the $4d_{3/2}$ and $4p_{1/2}$ states. The lower laser state then decays back to the ground state. In Fig. 8 are also shown the most important kinetic rates in Ni-like Er which are relevant for producing gain in the 64.8 Å laser line. A pump strength of 0.005 photons per mode, defined by Eq. (1), is assumed for the Al pump line. In Ni-like Eu five laser lines have been observed due to the collisional excitation and recombination processes. In Ni-like Er one would expect to see these five lines in the absence of the Al pump. With the Al pump present four of these lines should be significantly enhanced and a new laser line at 64.8 Å should also appear. The six laser lines, along with the calculated wavelengths of the laser transitions, are given in Table 3. The calculations were done with an Er ion density of 1.0 x 10^{19} ions/cc and an electron density determined by the ionization of the Er. The electron temperature was fixed at 700 eV while an ion temperature of 450 eV was used. Under these conditions the fraction of ions in the Ni-like ground state is 21.6 - 24.4% depending on the strength of the Al pump line.

3.2. Si-pumped Dy Laser

Using the laser line at 32.8 Å as an example, the process consists of H-like Si $2p_{3/2} \rightarrow 1s_{1/2}$ emission line resonantly photo-pumping an electron in the ground state of the Ni-like Dy ion to the $\overline{3s}_{1/2}\,4p_{3/2}(J=1)$ level, as shown in Fig. 9. This $\overline{3s}_{1/2}\,4p_{3/2}(J=1)$ level is the upper laser state which decays directly to the $\overline{3p}_{3/2}\,4p_{3/2}(J=1)$ level, which is the lower laser state. Unlike the collisionally pumped Ni-like lasers which lase between 4d and 4p electrons, in this scheme lasing occurs between the $\overline{3s}_{1/2}$ and $\overline{3p}_{3/2}$ electron holes. Therefore lasing can be thought of as taking place between holes in a closed M shell. This is equivalent to lasing from $3p_{3/2} \rightarrow 3s_{1/2}$. Even though this is a one electron process, it is between states with many equivalent electrons. The lower laser state then decays to several other singly excited states, principally the $\overline{3d}$ 4p levels.

Table 4. Laser transitions for Ni-like Dy				
Transition	λ(Å)	Gain(cm^{-1}) for n_λ =		
		0.002	0.005	0.008
$\overline{3s}_{1/2}\,4p_{3/2}(J=1) \rightarrow \overline{3p}_{3/2}\,4p_{3/2}(J=1)$	32.8	1.07	2.74	4.01
$\overline{3s}_{1/2}\,4p_{3/2}(J=1) \rightarrow \overline{3p}_{3/2}\,4p_{3/2}(J=2)$	33.2	1.21	3.22	4.75
$\overline{3s}_{1/2}\,4p_{3/2}(J=1) \rightarrow \overline{3p}_{3/2}\,4p_{3/2}(J=0)$	34.7		0.62	1.26
$\overline{3p}_{3/2}\,4p_{3/2}(J=1) \rightarrow \overline{3p}_{3/2}\,4s_{1/2}(J=1)$	94.1	2.40	4.57	6.21
$\overline{3p}_{3/2}\,4p_{3/2}(J=2) \rightarrow \overline{3p}_{3/2}\,4s_{1/2}(J=1)$	91.1	2.00	3.29	4.26
$\overline{3p}_{3/2}\,4p_{3/2}(J=0) \rightarrow \overline{3p}_{3/2}\,4s_{1/2}(J=1)$	81.4	2.86	3.70	4.34
$\overline{3s}_{1/2}\,4p_{3/2}(J=1) \rightarrow \overline{3p}_{1/2}\,4p_{3/2}(J=2)$	61.6	1.57	3.46	4.90
$\overline{3s}_{1/2}\,4p_{3/2}(J=1) \rightarrow \overline{3s}_{1/2}\,4s_{1/2}(J=1)$	90.9	2.21	4.82	6.80
$\overline{3s}_{1/2}\,4p_{3/2}(J=1) \rightarrow \overline{3s}_{1/2}\,4s_{1/2}(J=0)$	97.2	2.49	6.74	9.96

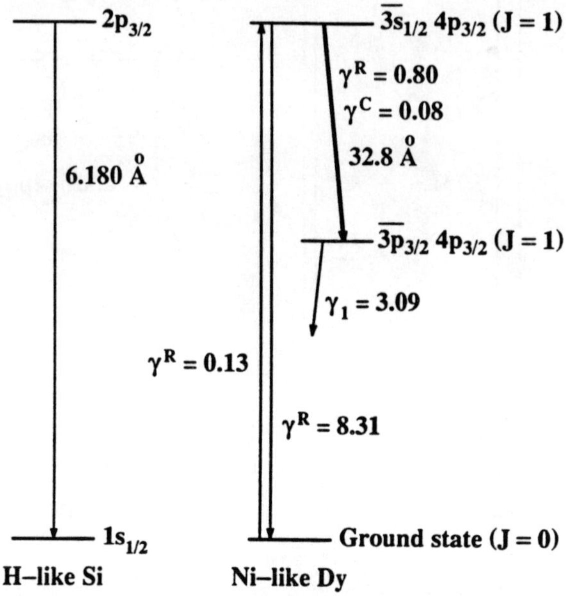

Fig. 9. Kinetic rates(psec^{-1}) relevant for the 32.8 Å Si-pumped Dy lasing transition with n_λ = 0.005.

Assuming an electron temperature of 650 eV and a Dy ion density of 1.0×10^{19} ions/cc, Table 4 shows the gains for several values of the pump strength.

3.3. New Class of Ni-like Lasers

The basic resonant photo-pumping scheme, using the 53.7 Å laser line as an example, is shown in Fig. 10. The process consists of the H-like Ar Ly-α emission lines resonantly photo-pumping an electron in the ground state to the $\overline{3d}_{5/2}\,6f_{7/2}(J = 1)$ level of Ni-like Re. This transition in Ni-like Re is the strongest 3 - 6 transition with an oscillator strength of 0.5. This 6f level decays directly to the $\overline{3d}_{5/2}\,4d_{5/2}(J = 0)$ level, which is the upper laser state. Lasing occurs between the 4d and 4p states. The lower laser state then decays back to the ground state. In Fig. 10 are also shown the most important kinetic rates in Ni-like Re which are relevant for producing gain in the 53.7 Å laser line. A pump strength of 0.001 photons per mode is assumed for the Ar pump line. The calculations were done with an Re ion density of 1.0×10^{19} ions/cc and an electron density determined by the ionization of the Re. The electron temperature was fixed at 700 eV while an ion temperature of 450 eV was used and the Ni-like fraction was set to 33%.

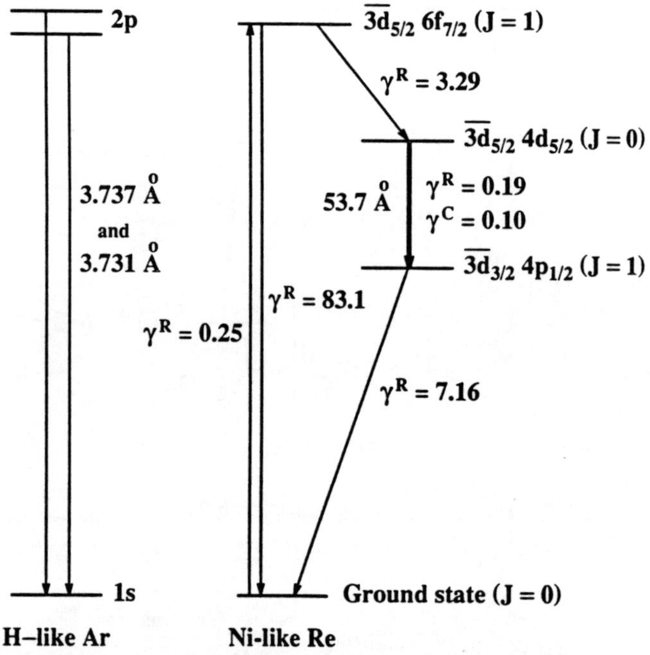

Fig. 10. Kinetic rates(psec^{-1}) relevant for the 53.7 Å Ar-pumped Re lasing transition with $n_\lambda = 0.001$.

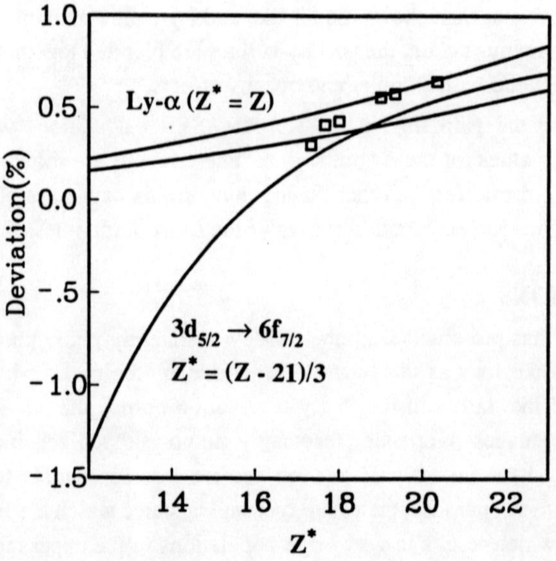

Fig. 11. Deviation of the Ly-α and Ni-like $3d_{5/2} \rightarrow 6f_{7/2}$ lines from 0.75 $(Z^*)^2$ Ry versus Z^*.

To understand the resonance, Fig. 11 plots the percent deviation of the Ly-α and Ni-like ground state(J = 0) → $\overline{3d}_{5/2}\,6f_{7/2}$(J = 1) lines from $0.75(Z^*)^2$ Ry where $Z^* = Z$ for the Ly-α lines and $Z^* = (Z\text{-}21)/3$ for the Ni-like lines. The solid curves are theoretical values but theory and experiment agree very well for the Ly-α lines, however there is a small offset between theory and experiment, shown by the boxes, for the Ni-like lines. Looking at the theoretical data, at $Z^* = 18$ the Ni-like line approaches closely to the lower Ly-α line, crosses it by $Z^* = 19$, and then tracks the Ly-α lines up to $Z^* = 23$. The boxes show the experimental data[41] which suggests a crossing

Table 5. Laser transitions for Ni-like Re

Transition	λ(Å)	Gain(cm^{-1}) for n_λ = 0.0005	0.0010	0.0020
$\overline{3d}_{5/2}\,4d_{5/2}$ (J = 0) → $\overline{3d}_{3/2}\,4p_{1/2}$ (J = 1)	53.7	2.0	3.4	5.2
$\overline{3d}_{5/2}\,4d_{5/2}$ (J = 0) → $\overline{3d}_{5/2}\,4p_{3/2}$ (J = 1)	64.7	4.6	7.8	11.9
$\overline{3d}_{5/2}\,4d_{5/2}$ (J = 2) → $\overline{3d}_{5/2}\,4p_{3/2}$ (J = 1)	70.4	2.0	3.4	5.2
$\overline{3d}_{5/2}\,4d_{5/2}$ (J = 1) → $\overline{3d}_{5/2}\,4p_{3/2}$ (J = 1)	73.3	9.0	15.1	23.0

about one Z lower and then shows the Ni-like and Ly-α line tracking almost exactly from $Z^* = 19 - 23$. In many spectra, the two Ly-α lines are blended together because of the line broadening associated with Doppler and opacity effects.

To estimate the gain for Ni-like Re, XRASER calculations were done assuming several different values for the Ar pump line. These results are shown in Table 5. Similar results should hold true for the other Ni-like ions in this new class of laser. This scheme has very large gains for very modest values of the Ly-α pump lines.

4. CONCLUSIONS

This paper has presented a number of new resonantly photo-pumped schemes using Ne-like and Ni-like ions as the lasants. In particular, the two Ne-like schemes and the new class of Ni-like lasers in which Ly-α radiation pumps the 3d → 6f transition offer tremendous advantages over other resonantly photo-pumped schemes suggested in the past in that they have higher gain per ion and are less susceptible to radiation trapping effects because the upper laser states are metastable states which are feed by strong radiative transitions which results in very large populations in the upper laser state. The lower laser state is fairly empty because it is a strongly allowed $\Delta n = 1$ transition with a spontaneous emission rate generally two orders of magnitude larger than the spontaneous emission rate for the $\Delta n = 0$ lasing transition. These three schemes depend completely on radiative processes and therefore can be fielded at lower electron density than is used for the collisionally pumped schemes currently fielded. This offers the additional advantage of lower index of refraction gradients which reduce the beam bending problems inherent in the higher density plasmas.

ACKNOWLEDGEMENTS

This work was performed under the auspices of the U. S. Department of Energy by the Lawrence Livermore National Laboratory under contract No. W-7405-ENG-48.

REFERENCES

[1] B. J. MacGowan *et al.*, "Progress Towards a 44-Å X-Ray Laser," *OSA Proceedings on Short Wavelength Coherent Radiation: Generation and Applications*, edited by Roger W. Falcone and Janos Kirz (Optical Society of America, Washington, DC, 1988), Vol. 2, pp. 2-10.

[2] D. L. Matthews *et al.*, Phys. Rev. Lett. **54**, 110 (1985).

[3] M. D. Rosen *et al.*, Phys. Rev. Lett. **54**, 106 (1985).

[4] T. N. Lee, E. A. McLean, and R. C. Elton, Phys. Rev. Lett. **59**, 1185 (1987).

[5] B. J. MacGowan *et al.*, Phys. Rev. Lett. **59**, 2157 (1987).

[6] B. J. MacGowan *et al.*, J. Opt. Soc. Am. **B 5**, 1858 (1988).

[7] S. Suckewer *et al.*, Phys. Rev. Lett. **55**, 1753 (1985).

[8] C. Chenais-Popovics *et al.*, Phys. Rev. Lett. **59**, 2161 (1987).

[9] P. Jaegle *et al.*, J. Opt. Soc. Am. **B 4**, 563 (1987).

[10] D. Kim *et al.*, J. Opt. Soc. Am. **B 6**, 115 (1989).

[11] M. A. Duguay and P. M. Rentzepis, Appl. Phys. Lett. **10**, 350 (1967).

[12] E. E. Fill, J. Phys.(Paris) Colloq. **47**, C6-117 (1986).

[13] E. J. McGuire *et al.*, J. Phys.(Paris) Colloq. **47**, C6-81 (1986).

[14] M. K. Matzen *et al.*, J. Phys.(Paris) Colloq. **47**, C6-135 (1986).

[15] P. D. Morley and J. Sugar, Phys. Rev. **A 38**, 3139 (1988).

[16] A. V. Vinogradov, I. I. Sobelman, and E. A. Yukov, Sov. J. Quant. Electron. **5**, 59 (1975).

[17] B. A. Norton and N. J. Peacock, J. Phys. **B 8**, 989 (1975).

[18] V. A. Bhagavatula, J. Appl. Phys. **47**, 4535 (1976).

[19] W. E. Alley, G. Chapline, P. Kunasz, and J. C. Weisheit, J. Quant. Spectrosc. Radiat. Transfer **27**, 257 (1982).

[20] J. P. Apruzese, J. Davis, and K. G. Whitney, J. Appl. Phys. **53**, 4020 (1982).

[21] J. P. Apruzese and J. Davis, Phys. Rev. **A 31**, 2976 (1985).

[22] J. P. Apruzese *et al.*, Phys. Rev. **A 35**, 4896 (1987).

[23] F. C. Young *et al.*, Appl. Phys. Lett. **50**, 1053 (1987).

[24] S. J. Stephanakis *et al.*, IEEE Trans. Plasma Sci. **16**, 472 (1988).

[25] R. C. Elton, T. N. Lee, and W. A. Molander, Phys. Rev. **A 33**, 2817 (1986).

[26] P. Monier, C. Chenais-Popovics, J. P. Geindre, and J. C. Gauthier, Phys. Rev. **A 38**, 2508 (1988).

[27] Joseph Nilsen, Opt. Comm. **72**, 371 (1989).

[28] Joseph Nilsen, Phys. Rev. **A 40**, 5440 (1989).

[29] Boris N. Chichkov and Ernst E. Fill, Opt. Comm. **74**, 202 (1989).

[30] Joseph Nilsen, "A Ne-like Cu laser resonantly photo-pumped by a He-like Mg line," UCRL-102608, Lawrence Livermore National Laboratory, Livermore, CA 94550 (1990). To be published in Optics Communications.

[31] Joseph Nilsen, "A Vanadium-pumped Titanium X-ray Laser," UCRL-102967, Lawrence Livermore National Laboratory, Livermore, CA 94550 (1990). To be published in Opt. Letts. **15** (July 15, 1990).

[32] N. Qi and M. Krishnan, Phys. Rev. Lett. **59**, 2051 (1987).

[33] Raymond L. Kelly, J. Phys. Chem. Ref. Data. **16**, Suppl. 1 (1987).

[34] N. J. Peacock, M. G. Hobby, and M. Galanti, J. Phys. **B 6**, L298 (1973).

[35] L. N. Ivanov, E. P. Ivanova, and U. I. Safronova, J. Quant. Spectrosc. Radiat. Transfer **15**, 553 (1975).

[36] V. A. Boiko, A. Ya. Faenov, and S. A. Pikuz, J. Quant. Spectrosc. Radiat. Transfer **19**, 11 (1978).

[37] H. Gordon, M. G. Hobby, and N. J. Peacock, J. Phys. **B 13**, 1985 (1980).

[38] P. L. Hagelstein and R. K. Jung, Atomic Data and Nuclear Data Tables **37**, 121 (1987).

[39] Joseph Nilsen, "Radiative-Hydro Modeling and Atomic Data Bases," *AIP Conference Proceedings 168 - Atomic Processes in Plasmas*, edited by Allan Hauer and A. L. Merts (American Institute of Physics, New York, 1988), pp. 51- 58

[40] G. D. Finn and D. Mugglestone, Monthly Notices Roy. Astronom. Soc. **129**, 221 (1965).

[41] N. Tragin *et al.*, Physica Scripta **37**, 72 (1988).

OBSERVATION OF GAIN AT 182Å IN WALL CONFINED CO2 LASER PRODUCED PLASMAS·

E. Miura, Y. Kitagawa, K. Sawai, K. Matsuo, H. Daido, K. Nishihara, Y. Kato, and S. Nakai
Institute of Laser Engineering, Osaka University, 2-6 Yamada-oka, Suita, Osaka, 565, Japan

A parylene(C_8H_8) hollow cylinder-type target was irradiated with a long pulse CO_2 laser (400J/50ns) to produce a confined soft x-ray laser gain medium. The amplification of Balmer-alpha line of hydrogen-like carbon at 182Å was observed. The time- and space-integrated gain coefficient of $1.5cm^{-1}$ was obtained from the axial to transverse intensity ratio. We also present the numerical simulation using the atomic physics code coupled with the one-dimensional hydrodynamic code to compare the experimental results. Based on the simulation, we propose the improved target which gives the long gain duration up to 20ns in FWHM and the large gain region up to 1.3mm in radius. These features offers us the advantages of constructing a soft X-ray laser cavity.

1. INTRODUCTION

Two successful approaches to soft x-ray lasers using laser produced plasmas are based on electron collisional processes from closed shell ions and recombination processes into highly stripped ions.[1] As for the recombination pumping scheme, amplified spontaneous emissions have been observed mainly in freely expanding plasmas [2-5], while a high gain coefficient has been achieved in a magnetically confined plasma column.[6] In freely expanding plasmas, although the rapid cooling due to the adiabatic expansion gives a high peak gain coefficient, the rapid decrease in the electron density gives the temporally and spatially limited gain coefficient. To produce the gain medium which has a long gain duration and large gain region is favorable to construct a soft x-ray laser cavity. To achieve the long gain duration, we should confine the gain medium plasma. If the plasma is able to be cooled rapidly without the rapid decrease in the electron density, the high gain coefficient is also expected. So we proposed a cylinder-type target to produce the confined gain medium using the recombination pumping scheme.[7] In this paper, we report the gain observation on the Balmer-alpha(Hα) line of the hydrogen-like carbon (CVI) using the cylinder-type target.

2. EXPERIMENT

The target structure and experimental arrangement are shown in Fig.1. A 30μm thick hollow parylene(C_8H_8) cylinder is 10mm in length and 3mm in diameter, having a 0.2μm thick parylene thin foil at the edge. A 0.5mm wide and 0.9mm long window is put on the wall to observe the transverse X-ray emissions. The long pulse CO_2 laser(400J/50ns) is focused on the thin foil, giving the peak intensity of $3X10^{12}W/cm^2$. The expected gain creation processes are described as follows. The plasma expands from both the thin foil and wall, producing the plasma column inside the cylinder. Without the rapid decrease in the electron density, the plasma column, which is confined by the wall, may be cooled down by the radiation and the heat transport to the wall after the laser irradiation,leading to the long gain duration.

To measure the gain coefficient of the CVI Hα line(182Å), the time- and space- integrated soft x-ray spectra from the axial and transverse directions were recorded on KODAK 101-07 films using two flat field type grazing incidence spectrometers with 0.3Å resolution. The characteristic curve of this film was determined with a laser plasma X-ray source.[2,8] The gain coefficient is estimated from the line intensity ratio of the axial and transverse directions.

The gain determination procedures are described as follows. The relative sensitivity ratio of the two spectrometers is measured using the rear side emission from the laser irradiated tilted thin foil target to give an identical source to the both spectrometers.[2] Since the transverse emission is observed through the view window, the observed volume ratio needs to be calibrated. It is assumed that the gain length is equal to the cylinder length. The gain coefficient is yielded from the following enhancement relationships:

$$E = [\exp(gL)-1]/gL, \qquad (1)$$

and

$$E = [I_{ax}(182)/I_{tr}(182)]/(R_s R_v). \qquad (2)$$

E, g, and L are enhancement factor, gain coefficient, and gain length, respectively. $I_{ax}(\lambda)$ and $I_{tr}(\lambda)$ indicate the line intensity of the wavelength λ of the axial and transverse direction. R_s is the relative sensitivity ratio of two spectrometers. Using the above method, it was determined that R_s was 0.23±0.08. R_v is the observed volume ratio of 5.2, which is determined from the target geometry. We can also estimate the gain coefficient using the reference line for the relative sensitivity calibration.[6,7] The CVI Hγ line(120Å) is chosen as a reference line. The enhancement factor is shown as follows:

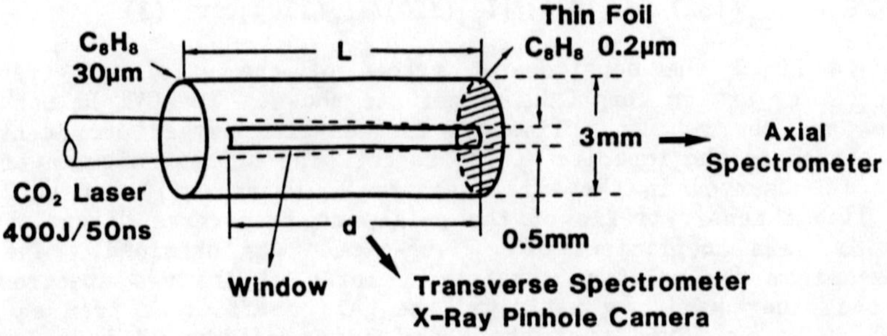

Fig.1 Target structure of cylinder-type target and experimental arrangement. The cylinder length L and the window length d are 1cm and 0.9cm. The window width is 0.5mm. The spectrum is observed from the axial and transverse directions with two spectrometers.

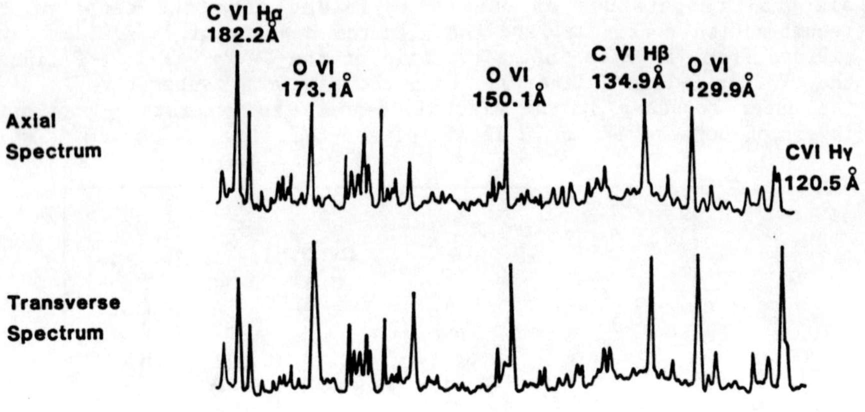

Fig.2 Axial and Transverse spectra in the region near the CVI Hα line(182Å). The strong emission of CVI Hα line is seen in the axial spectrum.

$$E = [I_{ax}(182)/I_{tr}(182)]/[I_{ax}(120)/I_{tr}(120)]. \quad (3)$$

In Fig.2 the densitometer traces of the axial and transverse spectra for an 1cm long C_8H_8 target are shown. The CVI Balmer series lines and the resonance lines of lithium-like oxygen are identified. The oxygen is the impurity of the target. The strong emission of CVI Hα line is observed in the axial spectrum. Using eq.(1) and eq.(2) from the line intensity ratio of the axial and transverse directions, the CVI Hα gain coefficient of $1.6^{+0.5}_{-0.7}cm^{-1}$ was obtained. The error arises from the relative sensitivity ratio of the two spectrometers. On the other hand, we estimated the gain coefficient from eq.(1) and eq.(3). It is noted that the third order of the CV $1s2p$-$1s^2$ line overlaps on the CVI Hγ line. Therefore we estimate the intensity of CVI Hγ line subtracting the overlapped component.[2] The gain coefficient was $1.4\pm0.2cm^{-1}$. The error arises from subtracting the overlap of the CV $1s2p$-$1s^2$ component. Since the two values, which are yielded from the different method, are consistent with each other, we obtain the definite gain coefficient of the CVI Hα line as $1.5^{+0.5}_{-0.8}cm^{-1}$.

The recombination continua and the dielectronic satellite lines are identified in the transverse spectrum of Fig.3, which was recorded using the grazing incidence spectrometer with 0.15Å resolution. We measured the electron temperature and density from this spectrum. The electron temperature of 54eV was yielded from the slope of the recombination continuum, and the electron density of $1.2X10^{19}cm^{-3}$ was yielded from the line intensity ratio of the CV $2p^2\ ^3P$-$1s2p\ ^3P$ line to the CV $2s2p\ ^3P$-$1s2s\ ^3S$ line.[9] When the electron temperature is 54eV, the upper boundary of the electron density to generate a population inversion between n=3 and 2 levels of the CVI is $3.2X10^{19}cm^{-3}$, where

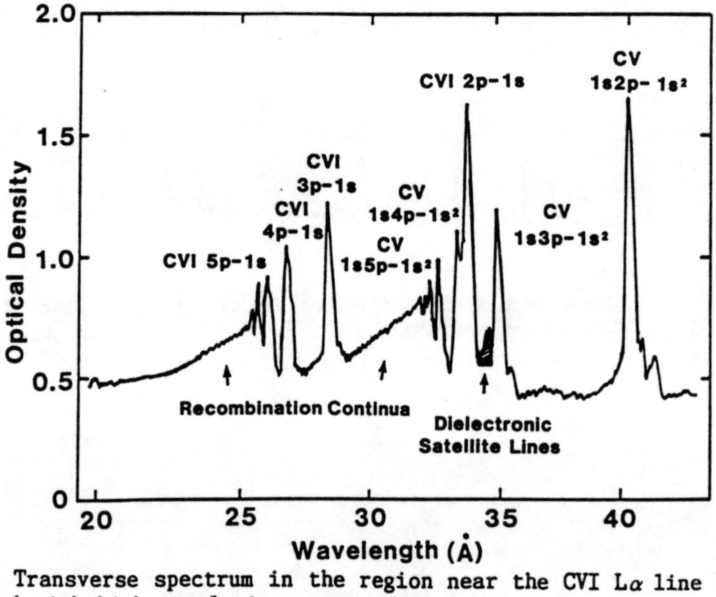

Fig.3 Transverse spectrum in the region near the CVI Lα line (33.8Å) recorded with high resolution spectrometer.

the radiative transition rate is comparable to the collisional de-excitation rate for the lasing transition. The measured electron density was satisfied with the this condition. However, since the measured electron temperature is higher than the suitable value to produce a large population inversion, the additional cooling such as a radiative cooling is necessary.

3. COMPUTER SIMULATION

We developed a simulation code using the atomic physics code coupled with one-dimensional hydrodynamic code. The details of this code have been given in our previous publication.[10] Figure 4(a) shows the flow diagram of the parylene target, which is calculated with the one-dimensional hydrodynamic code 'HISHO'.[11] The cylinder center is at 0mm and the initial position of the wall is at 1.5mm. The wall thickness is 30μm and the density is $1.1g/cm^3$. The laser energy is 400J/cm and the absorption fraction is 0.5. The laser pulse is Gaussian of 50ns duration in FWHM. It is assumed the laser heating stops at 60ns due to the closure by the critical density plasma. The rate equation of the carbon ions is calculated based on the collisional- radiative model. The opacity effect of the Lyman-alpha(Lα) line seems to be crucial to produce a population inversion between n=3 and 2 levels. However, the simulation results which include the opacity effect of the Lα line disagreed with the experimental results[12]. So, we neglect the opacity effect of the Lα line.

The spatial profile and time development of the CVI Hα gain coefficient for are shown in Fig.4(b). The gain occurs within 0.5mm radius of the cylinder after the laser heating stops at 60ns. The peak value is $0.15cm^{-1}$ and the duration is 3ns in FWHM. The time developments of the gain coefficient and the plasma parameters at the cylinder center are shown in Fig.4(c). The peak gain coefficient is very low. As described in Ref[10], in the case of Fig.4(c) the cooling time is too long to obtain the gain coefficient of $5cm^{-1}$. As shown in Fig.4(a) the hot plasma of the central region is gradually compressed after the laser heating stops. Therefore the ion and electron densities gradually increase in the recombination phase due to the adiabatic compression, resulting in the increase in the temperature. When the gain coefficient becomes peak, the electron density and temperature are $3X10^{19}cm^{-3}$ and 55eV, respectively. When the electron temperature is 55eV, the boundary electron density to produce a population inversion between n=3 and 2 levels of the CVI is $3.2X10^{19}cm^{-3}$ as discussed in section 2. The calculated value is nearly equal to this boundary value. Consequently, the increase in the electron density also limits the peak value and duration of the gain coefficent.

The measured gain coefficient is higher than the simulation result by a factor of 10. As shown in Fig.2, the resonance line of Li-like oxygen is identified. If the target material is $C_8H_6O_2$ to take into account of the oxygen impurity, the predicted gain coefficient is 0.5 cm^{-1}, which is closer to the experimental results. Since the dominant cooling mechanism is the radiative cooling as described in Ref[10], the oxygen impurity as a coolant seems to affect

(a)

(b)

(c)

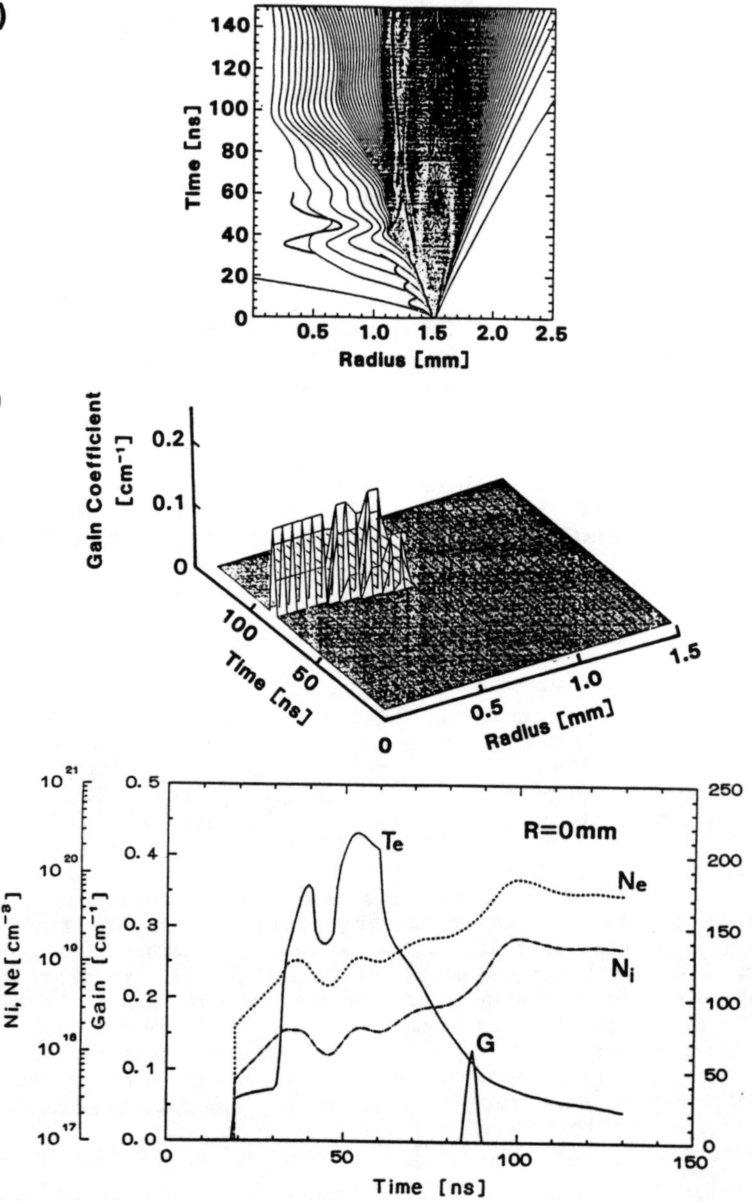

Fig.4 (a) Flow diagram in the cylinder. The dotted line is the critical density point. (b) Spatial profile of the CVI H line gain coefficient. (c) Time developments of the gain coefficient and the plasma parameters at the cylinder center. The parylene cylinder radius is 1.5mm. The wall thickness is 30um and the density is $1.1 g/cm^3$.

the increase in the gain coefficient. Further, the particle loss from the cylinder edge and view window also prevents the electron density exceeding the optimum value. As described above, the experimental results indicate the higher value than the simulation results.

The two problems discussed above should be solved to achieve the gain coefficient of $5cm^{-1}$. One is that the cooling time is too long. However, to achieve the rapid cooling of the plasma, it is effective to use the high Z impurity as a coolant, which has been described in Ref[10]. However, since the confined radiation in the closed cylinder returns the energy to the hot plasma, the plasma in the cylinder is not cooled effectively. So the radiation from the hot plasma should be removed out of the cylinder through the windows. Another problem is that the electron density exceeds to the optimum value in the recombination phase, which is caused by the adiabatic compression of the hot plasma.

By optimizing for the electron density, we obtained the improved gain profiles shown in Fig.5. The cylinder mass is reduced by a factor of 20 from the case of Fig.4. The wall thickness is 50μm and the density is $30mg/cm^3$, which is achieved using polystyrene foam. Figure 5 (a), (b), and (c) show the flow diagram, the spatial profile of the gain coefficient, and the time developments of the gain coefficient and the plasma parameters at the cylinder center, respectively. The gain occurs within 1.3mm radius of the cylinder, which is much larger than the case of the Fig.4(b). In Fig.5(a) the dotted line shows the critical density point. When the shock wave is propagated toward the wall, this point moves toward the wall at the same time. Therefore, the plasma volume, whose temperature is high enough to produce the fully ionized carbon, becomes larger, because the laser light is absorbed near the critical density point. Consequently, such a large gain region is achieved. After the laser heating stops at 60ns, the each mesh is fixed as shown in Fig.5(a), which implies that the electron and ion density is nearly constant. At the cylinder center, the electron density is kept to be $10^{19}cm^{-3}$, which is appropriate to obtain the CVI Hα gain. The gain duration up to 20ns in FWHM is achieved. The gain medium which has such a large gain region and long gain duration shown in Fig.5 is favorable to construct a soft x-ray laser cavity. Furthermore, it is presented that the spatial profile of the density in the cylinder is a waveguide one, which has the feature that the soft x-ray refraction is small.[10] This feature also offers us the advantage of constructing a soft x-ray laser cavity. If the optimization of the cylinder mass and the amount and species of the high Z impurity as a coolant is achieved, the gain medium, which has a peak gain coefficient of $5cm^{-1}$ and gain duration of 10ns, may be obtained.

4. SUMMARY

We proposed the cylinder-type target to produce the gain medium which had the long gain duration. The amplification of the CVI Hα line is observed. The gain coefficient is $1.5cm^{-1}$. The gain medium is satisfied with the condition to generate the population inversion between n=3 and 2 levels of the CVI. We presented the computer

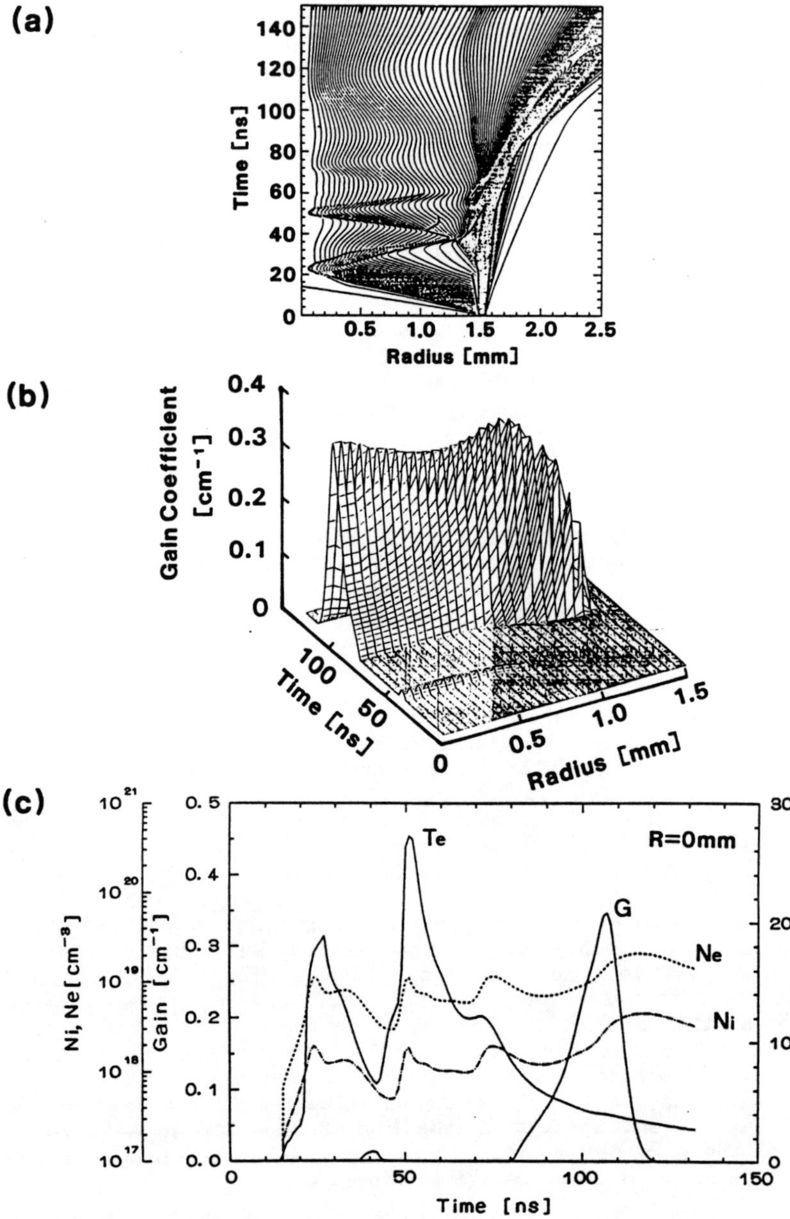

Fig.5 (a) Flow diagram in the cylinder. The dotted line is the critical density point. (b) Spatial profile of the CVI Hα line gain coefficient. (c) Time developments of the gain coefficient and the plasma parameters at the cylinder center. The cylinder radius is 1.5mm. The wall thickness is 50μm and the density is $30mg/cm^3$. The wall mass is reduced by a factor of 20 from the case of Fig.4. The improved gain profile is shown.

simulation results. The gain coefficient is limited by the long cooling time and the increase in the electron density. It is also shown that the electron density is able to be controlled by the cylinder mass. The long gain duration up to 20ns in FWHM and large gain production region up to 1.3mm in radius are predicted using the low density material like a polystyrene foam. These features offers us the advantages of constructing a soft x-ray laser cavity.

REFERENCES

[1] R.C.Elton, X-RAY LASERS, Academic Press, N.Y., (1990)p90.

[2] Y.Kato, E.Miura, T.Tachi, H.Shiraga, H.Nishimura, H.Daido, M.Yamanaka, T.Jitsuno, M.Takagi, P.R.Herman, H.Takabe, S.Nakai, C.Yamanaka, M.H.Key, G.J.Tallents, S.J.Rose, and P.T.Rumsby, Appl.Phys. B 50, (1990)247.

[3] C.Chenais-Popovics, R.Corbett, C.J.Hooker, M.H.Key, G.P.Kiehn, C.L.S.Lewis, G.J.Pert, C.Regan, S.J.Rose, S.Sadaat, R.Smith, T.Tomie, and O.Willi, Phys.Rev.Lett. 59, (1987)2161.

[4] P.Jaegle, G.Jamelot, A.Carillon, A.Klisnick, A.Sureau, and H.Guenou, J.Opt.Soc.Am. B4, (1987)563.

[5] J.F.Seely, C.M.Brown, U.Feldman, M.Richardson, B.Yaakobi, and W.E.Behring, Opt.Commun. 54(1987)299.

[6] S.Suckewer, C.H.Skinner, H.Milchberg, C.Keane, and D.Voorhees, Phys.Rev.Lett. 55 (1987)1753.

[7] E,Miura, H.Daido, Y.Kitagawa, K.Sawai, Y.Kato, K.Nishihara, S.Nakai, and C.Yamanaka, Appl.Phys.Lett. 55, (1989)223.

[8] H.Azuma, Private communication

[9] J.F.Seely, R.H.Dixon, and R.C.Elton, Phys.Rev.A 23,(1981)1437.

[10] H.Daido, K.Nishihara, E.Miura, and S.Nakai, J.Opt.Soc.Am. B7, (1990)266.

[11] M.Murakami and K.Nishihara, Jpn.J.Appl.Phys. 26, (1987)1132.

[12] A.I.Shestakov and D.C.Eder, LLNL Rep. UCRL-100485, (1989).

X-RAY LASER WITH ACTIVE COLD NON-PLASMA MEDIUM, CREATED BY THE DIRECT EXCITATION OF IONS DUE TO THE IMPACTS OF OSCILLATING ELECTRONS IN THE SUPERBRIGHT LASER FIELD

M.Yu. Romanovsky, I.N. Knyazev, and V.V. Korobkin
The General Physics Institute, 117942, Vavilov Street, 38, Moscow, USSR

Abstract

The physical situation when the superbright laser radiation (Intensity I > 10^{17}W/cm^2) is introduced into the cold gaseous medium is considered. Free oscillating electrons, arising due to the direct above threshold ionization, receive energies in keV region, ionize the top electrons of the ions and excite the first next electron which they can not ionize. Thus it is possible to realize, for example, the Ne- or Ni-like schemes for the X-rays coherent generation. The inner ionization is also possible, it leads to the holes in the electron structure of ions and the population inversion.

Such type of ions excitation provides several advantages as compared with the existing schemes: more narrow X-ray line, more gain and the opportunity to generate more short X-ray wavelength.

Introduction

All now existing lasers act in the following way: the optical laser pumping prepares hot plasma, and then either hot electrons excite ions under impacts, or rigid quanta, or free plasma electrons recombine with ions and therefore excite them (there are two types of recombination - three-body and dielectronic). All these ways include hot plasma as the necessary active medium (or the intermediate step for the recombination scheme). It is clear that the most part of pumping energy is distributed to all spectra of plasma quanta, electrons, and ions, not only to the necessary part of them. There is also the principal limit for the line width of X-ray stimulated coherent radiation due to the Doppler broadening, because plasma ions possess high hot velocities. Therefore it is very interesting to look for other ways to create the X-ray lasing [1]. We believe that the main of them is connected with the direct oscillating electrons excitation. These electrons appear if you introduce the superbright intensity optical laser field into cold gaseous (below we will show that target densities of an active medium are not necessary) medium. This field must produce the direct above threshold ionization, i.e. the maximal laser field strength must exceed so-called "inner-atomic" value - the field strength of atomic remainder in the point when the valence electron (or several top electrons) is situated. Thus, the original electrons appear due to this classical process. These electrons begin to oscillate in the pumping laser field and impact ions. Thus the second requirement for the pumping intensity arises - to provide the oscillation amplitudes larger than the electrons free track. This is more rigid requirement in the atmospheric density gases as compared with the first one but much more gentle in target experiments.

Ions can be ionized (including inner ionization) under these impacts. The Auger ionization is an additional channel of the free electrons multiplication - due to the inner ionization. Thus, the ionization includes both processes - impact and Auger, the last accelerates the main process of the direct impact ionization. These processes continue until oscillating electrons energies exceed the corresponding value of the ionization potential of one of the electrons in the ion. Since this moment, oscillating electrons can only excite the next (more rarely - one of the below laying) electron.

The electrons velocity distribution in time of collisions is, of course, non-Maxwell. Because the ions mass is much more than electrons one the ions practically rest during the electrons oscillations (the electrons oscillation amplitude $x_e \sim 10^{-5}$ cm under $I \sim 10^{17} W/cm^2$ and atmospheric pressure and the ions one $\sim 10^{-10}$cm - much less than the ions size). The ions distribution in space is homogeneous and it means that the electron can possess in time of collision the energy values among zero and the maximal oscillation energy, not more, which strongly differs from the Maxwell distribution.

Because the electrons velocities distribution strongly differs from the Maxwell, it is strongly non-equilibrium, therefore it is impossible to use the principle of detail equilibrium for the calculation of the rates of the direct and opposite processes, especially for the recombinations rates calculations.

This paper includes several separate chapters: the determinations of the collisions frequency and electrons velocities distribution function, the rates of all processes - ionizations, excitations, recombinations and opposite processes, and various schemes of X-ray coherent generation: Ne-like, inner shell transitions, recombination, and competition processes.

THE PUMPING STIMULATED COLLISIONS FREQUENCY. THE ELECTRONS VELOCITIES DISTRIBUTION FUNCTION.

It is impossible to investigate dynamics equations for all electrons in laser field. It is necessary to average these equations and to introduce physical values which describe correctly the model . Temperature can not be this value because the considered system is strongly unequilibrium.

Only two forces act on the electron between collisions. If m - the electron mass, e - the charge, $\bar{a}$ - the acceleration of the electron in the point $\bar{r}$ at time t, **E** is the laser electric field in this point, then

$$m\bar{a} = eE + MF_{\rho} \tag{1}$$

The first term in the right part is Lorenz force, F_{ρ} - the force, which is created by all other charges in the point of this electron. If $E \sim 10^{10}$ V/cm (laser beam intensity $I \sim 10^{18}$ W/cm^2) the maximum of oscillation velocity **v** is still much less than the light velocity **c**. The amplitude of oscillations without collisions and zero initial velocity of this free electron is $x \sim 2eE/m\omega^2$ ($\omega = 2\pi c/\lambda$, λ - laser wavelength) and if λ=1060 nm $x \sim 5*10^{-6}$ cm.

The electron free pass is $l \sim (n\sigma)^{-1}$, n is ions density, σ is collisions cross-section and for $n \sim 10^{19}$ cm^{-3} and $\sigma > 10^{-15}$ $l < x$, i.e. collisions exist, but $x < \lambda$ and the laser beam radius r_o is such that, at least, $r_o > \lambda$. Therefore $(E/\nabla_{\perp}E) >> x$; $\nabla_{\perp} = \{\partial/\partial x, \partial/\partial y\}$ and it can be expected in the first approximation that oscillations occur under constant laser field amplitude. Let us derive the expression for the number of electrons-ions stimulated collisions in such laser field.

We assume that the laser radiation polarization is linear and the laser beam propagates along z - axis. Oscillations occur under the constant field amplitude. Let us introduce the laser pulse envelope **f(t-z/c)**, the field amplitude E_m is the function of z in general case. The equation for the electron motion between collisions is:

$$\frac{d^2x}{dt^2} = \frac{eE_m(z)}{m} \cos(\omega t - kz) * f(t - z/c) \tag{2}$$

Here $k = 2\pi/\lambda$. Initial conditions in time $t = t_k$ - the time of the previous collision with number k are $x = x_k$ and $v = v_k$. What does occur under the collision? There occurs an instantaneous loss of the electron velocity with a parallel laser field vector. Transverse changes of the electrons velocity due to the attraction of electrons by ions also take place. The lost parts of velocities can have various values which, of course, are not more than the velocities of the electrons at the time of collisions. This means that the electrons motion becomes partially stochastic.

Let us choose the time moment t_0 and coordinate z such that the electron velocity is equal to zero (Fig.1). If f(t-z/c) is comparatively smooth ($\partial f/\partial t \ll \omega$), we have (not on the pulse front):

$$v \approx \frac{eE_m(z)}{m\,\omega} \sin[\omega(t-t_0)-kz] \tag{3}$$

and this expression is true up to the first collision. We remember that there are several collisions during laser optical period under such (and larger) laser field amplitudes. The amplitude of velocity decreases at the time of collision i.e. velocity becomes a jump higher but remains negative (point v_1 Fig.1). The same thing occurs at the time of the next collision (point v_2 Fig.1).

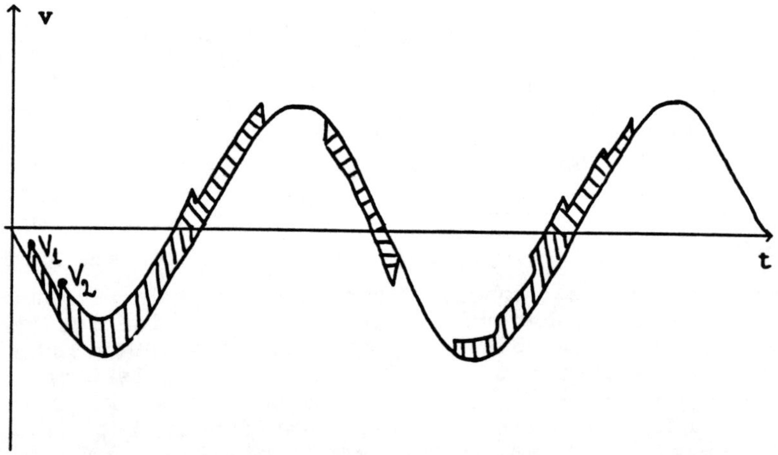

$\Sigma\Delta v$ Areas S: $S_{\text{≋}} + S_{\text{≋}} - S_{=} - S_{||} = 0$

Fig.1

The diagram of velocity consists of pieces of the sinusoid which are above "non collisions" solution (3) by the value Δv_1 at the v_1 point, by $\Delta v_1+\Delta v_2$ at the v_2 point, etc. As a result of several collisions, the velocity diagram crosses the abscissa earlier than in the non collisions case (3). Velocity losses after this crossing at time of collisions have already negative values. If we suppose that absolute values of velocity losses under collisions are identical (this means that impacting electrons excite or ionize the same ion shell, or more exactly - the same state, and the excitation energy or the energy of released electrons is always identical) the shaded area of the negative losses of velocity in Fig.1 is equal to the positive one's during several optical periods. If there are several values of velocity losses the total value is also leveling during several optical periods. The case when there are two loss values Δv_1 and Δv_2, $\Delta v_1=\Delta v_2/5$ is in Fig.2.

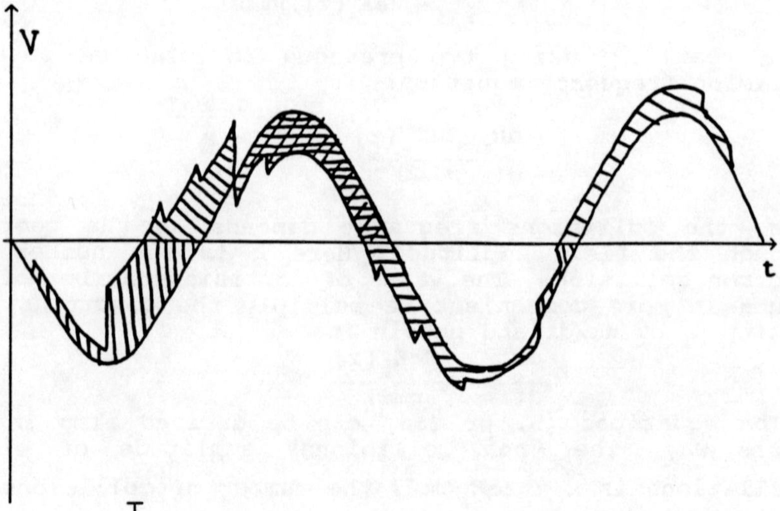

Δv_1 Δv_2 Areas S: $S_{≣} + S_{≋} - S_{=} - S_{||} = 0$

Fig.2

Here collisions when electron loses velocity Δv_2 occur five times more rarely than Δv_1 ones. Because triple collisions occur much more rarely than double ones, an electron can lose the velocity Δv_1 or Δv_2 but can not lose $\Delta v_1+\Delta v_2$, etc. Thus, the average contribution of the stochastic process to the velocity and the amplitude of oscillation is equal to zero during several optical periods. It is easy to obtain the same results by solving Equation (2) for each collision and by summing solutions.

Because the time interval between collisions $<<\omega^{-1}$, it is easy to obtain from (2) that the distance between k and k+1 collisions coordinates is

$$x_{k+1} - x_k \simeq (t_{k+1} - t_k)v^-_{k+1}. \quad (4)$$

Here v^-_{k+1} is the velocity right before the k+1 collision. Therefore

$$(x_{k+1} - x_k)/(t_{k+1} - t_k) \simeq v^-_{k+1}.$$

If we average this formula in space the left part will be $l\Delta k/\Delta t$ because the average distance between hard particles along electron trajectory is the electron free track length and Δk=1. The averaging of the right part must give "actual" value of the collided electrons velocity. It means that "actual" value of $\bar{v}_{k+1}$ is the average one of ordinate $v_m \sin\omega t$ within the optical half-period:

$$\overline{\bar{v}_{k+1}} = 2eE_m(z)/\pi m\omega$$

As a result, using two previous formulas we obtain the collision frequency equation:

$$\frac{dk}{dt} = \frac{2eE_m(z)}{\pi l m\omega} \qquad (5)$$

Thus, the collisions frequency depends on the coordinates through the field amplitude. Here k is the number of one electron collisions. The value of collisions number of a unit volume is more convenient we multiply the electrons numbers density n_e by dk/dt and obtain it:

$$n_e\frac{dk}{dt} = n_e\frac{2eE_m(z)}{\pi m\omega l} . \qquad (5a)$$

The equations (5) or (5a) can be derived also in a very simple way. The "non collisions" amplitude of electrons oscillations is $\bar{x} = 2eE_m/m\omega^2$, the number of collisions during the period of laser radiation is $2\bar{x}/T_{las}l$, or

$$dk/dt = 2eE_m(z)/\pi m\omega l$$

that coincides with (5).

Thus, we made simple estimation for the stimulated collisions frequency. Really, the exact averaging is necessary, because each electron can possess the velocity (3) in every point and therefore in the point of collision.

The first simplest case arises when $l \gg \bar{x}$ (or the stimulated collisions frequency is much more than laser pumping frequency). According to (4), we can consider the argument $\omega(t-t_o)-kz$ in the formula (3) as a random value with the homogeneous distribution function because ions are distributed homogeneously in space. If we write

$$v_m = \frac{eE_m(z)}{m\omega} ,$$

due to the general formula for the distribution function for random value y: $f(y)$, the distribution function for value $\phi = \phi(y)$ is $f[\phi^{-1}(y)]*\partial\phi^{-1}/\partial\phi$, here ϕ^{-1} is the opposite function for ϕ. Therefore, the distribution function for v is:

$$\phi(v) = \frac{2}{\pi(1-v^2/v_m^2)^{1/2}} \qquad (6)$$

This dependence is shown in Fig.3 also with the Maxwell distribution.

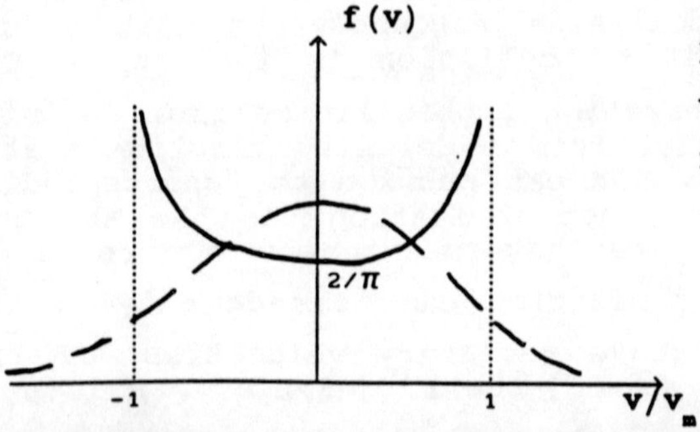

Fig.3. The electrons velocities distribution function under collision

——— - (6)

— — — - Maxwell ($T \sim mv_m^2$)

The absence of hot electrons (with velocities above v_m) determines the impossibility of the deep excitation and the presence of only one type of ions in the ions spectrum (which strongly differs from the hot plasma case).

The opposite situation arises when $\bar{x} \ll 1$. Here we can expect that the collisions process has Poisson distribution, the ratio $\bar{x}/l$ plays the role of the distribution parameter $\bar{n}$. The full distribution function is expressed from the complicated combination sin and the Poisson dependencies. Note, that the average value $\bar{v}$ which is calculated under function (6) coincides with the formerly introduced one (see(4)).

Thus, the electron distribution function in time of collisions strongly differs from the Maxwell one. It has simple view (6) in case of many collisions during one pumping optical period. Note, that the distribution function (6) depends on coordinates through E_m. The same condition exists for the Maxwell distribution with the changing temperature T.

Let us roughly describe the situation when the stimulated collisions distribution exists together with the Maxwell one and let us suppose that the temperature of the last one changes from $T \ll mv^2/2$ to $T \gg mv^2/2$. The first case ($T \sim 0$) means that the distribution (6) exists in pure view. If T increases, the sharp peak in (6) under $v=v_m$ remains high but final, wide and it decreases with the temperature grows. When T becomes $\sim mv^2/2$ the distribution looses the character (6) and now is like the Maxwell function. The exact determination of the full distribute function is the interesting separate problem.

THE IONIZATION AND EXCITATION

It is necessary to take into account that the density of free electrons n_e changes during the pumping laser pulse. Let us derive the equation for n_e.

The equation for electron density n_e will contain a source in the right part. This source has two terms. The first term is determined by the collision frequency in unit volume $n_e \partial k/\partial t$ - here the result of collision is ionization. The latter is the sum of ionization probabilities from definite shells, or, more correctly, from a definite electron state. The impacting electrons which can provide the corresponding process are necessary for such ionization. As the absolute value of electron velocity can have meanings from zero to $v_m = eE_m/m\omega$, that deep shell ionization can be made only by the part of electrons which have necessary velocities. If the ionization cross-section from j-shell (j=K,L,M,...) is σ_{ionj}, the first term in the electron density equation is the following sum:

$$\sum_{j=1}^{N_u} n_e^2 \langle \sigma_{ionj} v \rangle$$

Here the maximum deep ionization has number N_u, the averaging <> is made with the function (6).

The second term in this equation is the rate of the Auger-electrons appearance. An Auger process starts under each collision when a hole in electron configuration of ion appears. Therefore this term is proportional to the collision frequency of this ion $\partial \bar{k}/\partial t$ multiplied by ions density (it is the same sum like as in the first term), and corresponding probability of Auger-ionization γ_j°. It is evident that the frequency of ion-electron collisions and the frequency of electron-ion collision coincide in volume. Therefore this term can be written in the following way:

$$\sum_{j=1}^{N_u} \sum_{i \le j} \langle \tilde{\sigma}_{ionj} v \rangle n_e n \gamma_i^{\circ}$$

Here n is an ions density. Note that $\tilde{\sigma}_{ionj}$ can differ from σ_{ionj} due to the possible presence of above lying electrons under the inner shell ionization. It is necessary to remember that for the first ionization $\gamma_1^{\circ} = 0$. Finally, we obtain the following equation for the electron density:

$$\frac{\partial n_e}{\partial t} + \nabla(n_e \bar{V}) = \sum_{j=1}^{N_u} n_e \langle (\tilde{\sigma}_{ionj} * \sum_{i \le j} \gamma_i^{\circ} n + \sigma_{ionj} n_e) v \rangle \qquad (7)$$

Here V is the velocity of the electrons macroscopic collective motion due to the ponderomotive forces [2].

The initial condition of the equation (10) is $n_e = n_e^{(o)}$, where $n_e^{(o)}$ is the density of initial electrons. The number $n_e^{(o)}$ can physically include two terms. The first is determined by the exceeding of laser field amplitude above the value of "atomic" strength for few top electrons. The second term is determined by multiphoton processes on the laser pulse front.

Thus, we determine the electron density $n_e(\bar{r},t)$. According to (7), we can determine the full time of the ionization to the necessary force in a nonevident view:

$$\tau_{ion} = \sum_j \langle v[\sigma_{ionj} n_e(\tau_{ion}) + \tilde{\sigma}_{ionj} n \sum_{i \le j} \gamma_i^o] \rangle^{-1} \qquad (8)$$

The pumping laser pulse must have at least the duration $\tau_l \ge \tau_{ion}$ for the full necessary ionization if you try to prepare Ne- or Ni-like X-ray laser schemes. This - the ionization - stage coincides with the "heating" one in the "two-lasers" scheme.

The next - the excitation - stage is determined by the simple expression for the excitation time τ_{exc}:

$$\tau_{exc} = (\langle \sigma_{exc} v \rangle n_e)^{-1} \qquad (9)$$

Here we suppose, according to (7), that n_e can not be changed after the time moment τ_{ion}, if $\tau_{ion} < \tau_{plas}$, τ_{plas} is the time of plasma formation. This time for media in superbright laser pulses is not less than one picosecond [2]. The plasma absence determines the absence of hot electrons. The opposite value for the time (9) determines the rate of the X-ray laser working transition excitation.

Values (8) and (9) determine all necessary "stimulated" constant for Ne- or Ni-like schemes. For the inner shell excitation (due to the inner ionization), the excitation time of j-state is determined by the value:

$$\tau^i_{exc} = [n_e(t) \langle \tilde{\sigma}_{ionj} v \rangle]^{-1} \qquad (10)$$

The opposite value determines the rate of the inner shell excitation [2].

THE RECOMBINATION

Due to the absence of detail equilibrium, we ought to differ two stages of recombination - the recombination during the laser pumping pulse (electrons oscillate) and after the pulse. Of course, it is necessary to calculate directly the recombination rates for both these cases.

Electrons acquire the stochastic motion component under the collision, both the transversal and longitudinal part (from the field vector) [2]. The time when this motion becomes comparable with oscillations is τ_{plas} - the plasma formation time. But the stochastic part begin to increase with the first oscillation period (and the first collision). Therefore we can introduce a certain value which characterizes this process - it will be the temperature T. This temperature grows during the laser pulse. After the completion of the pulse, the stochastic motion part remains and at the first moment will be determined by the temperature

$T(\tau_1)$. Accordingly, the recombination during the pumping laser pulse will be determined by the regular oscillating motion.

Let us consider the situation, when an electron flies near the ion (see Fig. 4).

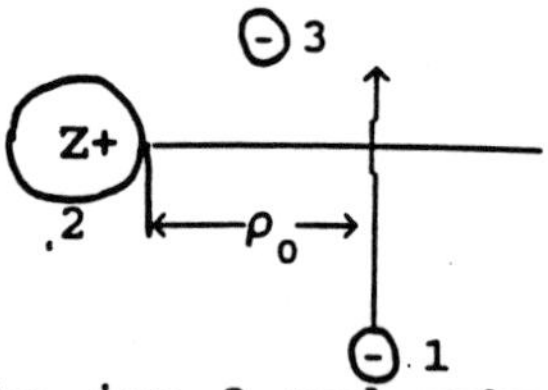

Fig.4. Electron 1 recombinates with ion 2. Electron 3 is the energy acceptor. ρ_o is the goal distance.

The electron 1 must be attracted by the ion 2 and makes compound with it. It means that the goal distance ρ_o must correspond the expression (Z is the ion charge):

$$\frac{Ze^2}{\rho_o} = \frac{mv_e^2}{2} = \begin{cases} e^2E^2/2m\omega^2 & t\le\tau_1 \\ T(\tau_1) & t\ge\tau_1 \end{cases} \tag{11}$$

Accordingly, the possibility of collision of one electron with the ion per one second is $nv_e\pi\rho_o^2$. This expression must be multiplied by the probability of the presence of another electron in the sphere with the ρ_o radius $n_e4\pi\rho_o^3/3$ for the determination of the recombination on the top levels of the ion in case $t>\tau_1$:

$$W_T^r = \frac{4}{3}\pi\rho_o^5nn_e\langle v_T\rangle_T = \frac{64\pi^2Z^6e^{10}n^2}{243(mT^9)^{1/2}} \tag{12}$$

Here $\langle v_T\rangle_T$ is the Maxwell distribution averaging of the heat electrons velocity v_T with the temperature $T(\tau_i)$.

In the case of the recombination to the deep empty states {j} with E_j >>T, we must change the possibility of the electron presence in the above mentioned sphere on the possibility of the energy E_j transmission to another electron $n_e\sigma_{ee}\rho_o$. Here ρ_o - the length of an interaction and, accordingly, the energy transmission, σ_{ee} is the cross-section of the energy transmission among two electrons Therefore we receive:

$$W_E^r = \pi\rho_o^3nn_e\langle\sigma_{ee}^E v_T\rangle_T = \frac{8\pi^2\sqrt{2}Z^4e^{10}}{27E_j(mT^7)^{1/2}} \tag{13}$$

For the calculations of the recombination on the top ions states $[E_j<T(t)]$ during the pumping laser pulse, we ought to change the possibility of the energy transmission among electrons on the same possibility with another v_e-oscillating one and change ρ_o [see (11a)].

We receive:

$$\tilde{W}^r_T = \pi\rho_o^3 n n_e \langle\sigma^T_{ee} v_e\rangle = \frac{2\sqrt{2}\pi Z^4 n^2 e^{10}}{(mE^5_{osc})^{1/2}\; T^2(t)} \qquad (14a)$$

Here E_{osc} is the maximal electrons oscillation energy $e^2E_m^2/2m\omega^2$.

The recombination rate to the deep states is:

$$\tilde{W}^r_E = \pi\rho_o^3 n n_e \langle\sigma^E_{ee} v_e\rangle = \frac{2\sqrt{2}\pi Z^4 e^{10}}{(mE^5_{osc})^{1/2}\; E_J^2} \qquad (14b)$$

Let us make several remarks about values (12)-(14). $E_J > 10^3$eV for the working transition wavelength in the Ne- or Ni-like schemes ~ 10-30 nm, Z~(20-40), $E_{osc} \sim 2E_J$ and for atmospheric pressures of the active media $\tilde{W}^r_E \sim 10^5-10^6\ s^{-1}$. If we roughly suppose for T the linear grows during the pumping laser pulse $T(t)= T(\tau_i)t/\tau_i$, we can estimate the current temperature by this expression. Accordingly, if $\tau_i<\tau_{plas}$, we receive with the same accuracy: $T(\tau_J)/\tau_i = T/\tau_{plas}$, T is the equilibrium electron temperature, and

$$T(t)= T\tau_i/\tau_{plas} \qquad (15)$$

Because T~ 100 eV, $\tilde{W}^r_T \sim 10^7-10^8\ s^{-1}$. Both values - $\tilde{W}^r_T$ and $\tilde{W}^r_E$ - can not be compared with the typical rates of the considered system - $10^{11}-10^{14}\ s^{-1}$ and we can conclude that during the pumping laser pulse we can neglect the recombination processes.

Because the low final temperature $T(\tau_i)$ is one of the advantages of the considered schemes, we ought to put that $\tau_{plas}>\tau_i$ (it is a usually fulfilling condition), therefore $T(\tau_i)<T$ in several times. It means, that W^r_T and W^r_E have the same order with the other rates in the system, and the recombination scheme which will begin to work after the pumping laser pulse completion must be considered.

THE MULTILEVELS SYSTEMS KINETICS. X-RAY LASERS NEAR-THRESHOLD CHARACTERISTICS

The full kinetics system for the electron configuration was described in [2]. Here we will consider three cases for multilevels systems: the inner ionization the direct electron excitation (Ne- or Ni-like), and the recombination case, when it is important - under $t>\tau_i$.

Let us consider the followings of inner-electron

ionization. When free electrons appear, two processes come into being - top electron ionization and inner one. Both processes change the ion Hamiltonian. The first result of them is the non-stationarity of electrons states in ion immediately after ionizations. It means that the electron that was situated directly above the appeared hole in ion electrons configuration is in nonsteady state and it rapidly transits into the steady state which can be excited. There are two ways of these transitions - radiative and Auger. The rate of Auger -Kronig-Coster processes is about 10^{15}-$10^{16}s^{-1}$ and usually much more than the corresponding radiative rates. Thus the first level of laser system is nonstationary. The second level is the steady excited state with electron wave function which is closely the same as the initial wave function of the "above-hole" electron. The third level is also a steady state which can be fundamental. There is a three-levels laser system:

n - nonstationary state,

2 - new excited steady state,

1 - steady state (fundamental) (Fig.5).

There are the following rates of transitions: W is a rate of the n-level excitation (note that it is not the 1-n excitation; in our conditions, the levels n, 1,2 appear only due to the inner ionization, so the population on the n-level goes from "above levels"). Γ_{n1} - a rate of the spontaneous transition n-1, Γ_{n2}- n-2, Γ_{21}- 2-1 ones, $V_{21}=V_{12}$- rates of stimulated transition, $\Gamma_{n2} \gg \Gamma_{n1}$. n_1, n_2, and n_n are the densities of states n, 1 and 2 and is described by equations:

$$\frac{dn_n}{dt} = W\times n - n_n(\Gamma_{n1}+ \Gamma_{n2}); \ \Gamma_{n2}+ \Gamma_{n1}= \Gamma_n,$$

$$\frac{dn_2}{dt} = - V_{21}(n_2- n_1) - n_2\Gamma_{21}+ n_n\Gamma_{n2}, \tag{16}$$

$$\frac{dn_1}{dt} = V_{21}(n_2- n_1) + n_2\Gamma_{21} + n_n\Gamma_{n1}.$$

The initial conditions $n_n= n_1= n_2= 0$, therefore

$$n_n= W\, n\, [1 - \exp(1 - \Gamma_n t)]/\Gamma_n.$$

Remember, that $W = <\tilde{\sigma}_{ion} v_e>n_e$ depends on t through $n_e(t)$. The system (16) does not include deexcitation terms because it is impossible for holes to decay this way [2]. The $\tilde{\sigma}_{ion}$ can also include the transitions to the empty top states. Near the threshold ($V_{21}<<\Gamma_{21}$), we receive for the population inversion $\Delta n = n_2- g_2n_1/g_1$:

$$\Delta n = \frac{Wn}{\Gamma_n} \{(1+\frac{g_2}{g_1})[1-\exp(-\Gamma_{21}t)] - \frac{g_2}{g_1}t(\Gamma_{n1}+ \Gamma_{n2})\} \tag{17}$$

Here g_1, g_2 are the multiplicity of levels 1 and 2

respectively. The formula (17) shows that the population inversion has nonstationary character; t=0 corresponds $\Delta n=0$, than (if $\Gamma_{n1} < g_2\Gamma_{21}/g_1$) Δn achieves the maximum:

$$\Delta n_m = \frac{Wn}{\Gamma_n \Gamma_{21}} \left[\left(1+\frac{g_1}{g_2}\right)\left(\frac{\Gamma_{21} g_1 - \Gamma_{n1} g_2}{g_1 + g_2}\right) - \frac{g_2(\Gamma_{n1} + \Gamma_{21})}{g_1 \Gamma_{21}} - \ln\frac{\Gamma_{21}(g_1 + g_2)}{g_2(\Gamma_{21} + \Gamma_{n1})}\right] \tag{18}$$

and achieves zero under $t \leq (g_1+g_2)/g_2(\Gamma_{n1}+\Gamma_{21})$. These expressions show that it is better to choose $g_2>g_1$ for X-ray laser working transition. It is clear that τ_1 can not be more than τ_{ion}, because the electrons above the hole are absent in opposite case.

The main advantage of this scheme is the opportunity for the X-ray laser to work into ground state and therefore to use the most part of ions "exciting" energy. The reabsorption can be avoided by the choice of the pumping pulse duration, which determines the force of the electrons collision ionization and therefore - the absence of reabsorption electrons in ions.

Let us make some estimations, taking Argon into consideration. The deepest possible working transition is 3s-2p. This situation arises when the $2p^6$ electron is pushed out of ion but at least one M-electron remains (better two or more electrons). For *Ar* we received X-ray wavelength of the main transition 3s-2p 5,1nm, $\Gamma_n \sim 10^{15}\ s^{-1}$, $\Gamma_{21} \sim 2{*}10^{13} s^{-1}$, W $\sim 0,9\ \alpha\ 10^{12} s^{-1}$ ($\alpha = n/2,67{*}10^{19} cm^{-3}$), $\Delta n/n \sim (10^{-2}-10^{-1})\alpha$.

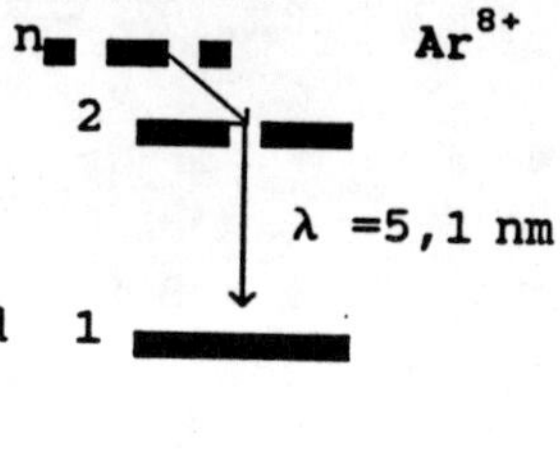

Fig.5. The levels system for "hole" laser. Ar ion lost seven upper electrons and upper $2p^6$ one. Levels 1(2p) and 2(3s) are the stationary state of ionized ion, the level n(3s) is an old state of atom or slightly ionized ion.

Of course, these calculations are correct if α does not exceed 10. The gain G:

$$G = \lambda^2 \Gamma_{21} \Delta n / 4\Delta\omega \tag{19}$$

also is determined by the X-ray linewidth $\Delta\omega$. There exist both Doppler, electron impacts, and the field broadenings. Because some temperature is present (see ch.3), it is Doppler case here with the effective temperature $T(\tau_1) = \tau_1 T/\tau_{plas}$. The temperature corresponds the Ar ionization energy ~ 400 eV T ~ 400/(15-20). If $\tau_1 \sim 500$ fs, $\tau_1/\tau_{plas} < 0,2$ and $T(\tau_1) \sim$ (5-6) eV, which determines the broadening $\Delta\omega$ =

$[8\pi^2 T(\tau_i)/\lambda^2 m_i]^{1/2} \sim 1,1*10^{13}\ s^{-1}$ (m_i is an ion mass) and G~ 870α cm^{-1}. Fig. 6 shows G as a function of t' and α if the pumping laser light intensity I≃ $2,3*10^{17}$ W/cm^2, using the KrF excimer laser λ = 248 nm.

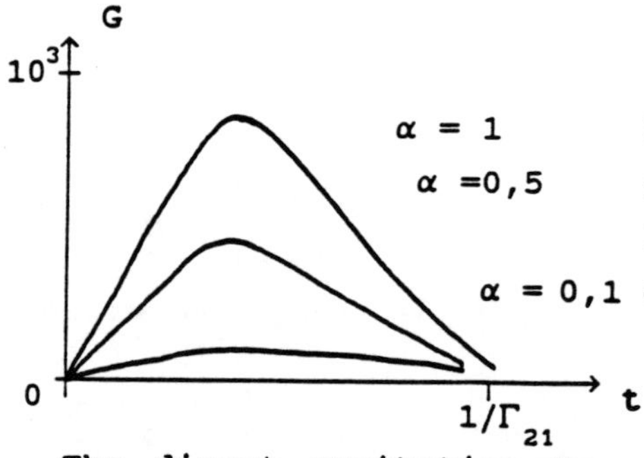

Fig.6. The dependence of the gain G from t and α for "hole" laser. Ar without seven upper electrons and upper $2p^6$ one.

The direct excitation Ne- or Ni-like schemes are actual during the pumping pulse after τ_{ion}. The nearthreshold equations system coincides with usual one for high-ionized ions, except the averaging of the excitation and deexcitation rates [3]. Here we will not write this system and discuss only the results.

First, the pumping pulse duration can not be shorter than the ionization time (8). Using the simple expressions for σ_{ion} and σ_{exc} [4], it is possible to obtain the result low estimation for τ_{ion}:

$$\tau_{ion} \simeq (2eE_{osc})^{1/2}/8\pi a_o^2 n I_H n \ln\kappa. \qquad (20)$$

Also the τ_{exc} of the state with the energy E_j:

$$W^{-1} = \tau_{exc} = (<\sigma_{exc} v_e> n_e)^{-1} = \frac{\sqrt{m/2}(E_{osc}-E_j)E_{osc}^2(2l_o+1)}{Zna_o^2 C I_H^2 E_j^{3/2} F(x)} \qquad (21)$$

Here a_o^2 is a Bohr radius, I_H= 13,6 eV, n is an average number of electrons in the subshell, lnκ is about 5 for Ne-like Se and about 4 for Ni-like Eu, C is the cross-section "amplitude" [4], l_o is the orbital quantum number of the excited electron. $F(x)$ is shown on Fig. 7, $x = (E_{osc}/E_j)^{1/2}$.

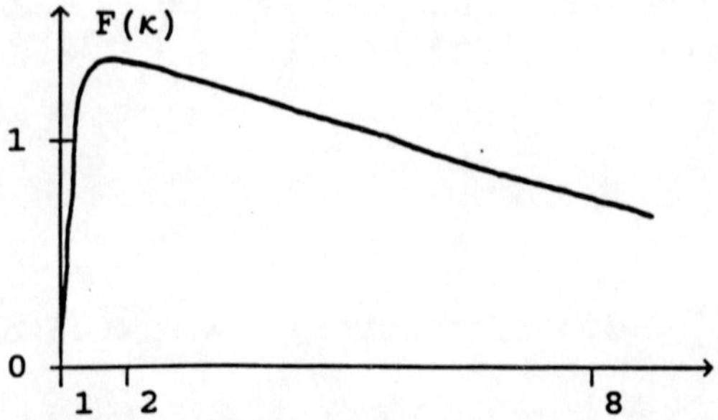

Fig.7. The dependence of the value excitation rate from $\kappa = (E_{osc}/E_j)^{1/2}$

Thus, if $\tau_{ion} < \tau_i < \tau_{plas}$, we have the stationary X-ray generation regime. Because $W(2p\text{-}3p) = W_1 \sim 9W(2p\text{-}3s)$ [4], the population inversion exists in every cases. The gain threshold value ($\Gamma_{21} >> V_{21}$), accordingly:

$$G = \frac{\lambda^2 \Gamma_{21} W n_o}{4\Delta\omega\Gamma_{21}} \left[\frac{1 - g_2\Gamma_{21}(W+W_1)/g_1\Gamma_{10}W}{1 + W/\Gamma_{21} + (W+W_1)/\Gamma_{10}}\right] \qquad (22)$$

Note, that the electron collision excitation cross-section has optimum value under E^o_{osc} and (20) must be calculated under this condition. Really, value G grows under W grows. The value (22) can not be strongly less due to the deexcitation process because it is calculated near the threshold and we can not expect the big population n_2. Here $\Delta\omega$ is also determined by the Doppler broadening with the temperature $T_e(\tau_i) = T_i(\tau_i)$.

Let us make several estimations. For Ne-like Se Z=24, the working transition 3p-3s, λ=20,6 nm, E_{osc} (provides the σ_{exc} maximum)= 2880 eV. Electrons with this energy can make the following ionization, and we choose for the receiving of only one type of ion - Se^{24+} $E_{osc} = E^{2p}_{ion}$ = 2540 eV. It means I= $2,4{*}10^{17}$W/cm^2, $F(x)$= 1,10 (F_{max}=1,34), $W_{ion} = 1/\tau_{ion}$ = $4,0\alpha{*}10^{10}s^{-1}$, $W_{exc} = 4,2\alpha{*}10^9 s^{-1}$ and $G \simeq 2,69{*}10^5\alpha^2 cm^{-1}$ ($\alpha<<1$), $G \simeq 7,55\alpha{*}10^5 cm^{-1}$ ($\alpha>>1$). If you possess the laser source with this intensity and pulse duration ~ 500fs, it means, that you must choose $\alpha>50$ (for $\tau_i > \tau_{ion}$). Therefore, last value G is only realized, $G \simeq 1,51{*}10^7 cm^{-1}$!

For Ni-like Eu^{34+} the working transition 4d-4p, λ=7,1 nm, to avoid the following ionization $E_{osc} = E^{3d}_{ion} \cong$ 1950 eV, $F(x)$= 1,03, I= $1.67{*}10^{17}$ W/cm^2, $W_{ion} = 1,23\alpha{*}10^{11}s^{-1}$, W = $3,04\alpha{*}10^{10}s^{-1}$, W_1(3d-4p)~ 0,15W, Doppler broadening with T_i ~ 5 eV and G= $2730\alpha^2$ ($\alpha<<1$), G=2410α ($\alpha>>1$). For the pulse with the duration about 500 fs α must be more than 15 and G is about $4{*}10^4$. It means that the practical realization of

Ni-like scheme is more convenient than the Ne-like, instead of the less gains. The gains for Ni-like Eu and Ne-like Se are shown on Fig. 8.

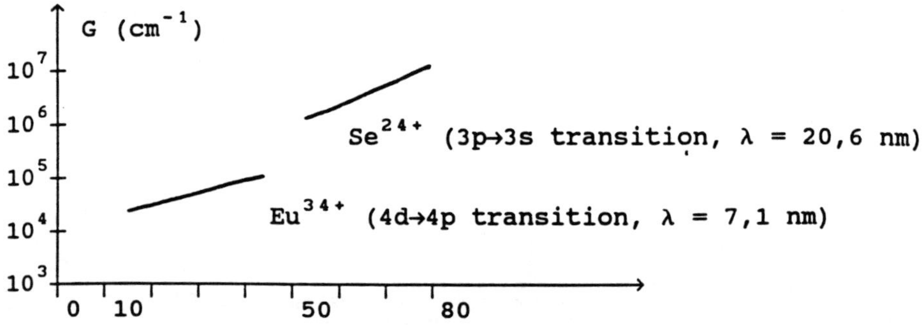

Fig.8. Gain factors G for Ne-like Se^{24+} and Ni-like Eu^{34+} - X-ray laser with the direct oscillating electrons excitation

The recombination laser really works if the rate of the low-level population is less than the radiation decay rate Γ_{10}. This condition limits the density of the active media and the total output. As we wrote above, the recombination scheme begin to work after $\tau > \tau_1$, for example, on the same Ne- or Ni-like ions. The gain of recombination scheme is the value (22), when W is changed to W_E^r (13), and $W_E^r < \Gamma_{10}$. For Ne-like Se $W_E^r = 1{,}29\alpha^2 * 10^{11} s^{-1}$, $G = 3{,}3\alpha^3 * 10^6 cm^{-1}$, $\alpha << 15$. According to the dependence W_E^r from E_j, there exists some opportunity to work to the ground state in H-, He- or Li-like schemes, when the recombination rate to the upper working level is several times more than the rate of it to the lower ground level.

ADVANTAGES. EXPERIMENTAL REMARKS

The pumping effect has a clear threshold character. First, the pumping laser electromagnetic strength must be more than inner atomic field in the point of one (or several) top electron(s). Secondly, the electrons oscillations amplitude can not be less than the electron free-track, which also connects the E_m and n.

Because all values - the deep of the ionization, ionizations times, excitations times - are directly connected with E, it strongly determines them. Therefore, it is necessary to determine the exact field behavior, because the pulse with such high field intensity is strongly distorted during the propagation. The pulse front propagates practically without distortions because the electrons reradiation propagates with the same velocity as the initial laser pulse and can not give the same contribution to the front amplitude as compared with the contribution to the another part of the pulse. Accordingly, the reradiation influences on the pulse back side stronger and extinguishes it . It is an example of "linear" selfaction.

Really, three values determine the type of X-ray lasing: τ_{ion}, τ_{plas}, τ_l. The first step is to choose the deep of the ionization and the pumping laser intensity, which are determined by τ_{ion}. I with the choosy medium density determines τ_{plas}. If $\tau_l < \tau_{ion}$, the active medium works in the regime of the inner ionization and you have the "hole" type laser. The point $\tau = \tau_l$ separates this laser type from the type with direct excitation in, for example, Ne-like scheme. This type exists during the pumping pulse, but not more than τ_{plas}. This type provides the most gains, but you loose in the emitting quanta energy. If $\tau_{ion} < \tau_l < \tau_{plas}$, the recombination scheme is actual. This scheme works under low active media densities. Fig. 9 show this distribution. If τ_l exceeds τ_{plas}, only usual X-ray laser schemes exist.

▓ -"hole" laser, █ -direct excitation laser,
░ -recombination laser

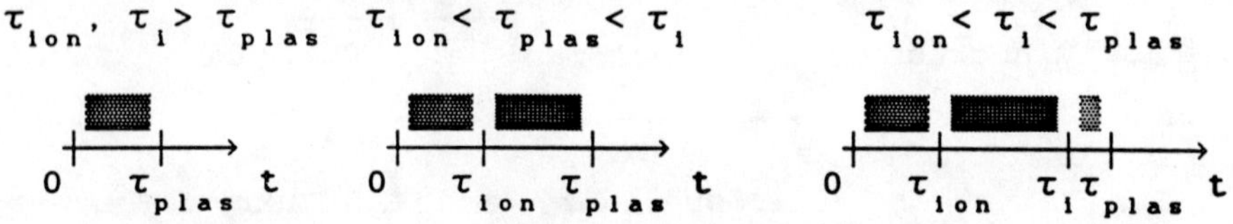

τ_{ion} - ionization time, τ_{plas} - tthe time of plasma formation, τ_l - pulse duration.

Fig.9. Time diagrams for various lasers types

What sizes of the active medium is it possible to achieve? For the "hole" type, it is clear that it is not necessary to light by the pumping laser the same volume of the gas more than time W^{-1} (for the sure inner ionization). Therefore we can transmit the laser beam on the neighbor volume of gas and so on. The length of the excited medium can be obtained as a result of this process. If the focal size of the laser beam is r_0, the laser pulse duration is $2\tau_l$, the sizes of the inverse population medium are $2r_0 \times 4r_0\tau_l W \times \delta^{-1}$. The last size - δ^{-1} is laser light length of absorption.

Now the 2 μm radius focal spot with intensities ~ 10^{19} W/cm^2 exists [5]. It means that the simple lighting of the active medium 10&100 μm is possible - it may the another way of the pumping.

The first way can be realized due to the rotative mirror or the electrooptics beam deviation, or by the axicon focusing.

The main advantage of the considered scheme is, of course, high gains. It occurs due to the high excitation rate and narrow line. Due to the upper limit in electron spectrum, it is possible to realize the pure medium with only one type of ions. Of course, there are the opportunities to work not only on the Ne- or Ni-like schemes, but on the every subshell transitions. The "hole" type laser and, may be, recombination one can provide the work the multilevels system to the ground state, what provides to use the most part of ions exciting energy for X-ray lasing. Also you can manage the sizes of the active medium.

REFERENCE

[1] V.V.Korobkin and M.Yu.Romanovsky. In book of Abstracts XIII Int.Conference on Coherent and Nonlinear Optics, Minsk, September 6-9, part 1, (1988),p.375.
V.V.Korobkin and M.Yu.Romanovsky. Sov.Phys. - Ac.of Sci.Reports. Physics Series, vol.53, No.4, (1989) p.808.

[2] M.Yu.Romanovsky and L.Sabirov. This issue.

[3] A.V.Vinogradov, I.I.Sobel'man, and E.A.Yukov. Sov. J. Quantum Electron., vol.7, (1977),p.32.

[4] L.A.Vainstein, I.I.Sobel'man, and E.A.Yukov. The atoms excitation and spectrum lines broadening. Moscow, (1979).

[5] C.K.Rhodes. This issue.

NEAR-ULTRAVIOLET RECOMBINATION LASER ON THE H-LIKE BE IV ION TRANSITION

B.A. Bryunetkin, V.M. Dyakin, and A.Ya. Faenov
NPO VNIIFTRI, Mendeleevo, Moscow Region, 141570, USSR

1. Introduction

The possibility of receiving stimulated radiation in short-wavelength region on multiply charged ions transitions in laser plasma has attracted the attention of many research groups during the last decade. Significant resources have been employed in the work in this direction and, at present laser installations with beam power up to several terawatt are being used to create plasma. Various schemes aimed at inversion reception were investigated, it was reported on the facts of amplification in H-, Li-, Ne- and Ni- like ions of various elements in the spectrum region 40-250 Å (see, for example /1-6/ and the works cited). However some peculiarities of mechanisms used to form active mediums are far from being clear at present, this fact is testified by the well-grounded criticism of some results obtained in works /7-11/. Besides, amplification was observed in a single- passage mode i.e. without using optical resonator and some reports on observing two- and three-passage amplification /12,13/ may still be regarded as a demonstration.

We believe that many principle matters may be clarified within the more limited parameters of heating radiation and respectively when investigating amplification in transitions of multiply charged ions with wavelengths in visible and near ultraviolet regions. It is also important that in this spectral region it is possible to employ sufficiently dense optical resonator and the question about the existence of amplification or generation on any transition is solved unambiguously.

The principle of stage-by-stage advance into the short-wavelength region was implemented for the first time in our works /14-16/ where we investigated amplification formation and generation on 3-4 transitions, Li-like ion Be II (λ =

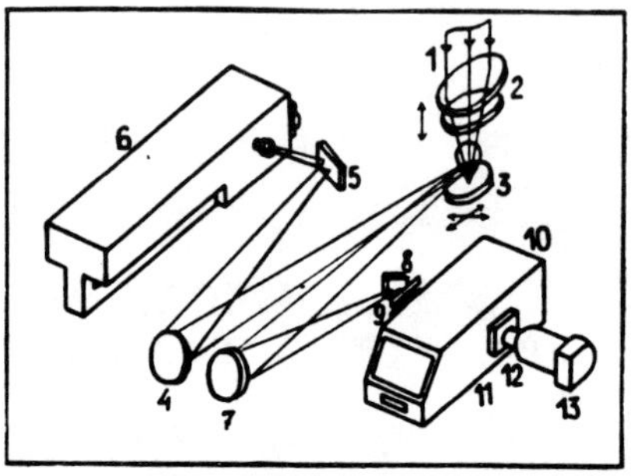

Fig.1.Placement of primary diagnostics used in our experiments.

436, 468, 527 nm) when laser plasma is expanding in the background gas and during its interaction with obstacles of various configuration. Working ion Be II is a single charged ion and therefore the transition to multiply charged ions arouses much interest.

This work reports on the results of recombining mechanism inversion research and generation observation on transition 4f - 5g in H-like beryllium in expanding laser-produced plasma.

II. Experimental setup

Experiments, the diagram is given in fig.1, were carried out by means of laser installations of two types: 1) Nd-glass laser, energy up to 50 J in the pulse with duration 3 ns on FWHM; 2) ruby laser with energy up to 10 J in the pulse with duration 8 ns. To produce plasma laser radiation (1) was focused by the lens (2) on the surface of a flat solid target made of beryllium (3), the flux density of heating radiation in all experiments described below was $q \approx (1 - 3)\,10^{11}$ W/cm^2. In the first spectroscopic channel the image of plasma flare was constructed with the help of mirrors (4,5) with reduction 1:2 in the input slit plane of 1-meter spectrograph (6) with dispersion 8 Å/mm and film registration. The spectrograph slit was positioned perpendicular to the target plane and spectra were recorded with the space resolution not worse than 0.1 mm in the direction of expansion axis of plasma flare. In the parallel channel plasma radiation was transferred to the monochromator input slit (10) by means of mirrors (7) and (8) which was provided with the space resolution diaphragm (9). Calibrated light filters (11) and plastic scintillator (12) can be installed on the monochromator output , further radiation was recorded by fast high-current photomultiplier (13). For this channel space resolution was ~ 1 mm, spectral ~2 nm, time ~ 3 ns. Both spectral channels were calibrated in the region 200 - 550 nm in relative units in accordance with standard methods. Special schemes of recording were used to investigate amplification and receive generation, their description is given in the respective sections.

III. Structure of plasma flare

In fig.2 one can see timely integrated photographs of beryllium plasma spectral region (a) and its representation in the light of separate spectral lines (b). It also contains the identification of transitions used for plasma parameters diagnostics. At the first stage we investigated velocities and angle characteristics of ion expansion of various ionization multiplicity. Velocity component which is perpendicular to a target surface, was determined by the shift of the maximum of the glow time bases in the ion lines

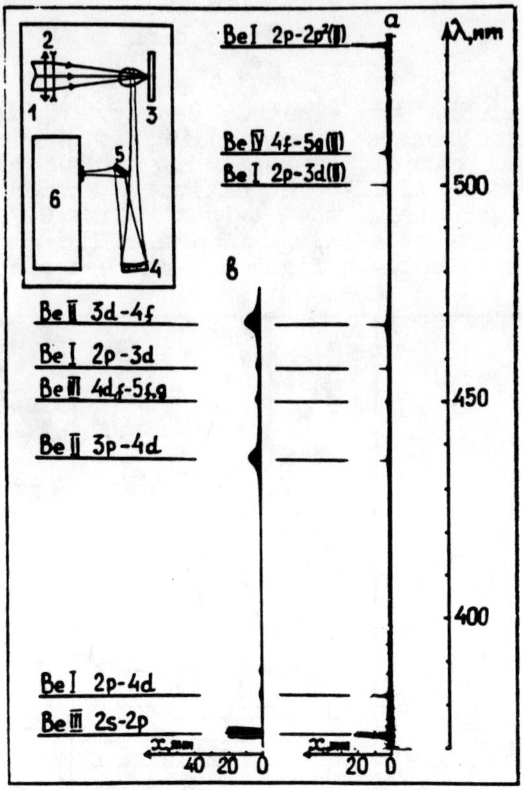

Fig.2. Time-integrated spectrograms (a) and spectroheliograms (b) of beryllium plasma (II-second order) in spectral region 200-500nm.

Be I - IV, recorded from the regions with dimensions 1x1 mm positioned at various distances from a target surface. For ions Be IV, III, II these velocities were $4\cdot10^7$, $1.3\cdot10^7$, $5\cdot10^6$ cm/s respectively. For further research it is important to point out that it is these groups of ions make the basic contribution in the timely integrated intensities of lines. However it is well known that ions of various multiplicity have different angle distribution and, respectively, various component of velocity which is parallel to a target surface and direction of observation. The availability of velocity component directed along the axis of observation leads to broadening of spectral line circuits (the so called the Doppler expansion broadening). According to the estimates this form of broadening predominated at long distances from the target (x > 0.7 cm in our experiments). This fact enables us to determine lateral (relative to expansion axis) velocities of ions with various multiplicity and assess respectively the angle of the expansion cone. Additionally the clearance angle of expansion cone is determined by spectroheliograms of plasma flare (fig.2b) and for ions Be IV, III, II it was respectively : 20°, 36°, 65°.

IV.Electron density

To determine the electron density N_e we employed line broadening parameters corresponding to the transitions with the principal quantum number being varied $n-n'=\Delta n=1$ ($n,n'=3$, 4, 5) in ions Be II- IV. The estimates showed that within the range interval from the target $0.1\leqslant x\leqslant0.7$ cm the major contribution to broadening of these lines is made by the linear Stark-effect. In fig.3 one can see spectral

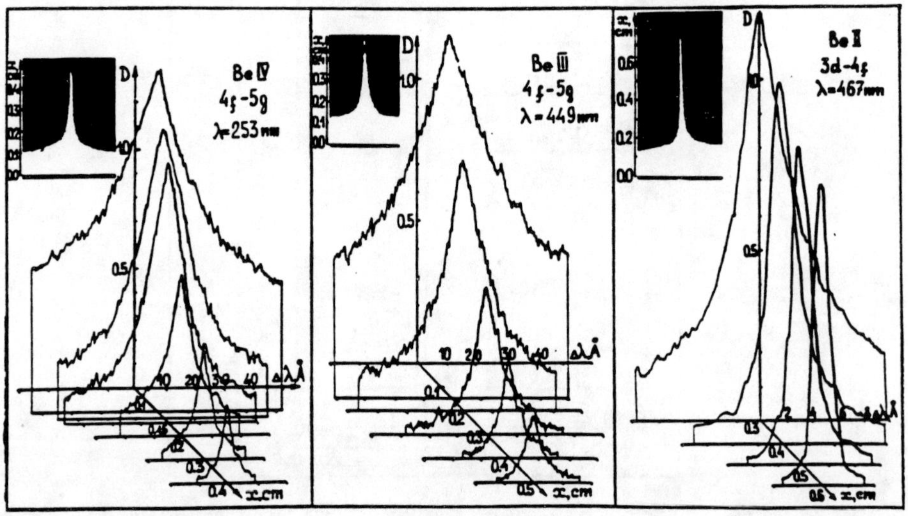

Fig.3.Space-resolved spectral lines of Be II-IV and it's densitograms.

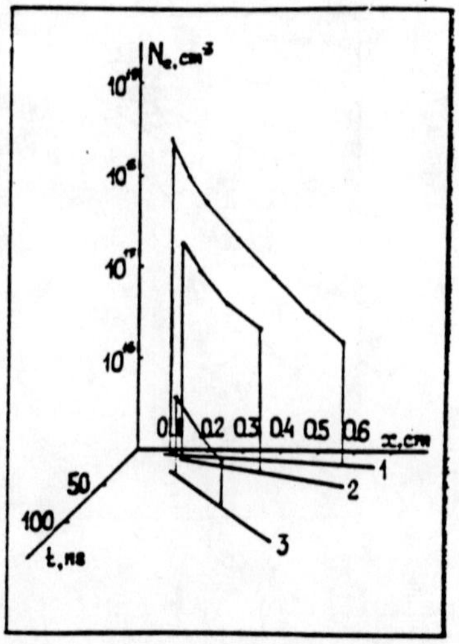

Fig.4.Electron density vs distance from target.

photographs (with spaced resolution) and densitograms of lines which were used for determining at various distances from the target. For transitions 4f-5g Be IV the computations of the line form /17/ were used to calculate the electron density, for transitions to Be III, II - the Holtzmark values of lines widths /18,19/ with the assumption that the levels are H-like. For all examined transitions there was also taken into account the contribution of the Doppler expansion and instrumental broadening . Profiles N_e at various stages of plasma flare expansion are given in fig.4 where curves 1, 2, 3 correspond to the groups of ions expanding with velocities $4\cdot10^7$, $1.3\cdot10^7$, $5\cdot10^6$ cm/s respectively.

V. Electron temperature.

In this work we were mainly interested in the possibility of producing inversion on the transitions 4f - 5g Be IV in the recombination regime, hence it was necessary to determine the electron temperature at exactly those moments of time when plasma is emitting at the designated transition. The method used to determine the electron temperature was as follows. On the basis of kinetic calculation /20/ with electron density values having been experimentally measured (curve 1, fig.4) we constructed the dependence of line intensities relationship I (2s -2p) to I (5g - 4f) of Be III ion under various values of electron temperature (fig.5a). The experimental values of intensities relationships I_{22}/I_{54} in every space point were taken at the instant of time corresponding to the maximum glow in the line 4f -5g Be IV i.e. when populations of the upper levels are

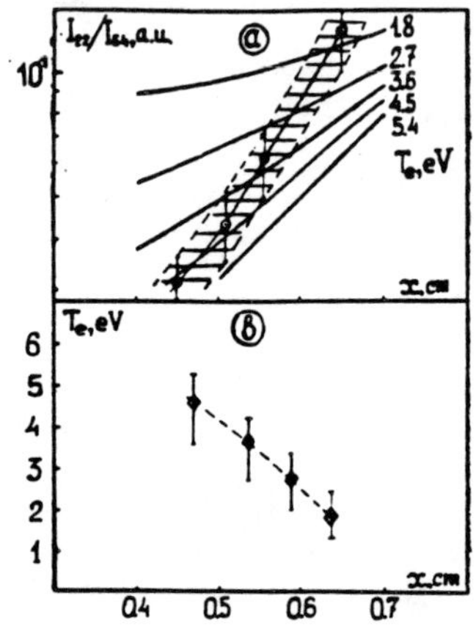

Fig.5.Results of calculation (a) and electron temperature (b) VS distance from target.

determined mainly by the recombination mechanism. It should be also taken into account that in general the measured intensity of the line 2s -2p Be III can strongly differ from the true as a result of radiation capturing effect. The estimates given with the account of experimentally measured width of this line showed that self-adsorption may be ignored at the distance from the target $x > 0.3$ cm. Therefore the values of electron temperature T_e (with an error of the experiment) obtained in such a way are given in fig.5b.

VI. Experimental measurement of the gain coefficient.

With respect to measurement results of plasma macroscopic parameters (section IV, V) the gain coefficient was estimated on the transition 4f -5g of Be IV ion: $5\cdot10^{-2}$ -10^{-1} cm^{-1} at the distance from the target $x \geqslant 0.4$ cm. The direct experimental determination of the gain coefficient is of great interest from the view point of conforming indirect estimates as well as for selecting optical resonator parameters for experiments in generation observation. This section describes experiments on alteration of the gain coefficient in expanding laser plasma by means of self-induced inspection.

In traditional methods of self-induced inspection to determine the gain (absorption) coefficient one mirror is usually used when there is no saturation. It is installed near the radiation source from the direction opposite to the recording device. In this case when recording is made in the distant zone the gain coefficient is determined in the following way:

$$g=(1/\ell)\ln[(A/R)(I_1/I_2-1)] \qquad (1)$$

where I_1, I_2 - radiation intensities with mirror being open or closed, ℓ -length of active medium, R - coefficient of a mirror reflection; A - factor~1, determined by the experiment geometry. In our case the values of the gain expected on the bases of the above mentioned estimates $g\ell \leq$ 1.1 and as it follows from (1) it is necessary to measure the relationship I_1/I_2 with accuracy better than 10%. Such precision is difficult to reach in the experiment, hence in this work the multipassage method was employed.

The experiment diagram is shown in fig.6a. The emission of ruby laser with the wave-front reversal (1) with energy up to 5 J and duration of 8 ns FWHM was focused in vacuum by the lens (2) on the surface of a hard target made of beryllium (3). The system of self-induced inspection included

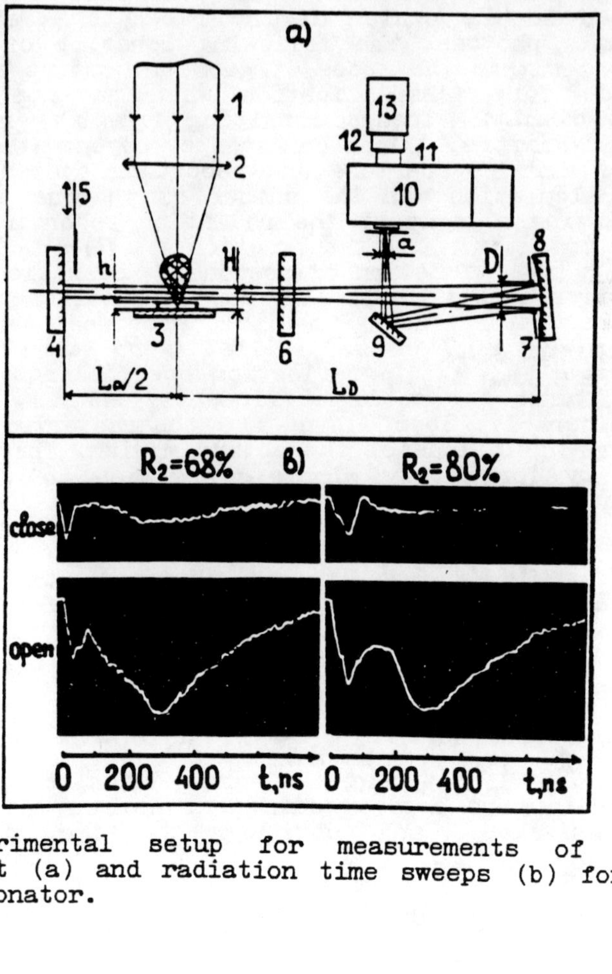

Fig.6.Experimental setup for measurements of the gain coefficient (a) and radiation time sweeps (b) for open and closed resonator.

two flat dielectric mirrors (4) and (6), its optical axis was parallel to the target surface. The mirror (4) could be overlapped by non-transparent screen (5). By means of spherical mirror (8) and flat mirror (9) plasma radiation was directed into the monochromator (10) with the input slit being in the mirror focus (8). The values of the system geometry parameters were as follows: L_r = 12cm, L_D = 180 cm, f = 80 cm, D = 8 cm, a = 0.6 cm, a = 0.4 cm. Radiation time sweeps were recorded at the wavelengths 253 nm with the rear mirror being open and closed by means of the PM-30 photomultiplier (13) and oscilloscope at the output of the monochromator (fig.6b).

Interpretation of these results is not that simple as in the case with the single passage diagram of self-induced inspection and demanded the development of the numerical model adequate to the experiment diagram. Calculation was carried out by means of the "large photons method" for real geometry of the experiment. It was considered that at the initial moment of time t_0 =0 the pin-point isotropic source positioned in the center of the resonator emits the pulse from n_0 photons. The following behavior of photons was described within the scope of geometry optics and allowance were made for: 1)amplification when passing through the active medium, 2) losses resulting from the reflection of resonator mirrors, 3)photon escape beyond the limits of resonator mirrors aperture and recording system. The result of the calculation was the number of photons which reached the receiving element of the radiation recording channel at the time moment t, i.e. the function $n_{ph}(t) = n_0 S_{ph}(t)$, where $S_{ph}(t)$ - function of response. Relaxation time τ of the function $S_{ph}(t)$ naturally depends on the quality of the system "resonator + active medium", provided that condition $R_1 R_2 \exp(-g\ell) \leq 0.7$ is fulfilled (were $g\ell$ - amplification per passage; R_1, R_2 - reflection coefficients of resonator mirrors) turns out to be sufficiently short: $\tau \cong 5$ ns, i.e.considerably less than the characteristic time of parameters variations of the active medium. This creates the opportunity for fairly simple transition to the situation when photon source in the resonator is not instantaneous. Actually let's assume that the emission function $n(t)$ (i.e. for the time period $t-\Delta t/2,\ t+\Delta t/2$ is emitted $\int_{t-\Delta t/2}^{t+\Delta t/2} n(t')dt'$ photons). Then the output signal may be presented as $n_{out} = \int n(t') S_{ph}(t-t')dt'$. Let's approximate $n_{out}(t)$ with respect to the interval $\Delta t \gg \tau$ in the presumption that over the time τ $n(t)$ is changing slightly:

$$\langle n_{out}(t)\rangle = (1/\Delta t)\int_{t-\Delta t/2}^{t+\Delta t/2} n_{out}(t')dt' = S_0 \langle n(t)\rangle \quad (2)$$

Here $S_0 = \int_0^\infty S_{ph}(\tau)d\tau$ depends on the parameters of the resonator and active medium but not on the function of photon emission and obviously the time, and $\langle n(t)\rangle$ averaged over the interval $\Delta t \gg \tau$ emission function. The value $S_0 = S_0(R_1, R_2, g\ell)$ was calculated in the program for various resonators with the most simple (δ-like) emission function. Using the introduced function S_0 let's write down

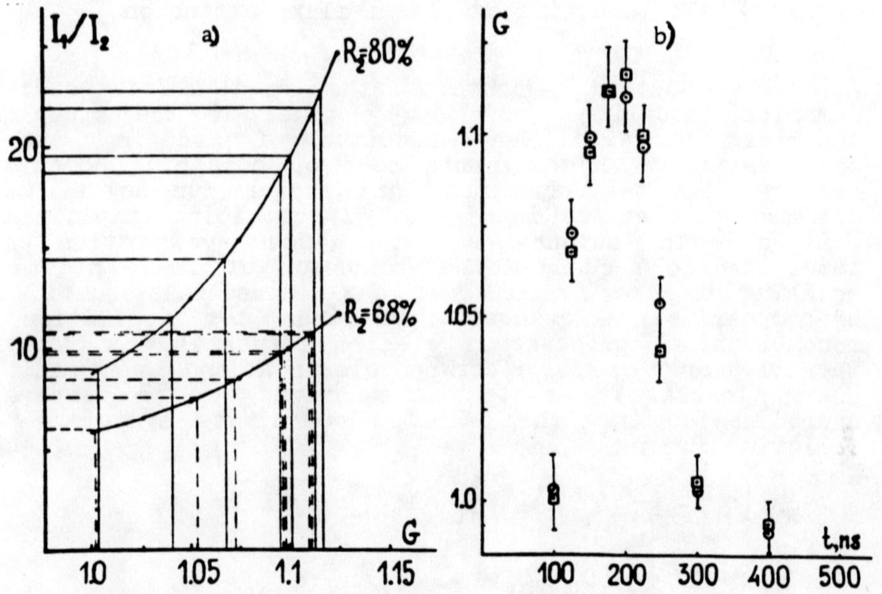

Fig.7.Calculated relation ships of output signals for open and closed resonator (a) and gain factor VS time (b).

the relationship of output signals when resonator is open and closed:

$$I_1(t)/I_2(t) = S_0(R_1, R_2, g\ell)/S_0(R_1=0, R_2, g\ell)$$

Computation results are given in fig.7a in the forms of dependencies of intensities with rear mirror (4) being open and closed from the gain factor $G = \exp(g\ell)$, with values of output mirror reflection coefficient R_2 =68% and 80%. Experimental values of this relationship at various moments of time were used to receive amplification factors values corresponding to these moments of time by means of the calculated curves, fig.7a. The results obtained in such a manner (with the experiment error) are given in fig.7b. Characteristic time of gain existence ~ 150nc, maximum of the gain factor G_{max} ~1.1, average value (FWHM) G_{avr} ≅1.06. With the length of the active medium ℓ ~1 cm it corresponds to the gain coefficient in transition 4f-5g Be IV (λ = 253 nm) g = (0.05-0.1)cm^{-1} that confirms the results of the indirect measurements.

The presented results show that if the active medium with length ℓ~1÷2 cm is created and sufficiently dense optical resonator is employed it is possible to achieve the generation on the transition 4f-5g Be IV.

VII. Modeling of plasma flare expansion

This section contains the results of conditions numerical modeling which took place in the experiments described above. The objective of modeling was the optimization of hydrodynamic code by comparing computation results with the experiment and search for active medium optimum parameters in terms of obtaining amplification dealing with advance to the shortwave region along isoelectronic sequence. The expansion of laser plasma was modeling by the system of one-dimensional equations of hydrodynamics. Allowances were made for ionization and recombination processes, electron thermal conductivity, thermal energy exchange between electrons and ions, radiation energy losses. The system of equations for two-temperature hydrodynamics in the Lagrange coordinate system has the following form:

$$\frac{\partial}{\partial t}\left(\frac{1}{\rho}\right)=\frac{\partial}{\partial s}(r^2 u)\ ,\ \frac{\partial r}{\partial t}=u$$

$$\frac{\partial u}{\partial t}=-r^2\frac{\partial}{\partial s}(p_e+p_i)$$

$$\frac{\partial \varepsilon_i}{\partial t}=-p_i\frac{\partial}{\partial s}(r^2 u)+\frac{1}{\rho}Q_{\Delta T} \qquad (3)$$

$$\frac{\partial}{\partial t}(\varepsilon_e+\varepsilon_{ion})=-p_e\frac{\partial}{\partial s}(r^2 u)-\frac{1}{\rho}(Q_{\Delta T}-Q_{rad})-\frac{\partial}{\partial s}(r^2 q_e)$$

where S - the Lagrangian mass coordinate determined by the relationship: $S=\int\rho(r')r'^2dr'$; t - time, r -Eiler coordinate; ρ - density; u - speed; p_e , p_i -pressure of electrons and ions; ε_e and ε_i - thermal energy of electrons and ions per a unit of mass; ε_{ion} - energy spent on ionization per unit of mass; $Q_{\Delta T}$ and Q_{rad} determine respectively the exchange of thermal energy between electrons and ions, radiation losses; $q_e=-\varkappa_e dT_e/dr$, $\varkappa_e$ - coefficient of electron thermal conductivity.

For electron and ion components it is expected that equation of ideal gas condition is fulfilled:

$$\varepsilon_e=p_e/(\gamma-1)\rho\ ,\quad p_e=N_eT_e \qquad (4)$$

$$\varepsilon_i=p_i/(\gamma-1)\rho\ ,\quad p_i=N_iT_i\ ,\quad \gamma=5/3$$

where N_e, N_i - concentration of electrons and ions; T_e , T_i - their temperatures. Charge composition of plasma is determined by the equation of the ion composition balance:

$$\frac{\partial}{\partial t}(N^{(z)}/N_i)=\Gamma^{(z)}\ ,\ z=1,2\ldots z_{nuc} \qquad (5)$$

Here $N^{(Z)}$ - ion concentration with spectroscopic symbol Z , Z_{nuc} - nucleus symbol, $\Gamma^{(Z)}$ - speed of ion appearance in a unit of volume as a result of ionization and recombination processes. Power of radiation losses Q_{rad} is determined by the charge composition, temperature and electron density. We assume that plasma is thin for its own radiation.

The solution method (3) - (5) was taken from the work /21/ and is based on splitting physical processes. Let's assume that at the moment of time t_0 we know values of all quantities.The calculation of these values at the time moment $t_0+\tau$ is carried out in three stages. At the gas dynamics stage we calculate gas movement under pressure forces and variation of its inner energy owing to the operation of pressure forces. In doing so we consider thermal conductivity processes, e-i exchange, ionization and recombination are frozen. At the second, thermal stage it is assumed that there are no gas dynamics movement, ionization and recombination processes and temperature variations of electrons and ions due to e-i exchange and electron thermal conductivity are calculated. When solving the equations at the thermal stage, values obtained after the gas dynamics stage are regarded as the initial condition. At the third kinetic stage the variation of plasma charge composition owing to the process of ionization and recombination and temperature variation of electrons due to the non-elastic collisions are taken into account. The processes of electron thermal conductivity, e-i exchange are assumed to be frozen as well as the absence of gas dynamics movement. Ionization and recombination speeds included in the right part of the equation (5) as well as power of energy expenditure for ionization and recombination were determined from the tables /22/ by means of interpolation.

The following calculated plasma characteristics were compared with the experiment:

a) electron density at a distance x from a target averaged over time with the weight of line intensity 5g-4f Be IV:

$$N_e(x)=\int_0^t N_e(x,t)I_{54}(x,t)dt\Big/\int_0^t I_{54}(x,t)\,dt$$

b) time dependencies of line intensity 5g-4f Be IV $I_{54}(t)$ in the fixed points x ;

c) maximum value of intensity line 5g-4f Be IV in the point x: $I_{54}^{max}(x)$;

d) relationship of lines intensities I(2p → 2s) and I(5g -4f) Be III at the moment of maximum $I_{54}(x,t)$ as the x function.

As is seen from the results presented in fig.8 by and large the calculation correctly describes the experimental situation. However, it should be pointed out that the calculated profile $N_e(x)$ is dropping slightly slower than that in the experiment. This event is evidently linked with multidimensional character of plasma expansion and it is this occurrence that brings about the difference of calculated and experimental values of the most sensitive to the plasma parameters the relationship of intensities I_{22}/I_{54} Be III ion.

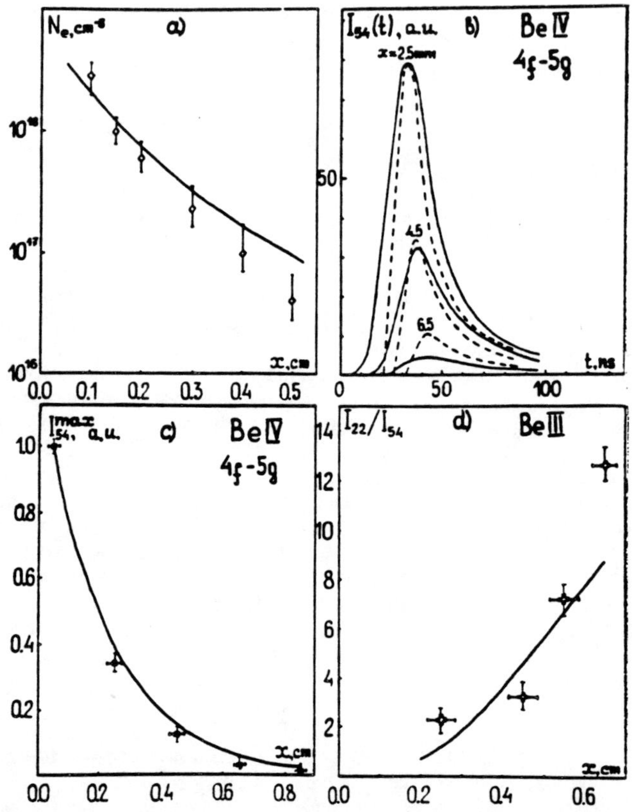

Fig.8.Plasma parameters computation results and it's comparison with experiment (see text).

Therefore the calculation shows that plasma parameters at a distance $x \geqslant 0.4$ cm are close to the optimum in terms of obtaining maximum inversion in transition 4f - 5g Be IV and gain coefficient can reach ~ 0.1 cm^{-1}.

VIII. Generation at the wavelength 253 nm.

In the experiment on generation observation, with the diagram shown in fig.9, plasma was produced inside the optical resonator made by the mirrors (1), (2) when focusing radiation of Nd-laser (3) on the surface of a solid target made of beryllium (4) into the system comprising 5 points positioned on the straight line with 8 mm interval. The resonator length was ~20cm, transmission coefficients of mirrors T_1 = 0.1% and T_2 =0.3% at the wavelength 253 nm. The output radiation by means of the quarts lens (7) with

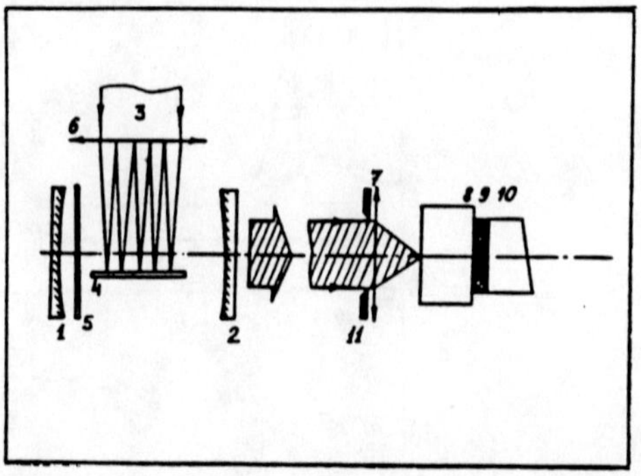

Fig.9.Experimental setup for observation of generation.

f = 10 cm was directed to the monochromator (8) and was recorded at the output by the high-current PM (10). The diaphragm (11) was placed before the lens with the variable diameter form 2 to 16 mm, the distance from the output mirror of resonator was L ~ 150 cm. The output slit of the monochromator was in the lens focus, so that radiation was recorded only at the flat angle $\leqslant 10^{-2}$ rad near the resonator axis.

The typical pulse form of the generation and dependence of its power from length of the active medium are given in fig.10. Experiments were carried out with the fixed density of the flux of heating radiation in the focusing points ($3 \cdot 10^{11}$ W/cm^2), length of the active medium was varied by changing the number of plasma clusters being produced. The linear growth of generation power P_{gen} was observed depending on the length of the active medium that qualitatively corresponds to the quasistationary generation in the conditions of the experiment (life time of inversion $\tau_{inv} \sim 100$ ns is much longer that the life time $\tau_r < 1$ ns of photon in the resonator). As the length of the active medium could be varied only discretely, the threshold of the generation in our experiments was determined by extrapolation of data, fig, 10b. The values of the threshold length of the active medium $l_{thr} \cong 0.11$ cm obtained by this method can be used to estimate the threshold gain coefficient $g_{thr} \sim (4\text{-}5) \cdot 10^{-2} \text{cm}^{-1}$, that agrees with the results of direct measurements, par.VI. The measurement of divergence was made by changing the diaphragm diameter (7, fig.9), generation signal stops growing with diameter ~ 0.6 cm that corresponds to the divergence ~ $4 \cdot 10^{-3}$ rad.

Hence, in this work the processes of formation recombination parameters of laser active medium in transition 4f-5g Be IV were investigated in expanding laser plasma, the generation was obtained for the first time in this transition by means of optical resonator. In part the obtained were presented in works /23-25/.

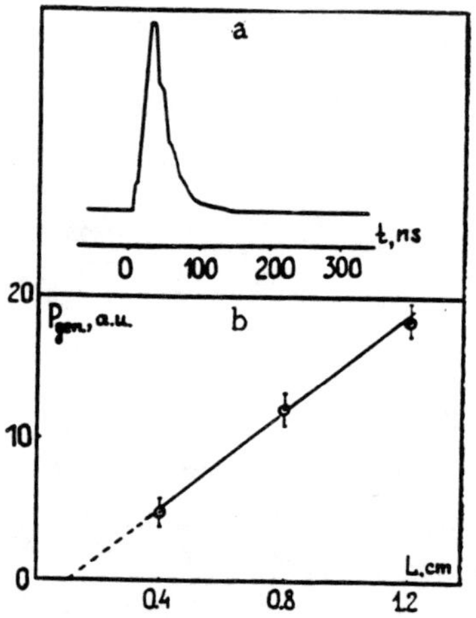

Fig.10.Pulse form of the generation (a) and it's power vs length of the active medium (b).

Basic parameters of the active medium and generation:
Wavelength - 253 nm
Gain coefficient - $(5-10)\cdot 10^{-2}cm^{-1}$
Generation energy - 50-100 mJ
Pulse duration - 50-60 ns
Power - 10^3 W
Divergence - 4 mrad

IX. Conclusion

In conclusion let's formulate our findings which proceed from the results of this work:

- recombination mechanism for inversion production in expanding laser plasma was implemented in the experiment;
- for the first time the recombination laser was launched in transition of multiply charged ion by means of optical resonator;
- transitions between the highly-excited levels (n= 3,4,5) are preferable for obtaining generation in the recombination regime as they make it possible to eliminate effects connected with radiation reabsorption;
- recombination mechanism is a very promising direction for creating short-wave region lasers on transitions of multiply charged ions.

REFERENCES

[1]C.J.Keane et al. J.Phys.B: 1989, v.22, N21, p.3343-3362
[2]C.H.Skinner et al. IEEE Trans. of Plasma Science, 1988,

v.16, N5, p.512-519
[3]C.L.Lewis et al. Plasma Phys.and Contr.Fusion, 1988, v.30, N1, p.35-44.
[4]B.Yaakobi et al. IEEE Trans. of Plasma Science, 1988,v.16, N5, p.505-511.
[5]G.Jamelot et al. IEEE Trans. of Plasma Science, 1988,v.16, N5, p.497-504.
[6]P.R.Herman et al. IEEE Trans. of Plasma Science, 1988, v.16, N5, p.520-528.
[7]J.F.Selly et al. Opt Commun., 1985, v.54, N5, p.289-294
[8]J.P.Appruse et al. Phys.Rev.Lett., 1985, v.55, N18, p.1877-1879.
[9]V.I.Derzhiev, A.G.Zhidkov, S.A.Mayorov, S.I.Yakovlenko.DAN USSR (in Russian), 1987, v.294, N3, p.588-591.
[10]V.I.Derzhiev, A.G.Zhidkov, S.A.Mayorov, S.I.Yakovlenko. Kvantovaya Elektronika (in Russian), 1988,v.15,N2, p.412-421.
[11]A.V.Gulov, V.I.Derzhiev et al. J.Phys.B., 1990,v.23, N5, p.L67-L70.
[12]N.M.Ceglio et al. Applied Optics, 1988, v.27,p.5022-5025.
[13]N.M.Ceglio et al. Opt.Commun., 1989, v.69, N3,4,p.285-288
[14]V.A.Boiko, B.A.Bryunetkin, F.V.Bunkin etc. Preprint FIAN, (in Russian) N237, M., 1982.
[15]V.A.Boiko, B.A.Bryunetkin, F.V.Bunkin etc. Kvantovaya Elektronika (in Russian), 1983, vol.10, N5, p.901-903.
[16]V.A.Boiko, B.A.Bryunetkin, F.V.Bunkin etc. Kvantovaya Elektronika (in Russian), 1985, vol.12, N11, p.2340-2341.
[17]V.I.Derzhiev, A.G.Zhidkov, A.I.Magunov, S.I.Yakovlenko. Preprint IOFAN (in Russian), N56, M., 1986, 37 p.
[18]H.R.Griem Astroph.J., 1960, vol.132, p.883-893
[19]C.F.Hooper.Phys.Rev., 1968, vol.165, p.215-222
[20]I.Yu.Skobelev, S.Ya.Hahalin, S.I.Yakovlenko. Plasma Physics (in Russian), 1987, vol.13, N7, p.872-875.
[21]S.A.Mayorov. FVM (in Russian) and MF, 1985, vol.26, N7.
[22]I.Yu.Skobelev, S.Ya.Hahalin, S.I.Yakovlenko. Preprint IOFAN (in Russian) N32, M., 1985.
[23]B.A.Bryunetkin, etc. Pis'ma Zh.Tech.Fiz. (in Russian), 1986, vol.12, issue 10, p.613-618.
[24]B.A.Bryunetkin, etc. Kvantovaya Elektronika (in Russian), 1989, vol.16, N3,p.407-409.
[25]B.A.Bryunetkin, etc. Zh.Tech.Fiz. (in Russian), 1989, vol.59, issue 11, p.110-112.

EXCITATION OF ^{235}U NUCLEI ON THE ISOMETRIC LEVELS IN A HOT PLASMA AS A STEP IN INVESTIGATION OF THE POSSIBILITY OF COHERENT GAMMA-RAY GENERATION

R.V. Arutyunyan, V.Yu. Baranov, L.A. Bol'shov, V.D. Vikharev, S.A. Dorshakov, V.V. Kol'tsov, V.A. Kornilo, A.A. Krivolapov, D.D. Malyuta, G.A. Polyakov, A.A. Rimsky-Korsakov, V.P. Smirnov, and E.V. Tkalya

Branch of the Kurchatov Atomic Energy Institute, 142092, Troitsk, Moscow Region, USSR

The problem of nuclear excitation in the hot and dense plasma is definitely actual due to some possible scientific and technological applications of nuclei with the low isomeric level and nuclei which have a short-living level near a long-living ($T_{1/2}$ >1s) isomeric level. The following problems are of the great interest in this line:

- investigation of nuclear excitation mechanisms in plasma;
- creation of a system with inverse nuclear level population that is necessary to materialize a γ-laser concept;
- study of physical processes in the system of excited isomeric nuclei;
- acceleration of isomers decay in the intense external field;
- development of new methods for the plasma diagnostics, etc.

The isomeric state of ^{235}U was until recent years the unique known nuclear level with the very low transition energy. The energy is 76.8 eV and the half-life is 26 min.

It is assumed that the nuclear excitation in a hot plasma can be attributed to the inverse internal electron conversion mechanism [1], the inelastic electron scattering by the nucleus [2], and also the resonant absorption of the plasma emission [3].

Morita [4] pointed out another theoretical mechanism named "Nuclear Excitation by Electron Transition" (NEET). An essence of NEET for ^{235}U isotope is the resonance between nuclear and atomic energy levels. If an electron hole is produced in the $5d_{3/2}$ level and an electron transition a nuclear excitation. In order to create the electron hole in the $5d_{3/2}$ level of ^{235}U, one can use a hot plasma (for example, a laser-produced uranium plasma).

The results of the first experimental work on ^{235}U nuclear excitation in a plasma produced by a TEA CO_2 laser pulse (1 J in 100 ns) at λ=10.6 μm with a repetition rate of 0.5 Hz, was reported by Izawa and Yamanaka [5]. A laser beam was focused by a Ge lens on a metallic natural (0.71% ^{235}U) uranium target in a vacuum chamber.

A number of produced isomers was estimated to be $N^* \approx 10^3$ nuclei/laser pulse. The nuclear excitation rate value of $\langle\sigma_N \cdot v_e\rangle = 1.4 \cdot 10^{-20}$ $cm^3 \cdot s^{-1}$ was obtained, where σ_N is the NEET cross-section and v_e is the electron velocity. Brackets indicate the averaging over the Maxwell distribution of the electron velocities. A value of $\sigma_N \sim 10^{-29}$ cm^2 was estimated by using a value of electron temperature of the plasma $T_e \sim 100$ eV.

This cross-section is one order of magnitude smaller than calculated by Okamoto [6]. However, a more correct calculation [7] demonstrated that the NEET probability evaluation in [6] overestimates the cross-section value by six orders of magnitude as compared to the realistic value.

A nuclear excitation of ^{235}U due to ionization of the atoms in uranium foil by continuous electron beam (E_e=600 eV, i=20 mA, t=30 min) was investigated by Zhudov et al. [8]. Enrichment degree of $^{235}UO_{2+x}$ (0<x<1) target was 99.995%. Excitation cross-section $\langle\sigma_N\rangle \sim 10^{-32}$ cm^2 was obtained, but the signal-to-noise ratio was very small. In this case, half-lives of the decays were defined to be $24 < T_{1/2} < 144$ min for the first and $6 < T_{1/2} < 30$ min for the second experiments.

Thus there is a serious discrepancy between theoretical evaluations made in the frames of known mechanisms and measured cross-section values obtained in not-too-numerous experiments. Therefore a new experimental research is needed at first to confirm the principle possibility of nuclear excitation effect in plasma.

Results of a nuclear excitation of ^{235}U from the ground state to the isomeric state in a plasma produced at the surface of solid targets by the powerful laser irradiation or the relativistic electron beam (REB) are presented in this report.

LASER EXPERIMENT. A scheme of the interaction chamber for this experiment is shown in Fig.1a. We used a TEA CO_2 laser with pulse energy 5 J; pulse width, 200 ns; wavelength, 10.6 μm, and repetition rate, 10Hz. A target was made of uranium ceramics (UO_2) enriched with ^{235}U up to about 6%.

The laser beam was focused normally on the target (5) by a NaCl lens (1) in the vacuum ($p_1 \sim 10^{-2}$ Torr) chamber (3). The NaCl plate was used as entrance window (2) of the chamber. Maximum intensity of the laser irradiation was approximately 10^{11} Watt/cm^2.

The parameters of the laser-produced plasma were not measured directly. The estimation of electron density ($n_e < 10^{19}$ cm^{-3}) and moderate temperature ($T_e \sim 10$ eV) was based on previous experience.

A vaporization of the target occurs from the surface and is accompanied by ejection of material. This material consists the neutral atoms, ions, and electrons of the plasma and also some macroscopic particles of the uranium ceramics. To cut off these particles, a copper diaphragm (4) (transparence of about 10% for vapors) was used .

Uranium ions and neutral atoms were collected on copper foil plates (Cl1 and Cl2). The diameter of collectors was 20 mm, the distance between collectors was 20 mm, the distance between collectors and the uranium target was 50 mm.

Measurements of number of isomers ^{235m}U was based on registration of internal conversion electrons accompanied by characteristic decay like in [7] and was carried out in the following way. After irradiation of the target by either laser beam or REB, a collector coated by evaporated uranium, was transferred into the second diagnostic vacuum chamber by the pivot through the special port without

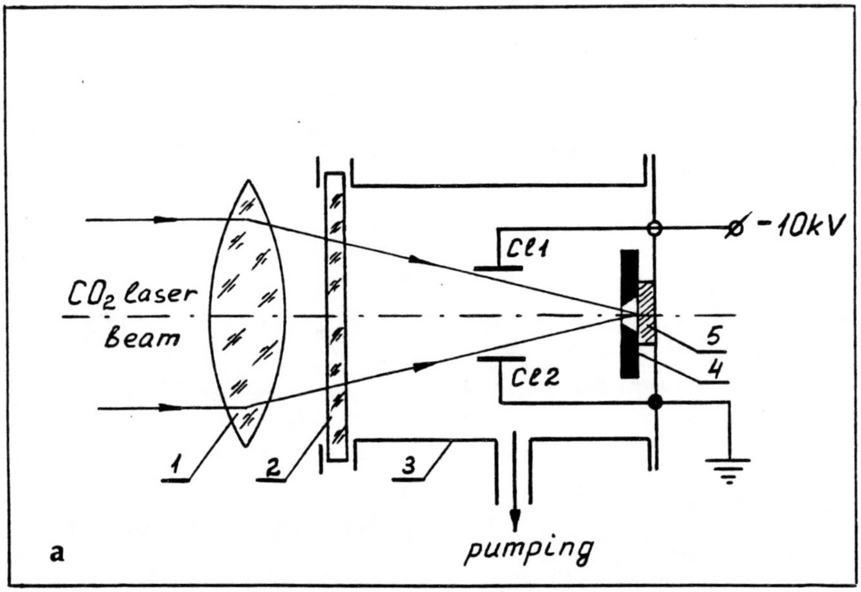

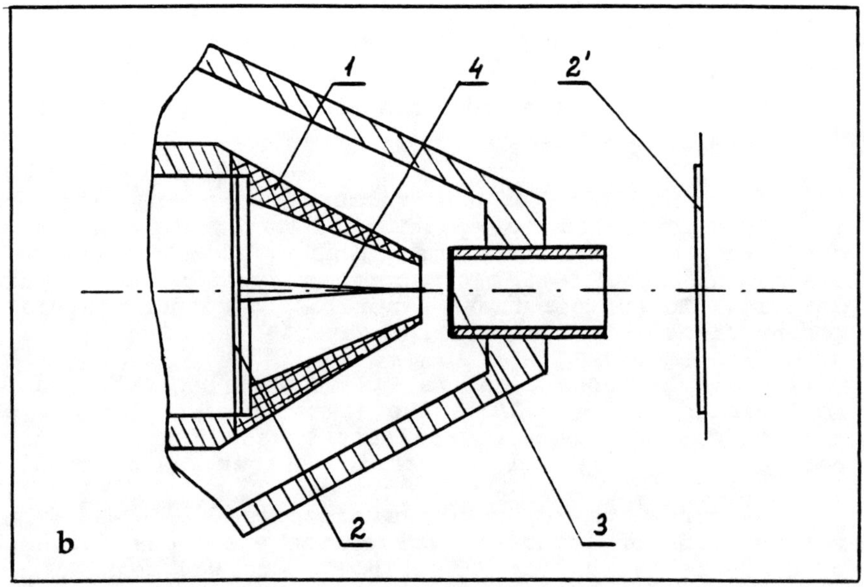

FIG.1. A scheme of interaction chamber for laser experiment (a); A diode construction of "TRITON" accelerator for the experiment with REB (b).

degrading the vacuum ($p_2 < 10^{-5}$ Torr) inside. Measurement of the electron counting rate from the collector was started within 5-6 min after laser or REB irradiation of the target. The system with different chambers for irradiation and measurement has an essential advantage as compared with single-chamber system described in [7], namely the electron multiplier operates in better and unchanging conditions all the time of an experiment.

As the detector we used a channel electron multiplier with preamplifier and special counter. The system operated in electron counting mode. The distance between the collector surface and multiplier entrance was nearly 15 mm. The grid and the multiplier entrance were grounded, the collector was kept at the potential of -300V. This improves collection of very soft electrons and an efficiency of the counting system. We assume its value to be $\eta \approx 50\%$. The detector had the about 10÷20 counts/min background.

The whole system was calibrated in absolute units of uranium isomers with the aid of collecting the recoil atoms from the α-decay of a very thin plutonium-239 source.

The number of uranium atoms on the collector was measured by the α-spectrometer, which consists of a semiconductor detector, an electronic amplifier and a multichannel analyser NOKIA LP 4900B.

After obtaining the experimental dependence of the counting rate as a function of time, the mathematical processing was carried out. This processing was essentially the minimization of the sum of deviations squares of the experimental data from the dependence

$$n(t)=A\cdot\exp(-t\cdot\ln2/T_{1/2})+C, \quad (1)$$

where A is the initial counting rate of detection system, $T_{1/2}$ is half-life of the ^{235}U isomers, C is the background. Constants A, $T_{1/2}$, C and their standard deviations were determined.

In the laser experiment, several series of the irradiation of the uranium-containing targets were made. After each series Cl1 collector was transported into the diagnostic chamber. A decrease of the counting rate of the detector with time was not observed in every measurement. The constant counting rate was caused by α-emission from the ^{234}U, ^{235}U and ^{238}U atoms found on the collector. The absence of the excitation effect of the ^{235}U isomeric level is supposed to be due to low electron temperature of plasma. An excitation of isomers ^{235m}U apparently has demonstrated the threshold character and the best efficiency is reached for $T_e \approx E_{ie} = 76.8$ eV. As a result, the upper bound estimation for the cross-section of excitation process $\sigma_e < 10^{-32}\,cm^2$ was obtained.

EXPERIMENT with REB. This experiment was carried out at "TRITON" high-current accelerator with 500 keV electron energy, 150 kA beam current, 30 ns pulse duration. In comparison with the laser experiment mentioned above and also [7], here a plasma with a slightly higher value of $n_e \cdot \tau = 10^{12} \div 10^{13}\ cm^{-3} \cdot s$ was expected along with essentially larger volume occupied by a plasma formed from uranium samples (containing nearly 100% of ^{235}U). All these circumstances should lead to the production of the number of isomers per pulse approximately four orders higher than in [7].

The uranium target was placed in the diode of the accelerator. A diode construction is shown in Fig.1b. A hollow polyethylene cone was used as a cathode (1) of the diode. A copper foil collector (2) was placed inside the cathode. The uranium vaporized from the target-anode (3) was deposited on the collector. In series of the experiments a collector (2') was placed near the anode. A stainless steel needle (4) creating favorable beam focusing conditions was situated at the cathode axis. The foils with different ^{235}U isotope content and mentioned above uranium ceramics examples was used as the targets. Characteristics of the foil targets are summarized in the table 1.

For the targets of all types, several tens of experiments has been made. But only in experiments with A- and D- targets the decrease of number of particles hitting the detector was observed.

The following circumstances support the assumption that ^{235m}U have been produced in the plasma:

a) A well-known half-life of ^{235m}U ($T_{1/2}$=26 min) is within the trust interval of $T_{1/2}$ obtained by mathematical processing of experimental data with the aid of (1).

Table 1. Characteristics of the foil targets.

Type of target	Uranium, mg/cm^2	Al-plate, mg/cm^2	Isotope content, %		
			U-234	U-235	U-238
A	20	10	0,8	93,4	5,8
B	20	10	-	0,6	99,4
C	190	-	0,03	~2,3	97,7
D	1,5	1,6	0,002	99,99	0,002
E	1,5	1,6	0,003	0,003	99,99

b) A direct dependence between the quality of electron beam focusing at the foil target and following decay curve registration was marked in the experiments.

c) For the B-, C- and E-targets (contained less than 90% ^{235}U), a detector counting rate decrease with time has never been observed in the experiments.

To diminish the error of half-life definition when the signal-to-background ratio is smaller than 1, a summation of experimental data received in the identical conditions was used in our case. For example, a sum of four decay curves (D-targets) and a following mathematical processing (see Fig.2) gives the value of $T_{1/2}$=24.6±6.8 min, which is fairly close to 26 min.

A input and measuring procedure for the some collectors was repeated 10-15 hours after the target irradiation. A counting rate in this case was not independent of time.

Thus, assuming that the produced decay curves characterize the ^{235}U nuclear excitation process in the plasma, the cross-section $\sigma_{\bullet}$ can be estimated. The energy α-spectrums from collector surfaces were taken for this purpose. It gave a possibility of the average thickness of collector layer definition (δ) by the certain isotope composition of the uranium. We assumed efficiency of the counting system η=50% and a range of conversion electron in the substance λ=50Å. The total number of isomers in our experiment is given by

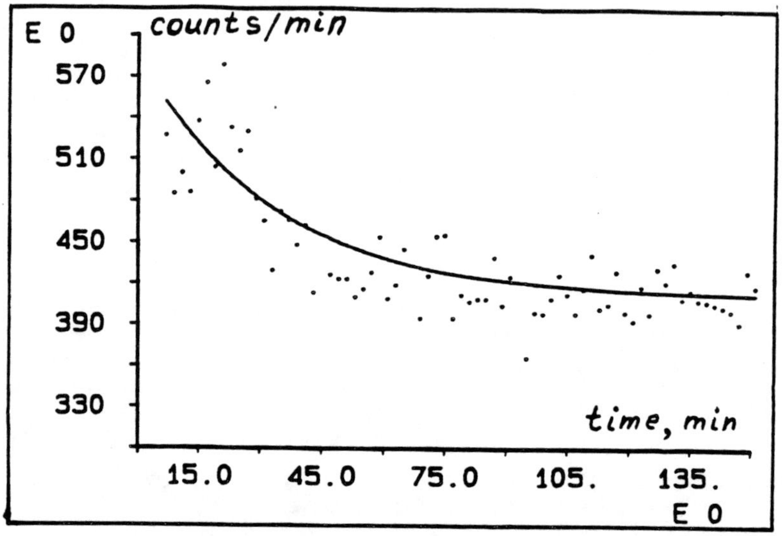

FIG.2. A sum of four experimental decay curves and a result of the mathematical processing.

$$N^* \cong \frac{A \cdot T_{1/2} \cdot \delta}{\eta \cdot \lambda \cdot \ln 2} \tag{2}$$

and excitation efficiency

$$\xi = \frac{N^*}{N}, \tag{3}$$

where N is the total number of ^{235}U nucleus.

Eqs.(2,3) give the maximum value of $\xi = 8.5 \cdot 10^{-13}$ for the experiments with A-type targets. This value is one order of magnitude smaller than for D-type targets. Probably this fact can be explained by more efficient deposition of aluminium atoms at the collector because here a collector was placed in the position 2' (see Fig.1b). The value of $\sigma_{\bullet}$ thus obtained is then used to calculate ξ using the formula:

$$\xi \approx n_e(T)\cdot\tau\cdot\exp(-\Delta E/T)\cdot(2\Delta E/m_e)^{1/2}\cdot\sigma_e \quad . \qquad (4)$$

Here T is the plasma cloud temperature (~ 20 eV), τ is the plasma life time (30ns), ΔE=76.8 eV, m_e is the electron mass, $n_e\tau = 10^{12} \div 10^{13}$ $cm^{-3}\cdot s$. An estimation of the excitation cross-section $\sigma_e \sim 1\cdot 10^{-32} \div 1\cdot 10^{-31}$ cm^2 was obtained. The value of σ_e is at least two to three orders of magnitude less than the one in well-know experiments of Izawa and Yamanaka [7] *and* very close to the results of the work [8].

REFERENCES

[1] V.I.Gol'dansky and V.A.Namiot, J.Nucl.Phys. v.33, N2, (1981) p.319.

[2] D.P.Grechukhin and A.A.Soldatov, Preprint IAE-2976, (1978) I.V.Kurchatov Institute of Atomic Energy, Moscow.

[3] R.V.Arutyunyan , L.A.Bol'shov, V.F.Strizhov, E.V.Tkalya, Preprint IAE-4864/2, (1989) I.V.Kurchatov Institute of Atomic Energy, Moscow.

[4] M.Morita, Progr.Theor.Phys. v.49, N5, (1973) p.1574.

[5] K.Okamoto, J.Nucl.Sci.Techn. v.14, N10, (1977) p.762.

[6] D.P.Grechukhin and A.A.Soldatov, Preprint IAE-2706, (1976) I.V.Kurchatov Institute of Atomic Energy, Moscow.

[7] Y.Izawa and C.Yamanaka, Phys.Lett. v.88(B), (1979) p.59.

[8] V.I.Zhudov, V.M.Kulakov, B.V.Odinov, A.D.Panov, In book: Precision Measurements in Nuclear Spectroscopy, Vilnius, (1984) p.109.

OPTICALLY PUMPED X-RAY FREE ELECTRON LASER AS A TWO-LEVEL QUANTUM GENERATOR WITH FULLY INVERTED ACTIVE MEDIUM

E.M. Belenov, S.V. Grigoriev, A.V. Nazarkin, and I.V. Smetanin

P.N. Lebedev Institute of Physics, Academy of Science, Leninsky Prospect 53, Moscow, 117924, USSR

1. INTRODUCTION

One way to create an X-ray free electron laser (FEL) is connected with using of counter laser wave as an undulator field (so called dynamic wiggler /1-3/). The dynamic wiggler FEL is characterized by relatively small energy of the used electrons (1-10 MeV). In order to achieve high values of X-ray gain it is necessary to use rather monoenergetic electron beams ($\Delta\gamma/\gamma \lesssim 10^{-4}$) and the wigglers with a large number of periodic elements ($N \gtrsim 10^4$) /2/. In such "optically pumped" X-ray FEL the energy of X-ray quantum can exceed the induced emission (absorption) linewidth. Adequate description of the laser operation in this case can't be done in classical terms, and a quantum approach is necessary. This statement is explained below.

The amplification in dynamic wiggler FEL can be interpreted as a stimulated Compton back-scattering /1/. When the pump frequency ω_i and generated frequency ω_s are fixed, the energy and momentum conservation laws define the energy of emitting (absorbing) electrons /4/:

$$\mathcal{E}_{e,a} = \mathcal{E}_c \pm \tfrac{1}{2}\hbar\omega \quad , \qquad (1.1)$$

where $\mathcal{E}_c = mc^2(1 - \omega^2/c^2K^2)^{-1/2}$ is the energy corresponding to classical condition of electron synchronism with a combination wave, $\omega = \omega_s - \omega_i$, $K=K_s+K_i$; $K_{s,i}$, are the wave numbers of pumping (i) and generated field ($\omega_i \ll \omega_s$).

Absorption and emission lines have its centres at the energies $\mathcal{E}_a$ and $\mathcal{E}_e$,respectively. They can have both homogeneous (because of finite interaction time) and inhomogeneous broadening (because of energy spread of the electron beam).The homogeneous and inhomogeneous linewidths are

are given by

$$\Delta\mathcal{E}_{hom} = \mathcal{E}(4N)^{-1}, \quad \Delta\mathcal{E}_{inhom} = \mathcal{E}\,\Delta\gamma/\gamma \ , \tag{1.2}$$

where γ is electron relativistic factor; $\Delta\gamma/\gamma$,relative energy width of the electron beam. Total linewidth is determined by

$$\Delta\mathcal{E} = max\{\Delta\mathcal{E}_{hom}, \Delta\mathcal{E}_{inhom}\}. \tag{1.3}$$

The quantum properties of FEL are characterized by a parameter

$$\eta = \hbar\omega/\Delta\mathcal{E} \ . \tag{1.4}$$

In the case of $\eta \ll 1$, taking place in magnetic undulator FEL, the emission and absorption lines are almost in coincidence.

The difference of competing processes' probability occurs smaller in comparison with the probability of each of the processes. This difference may be calculated by expansion of the quantum probabilities into a series with expansion parameter η . The first term of expansion coincides with the results of classical theory /4,5/. Thus, classical theory describes FEL operation adequately in the limit $\eta \ll 1$.

The generation frequency increase makes classical description impossible. An exceed of the "quantum threshold" ($\eta > 1$) can actually be reached in the X-ray FEL pumped by a counter laser wave. At $\mathcal{E} \simeq 3$ MeV, $\Delta\gamma/\gamma = 5.10^{-5}$, $N = 10^4$ the given condition is fulfilled for $\hbar\omega_s \gtrsim 100$ eV, and the pump frequencies lie in the IR or visible range.

In this work X-ray FEL operation in quantum limit $\eta \gg 1$ is considered. Quantum limit is characterized by absolute separation of emission and absorption lines. Below it is shown that in this case the X-ray FEL appears to be two-level quantum generator with the properties analogous to those of the lasers in discrete spectrum. It should be noted that the "two-level" FEL is of interest from the viewpoint of energy output. In fact, the electron beam injected into the region of interaction with the energy close to the energy of the upper "working" level, $\mathcal{E}_e$ represents a fully inverted medium, whereas in the classical limit, $\eta \ll 1$, only partial inversion is realized, since both working levels are pumped simulataneously. We will also show the possibility of 100% output of inversion per passage under quantum FEL operation. This allows one to use small reflection mirrors in such a FEL.

2. TWO-LEVEL APPROXIMATION FOR ELECTRON-FIELD INTERACTION IN OPTICALLY PUMPED FEL

Tovalidate two-level model of FEL operation we consider the interaction of the electron beam propagating in +z-direction with plane monochromatic waves described by vector-potentials:

$$\vec{A}_s(z,t)=\frac{\vec{e}_s}{2}A_s e^{i(\omega_s t-\kappa_s z)}+c.c.,\ \vec{A}_i(z,t)=\frac{\vec{e}_i}{2}A_i e^{i(\omega_i t+\kappa_i z)}+c.c. \quad (2.1)$$

where $A_{i,s}$ are the amplitudes of the pumping (i) and generated (s) fields; $\vec{e}_{i,s}$ are unit orts lying in xy plane, $(\vec{e}_i,\vec{e}_s)=1$. The wave function Ψ of the electron satisfies the Klein-Gordon equation /4/:

$$\hbar^2\frac{\partial^2\Psi}{\partial t^2}-\hbar^2c^2\frac{\partial^2\Psi}{\partial z^2}+2e^2\vec{A}_s\vec{A}_i\Psi+m^2c^4\Psi=0. \quad (2.2)$$

The electron interaction with the fields proceeds within the electron flight time $\tau_{in}\simeq L/c$ through the interaction region of the length $L=N\lambda_i$. Before the interaction ($t<0$) the electron has momentum p and energy $\mathcal{E}(p)=(m^2c^4+p^2c^2)^{1/2}$ and is described by de Broglie wave fucntion

$$\Psi\sim\exp\left\{\frac{i}{\hbar}(pz-\mathcal{E}(p)t)\right\}. \quad (2.3)$$

When the filds are switched on ($t>0$) the dynamics of the electron obeys the equations which follow from (2.2):

$$i\dot{a}_n=g_n a_{n-1}e^{-i\Delta_n t}+g_{n+1}a_{n+1}e^{i\Delta_{n+1}t},\quad a_n(0)=\delta_{n0}, \quad (2.4)$$

where a_n is the amplitude of theprobability of the electron state with the momentum $p_n=p+n\hbar\kappa$ and energy $\mathcal{E}_n=(m^2c^4+p_n^2c^2)^{1/2}$, $n=0,\pm1,\pm2,\ldots$, $g_n=\frac{e^2}{\hbar}A_iA_s(\mathcal{E}_{n-1}\mathcal{E}_n)^{-1/2}$.

Equation (2.4) describes the excitation of the unbounded-level anharmonic oscillator by the field of the frequency ω. Frequency tuning out from the resonance with the neighbouring levels can be given by:

$$\Delta_n=\omega+\hbar^{-1}(\mathcal{E}_{n-1}-\mathcal{E}_n)\simeq-8\pi N\frac{\mathcal{E}(p)-\mathcal{E}_e+n\hbar\omega}{\mathcal{E}_c}\frac{1}{\tau_{in}}. \quad (2.5)$$

Let the initial electron distribution over the energy be formed near the resonance value $\mathcal{E}_e$ (see (1.1)). Then from the condition $\hbar\omega\gg\mathcal{E}(4N)^{-1}$ it follows that the phase in (2.4) at $n\neq0$ is much higher than 2π for $t\lesssim\tau_{in}$.

The respective exponential terms oscillate rapidly in time, and can be omitted. Besides, the condition $\hbar\omega\gg\mathcal{E}\Delta\gamma/\gamma$ means that the above statement is true for all electrons of the beam. It is easily seen that only two amplitudes with n=0,-1 remain out of the infinite chain of amplitudes ,
and we come to a system of two equations:

$$i\dot{a}_0=g_0a_{-1}e^{-i\Delta_0 t},\quad i\dot{a}_{-1}=g_0a_0e^{i\Delta_0 t}. \quad (2.6)$$

3. EQUATIONS OF TWO-LEVEL FEL DYNAMICS

An experimental scheme of optically pumped FEL has been proposed by Gea-Banacloche et al. /2/. According to /2/ the electron-field interaction region is concentrated in the pump beam focal volume. The electron bunch and X-ray pulse, passing through the interaction region, are assumed to be synchronized with the pump pulse appearance.

Due to the absence of a rather high reflecting X-ray mirrors the value of the gain for one passage must be by an order of unit. It means that the generated field amplitude and phase variations along the interaction region are considerable. Thus, the dynamics of amplification in the considered case can be described only within the framework of selfconsistent approach, taking into account spatial and temporal effects.

Neglecting transverse structure of the electron and the wave beams, let us, consider the interaction of the electron bunch with the fields in one-dimensional geometry.

The electron distribution in electron bunch is described not by single De Broglie wave (2.3) but by a wave superposition, representing a wave packet.

In quantum limit, $h \gg 1$, the wave function of the electron, interacting with the fields will naturally be a sum of two wave packets. Each packet presents one of two resonant states of the electron:

$$\Psi(z,t) = \left(\frac{mc^2}{2\mathcal{E}_0}\right)^{1/2} a_0(z,t)\, e^{\frac{i}{\hbar}(p_0 z - \mathcal{E}_0 t)} + \left(\frac{mc^2}{2\mathcal{E}_{-1}}\right)^{1/2} a_{-1}(z,t)\, e^{\frac{i}{\hbar}(p_{-1} z - \mathcal{E}_{-1} t)} . \qquad (3.1)$$

The slowly varying amplitudes a_0, a_{-1} of the wave function (3.1) in the following way are connected with spacial-temporal electron bunch density distribution n_e :

$$n_e(z,t) = i\frac{\hbar}{mc^2}\left\langle \Psi^* \frac{\partial \Psi}{\partial t}\right\rangle + c.c. = |a_0(z,t)|^2 + |a_{-1}(z,t)|^2, \qquad (3.2)$$

where the averaging is implied to be made during the time much longer than ω^{-1}. Before the interaction ($t \to -\infty$) the wave function (3.1) is given by single resonant state: $a_{-1}(z,t) = 0$, and $a_0(z,t)$ is described by a correlation:

$$|a_0(z,t)|^2 = n_e(z - v_0 t), \qquad (3.3)$$

where $v_0 = c^2 p_0/\mathcal{E}_0$ is the wave packet group velocity; $n_e(z)$, the electron density spatial distribution ($\int dz\, n_e(z) = N_e$, N_e is total number of electrons of the bunch per unit cross-section).

By substitution (3.1) into (2.1) one can obtain the following equation for a_0, a_{-1} :

$$\frac{\partial a_0}{\partial t} + v_0 \frac{\partial a_0}{\partial z} = -\frac{i}{4\hbar} \frac{e^2 A_i A_s^*}{(\mathcal{E}_0 \mathcal{E}_{-1})^{1/2}} e^{-i\Delta_0 t} a_{-1} , \quad (3.4)$$

$$\frac{\partial a_{-1}}{\partial t} + v_{-1} \frac{\partial a_{-1}}{\partial z} = -\frac{i}{4\hbar} \frac{e^2 A_i^* A_s}{(\mathcal{E}_0 \mathcal{E}_{-1})^{1/2}} e^{i\Delta_0 t} a_0 ,$$

where $\Delta_0(\mathcal{E}_0) = \frac{\omega_s}{mc^2\gamma^3}(\mathcal{E}_0 - \mathcal{E}_e), \quad v_{-1} = \frac{P_{-1}c^2}{\mathcal{E}_{-1}}$.

Note that the electron mass shift, induced by the pump field, is assumed to be compensated and is not taken into account. The possibilities of such a compensation have been suggested in /2/. It's easily seen that above obtained system of Eqs.(3.4) has an integral

$$\int_{-\infty}^{\infty} (|a_0|^2 + |a_{-1}|^2) dz = N_e \quad (3.5)$$

which means total charge conservation.

For subsequent analysis it's convenient to make a substitution

$$\tilde{a}_0 = a_0 , \quad \tilde{a}_{-1} = a_{-1} e^{-i\Delta_0 t} , \quad (3.6)$$

and Eqs.(3.4) can be written:

$$\frac{\partial \tilde{a}_0}{\partial t} + v_e \frac{\partial \tilde{a}_0}{\partial z} = -\frac{i}{4\hbar} \frac{e^2 A_i A_s^*}{(\mathcal{E}_e \mathcal{E}_a)^{1/2}} \tilde{a}_{-1} \quad (3.7)$$

$$\frac{\partial \tilde{a}_{-1}}{\partial t} + v_a \frac{\partial \tilde{a}_{-1}}{\partial z} = i\Delta_0 \tilde{a}_{-1} - \frac{i}{4\hbar} \frac{e^2 A_i^* A_s}{(\mathcal{E}_e \mathcal{E}_a)^{1/2}} \tilde{a}_0 .$$

Because of $h >> 1$, in Eqs.(3.7) we have neglected the difference of the group velocity values v_0, v_{-1} from the resonant ones v_e, v_a. It's not the case for phase mismatch Δ_0 ,whose dependence on $\mathcal{E}_0$ is essential.

To make the consideration selconsistent it's necessary to supplement Eqs.(3.7) with wave equation:

$$\left(\frac{\partial^2}{\partial z^2} - \frac{1}{c^2}\frac{\partial^2}{\partial t^2}\right)\vec{\mathcal{A}}_\perp = -\frac{4\pi}{c}\vec{j}_\perp , \quad \vec{\mathcal{A}}_\perp = \vec{\mathcal{A}}_i + \vec{\mathcal{A}}_s , \quad (3.8)$$

where transverse current density $\vec{j}_\perp$ is determined by

$$\vec{j}_\perp = -\frac{2e^2}{mc}\vec{A}_\perp \Psi^* \Psi . \tag{3.9}$$

When passing in Eqs.(3.8),(3.9) to slowl varying amplitudes we can easily come to equation of generated field dynamics

$$\frac{\partial A_s}{\partial z} + \frac{1}{c}\frac{\partial A_s}{\partial t} = i\,\frac{2\pi c e^2 A_i}{\omega_s(\mathcal{E}_e\mathcal{E}_a)^{1/2}}\int_0^\infty d\mathcal{E}_0\, f(\mathcal{E}_0)\,\tilde{a}_0^*\tilde{a}_{-1} . \tag{3.10}$$

In right-hand part of Eq.(3.10) $\dot{j}_\perp$ is averaged over the initial energy distribution $f(\mathcal{E}_0)$, $\int_0^\infty f(\mathcal{E}_0)d\mathcal{E}_0 = 1$.

As the generated field energy is extracted mainly from the energy of the electron beam, in all the cases of practical importance the pump field exhaustion can be neglected.

One can see that Eqs.(3.7), (3.10) are analogous to Maxwell-Bloch equations (M-B Eqs.), which describe resonant field interaction with a medium of two-level atoms /6/. The only difference of Eqs.(3.7), (3.10) from M-B Eqs. is that the former are written for the moving particles.

4. AMPLIFICATION IN QUASISTATIONARY CASE

The main features of "Two-level" FEL operation can be most easily studied for quasistationary amplification, which occurs when the electron bunch and the generated and pump pulses have the lengths much greater than the one of the interaction region:

$$L_B,\ L_s,\ L_i \gg L . \tag{4.1}$$

Interaction region length L is determined by Rayleigh range of the pump beam /2/:

$$L \approx 2K_i a_i^2 , \tag{4.2}$$

where a_i is pump beam waist in the lens focus.

Under condition (4.1) temporal derivatives in Eqs.(3.7), (3.10) can be neglected. Let the electron beam be rather monoenergetic, and the following condition is fulfilled:

$$\Delta\mathcal{E}_{hom} > \Delta\mathcal{E}_{inhom} . \tag{4.3}$$

This means that phase mismatch in Eq.(3.7) can be omitted, as the result equations are easily integrated. By taking into account boundary condition at the interaction region input (z=0), one can get

$$\tilde{a}_0(z) = \tilde{a}_0(0)\cos\chi(z) , \quad \tilde{a}_{-1}(z) = -i\tilde{a}_0(0)\sin\chi(z) , \tag{4.4}$$

where $\chi(z) = (\mu/c\hbar)\int_0^z E_s(z')dz'$, $\mu = \frac{e^2 E_i}{m\omega_i^{1/2}\omega_s^{3/2}}$

In Eqs.(4.4) we use the amplitudes of electric fields $E_{i,s}$ instead of vector potentials. The parameter μ, by anology with two-level particles, plays the role of a dipole momentum of the transition.

By substituting (4.4) into (3.10) we come to equation for $\chi(z)$:

$$\frac{\partial^2 \chi}{\partial z^2} = \alpha^2 \sin \chi , \quad (4.5)$$

where $\alpha = \left(\frac{2\pi n_e e^4 E_i^2}{\hbar\omega_s^2 \omega_i m^2 c^2}\right)^{1/2}$ is linear amplification index.

Let us introduce the maximum intensity that the beam of "two-level" electrons can radiate: $I_B = n_e c\hbar\omega_s$. Then the solution of Eq.(4.5) yields the following connection between input intensity of generated field and output one:

$$I_s(L) = dn^{-2}(\alpha L \varkappa^{-1}, \varkappa) I_s(0) , \quad (4.6)$$

where $dn(x,y)$ is elliptic Yacobi function, $\varkappa = (1 + I_s(0)/I_B)^{-1/2}$. Periodic dependence of output I_S on L (see(4.6)) is due to electron-field energy exchange. When the interactio on length is equal to

$$L_m = \alpha^{-1}\varkappa \mathcal{K}(\varkappa)(2m+1), \quad m = 0,1,\ldots, \quad (4.7)$$

Is(L) has the maximum value

$$I_s(L_m) = I_s(0) + I_B , \quad (4.8)$$

where $\mathcal{K}(y)$ is a complete elliptic integral. From (4.8) it follows that just at the interaction length L_m all the energy stored in the upper electron level is emitted into the field. Note that condition (4.7) is analogous to the condition when the coherent pulse propagating in two-level amplifier /6/ takes all the energy stored in the medium.

Optimal FEL operation can be achieved when $L=L_o$, and maximal energy output is observed under condition:

$$\alpha^{-1}\xi \mathcal{K}(\xi) = L_o , \quad \xi = (1 - R^2)^{1/2} , \quad (4.9)$$

where R is reflection index of X-ray mirror.

Let us make some estimates. If pump radiation of the wavelength λ_i=1.06μ is focused into a spot with the radius equal to the one of the electron beam $r_B = 10^{-1}$cm, then the active region length is equal to L = 10 cm (this requires the electron energy $\mathcal{E} \approx 2.5$ MeV), the reflection index and the electron beam density are equal to R=0.5, and $n_e = 3 \cdot 10^{12}$ cm^{-3}, then the required pump intensity in the active region : $I_i \approx 10^{15}$ W/cm^2.

The X-ray intensity I_s and the total power W_s are equal to 10^6 W/cm^2 and 100 W, respectively.

CONCLUSION

We would like to emphasize that X-ray FEL, operating in quantum limit, demonstrates the properties analogous to the ones of the lasers in discrete spectrum. The possibility of FEL generation with full inversion output indicates the connection of FEL dynamics with coherent effects, such as nutations, selfinduced transparency, and so on. Finally, the above estimations show that energy output of optically pumped FEL is not inferiour to that ofX-ray undulator sources / 7/. Moreover, the electron energy should not be as high as GeV.

REFERENCES

1. R.H.Pantell, G.Soncini, H.E.Puthoff. - IEEE J. Q.E., vol.4, p.905 (1968).
2. J.Gea-Banacloche, M.D.Scully, G.T.Moore et al. - IEEE J.Q.E., v.23, N 9, (1987).
3. J.C.Gallardo, R.C.Fernow, R.Palmer et al. IEEE J.Q.E., v.24, N 8 (1988).
4. M.V.Fedorov. Progr. in Quant.Electr., v.7, p.73 (1981).
5. F.A.Hopf, G.T.Moore, M.O.Scully et al. - Physics of Quant. Electr., v.7. Free electron generator of coherent radiation. Addison-Wesley, 1980.
6. E.M.Belenov, P.G.Kryukov, A.V.Nazarkin et al. -JOSA, v. B-5, p.946 (1988).
7. D.Attwood, K.Halbach, Kwang-Je Kim. - Science, v.228, p. 1265 (1985).

FIELD-INDUCED NARROWING OF AUTOIONIZATION ATOMIC STATES AS A MEANS OF CREATING INVERSE POPULATIONS

S.A. Kotochigova
State Optical Institute, 199034, Leningrad, USSR

Abstract

We discuss the possibility of narrowing the atomic autoionization states via their resonance mixing in a field. The results of ref.[1] show that, in contrast to the mixing of isolated states [2], with mixing of multiplets one may expect substantial narrowing of autoionization states owing to their intersection with bound electron states.

1. Introduction

In this paper we discuss the mechanism of laser–induced narrowing of autoionization atomic states as a means of creating the inverse population of these states and obtaining short–wave laser amplification. This mechanism is based on the mixing of autoionization states with those of the discrete spectrum that are resonant to the frequency of the laser radiation. A consistent theoretical examination of the mixing of multiplet terms in alkaline earth atoms has been made in refs.[1] and [3]. Here we will consider the principles of the calculation method and the concrete schemes of realization for the mixing of multiple resonances in the barium atom spectrum.

The first suggestion to use multiple resonances between discrete and autoionization states for narrowing the latter in a field of laser radiation has been made by Geller and Popov [2]. This mechanism of narrowing is based on the interference between the decay channels of an ionization state associated with a spontaneous decay into continuum and with an induced transition to a discrete level with subsequent transition into continuum. In this representation, the width of the autoionization state has the form [2]

$$\Gamma_A^{\,f} = 2\pi |V_i \mathscr{E} + W_A|^2 \qquad (1)$$

where V_i and W_A are the matrix elements of the photo– and autoionization transitions, respectively (Fig.1) and $\mathcal{E}$ is the electromagnetic field strength. It is evident that, according to (1), the substantial narrowing of the level can be achieved at approximately the same values of $|V_i\mathcal{E}|$ and $|W_A|$ and opposite signs of V_i and W_A. It is important that there be only one channel of spontaneous decay, since the existence of two such channels will lead to additional broadening effects. We would like to emphasize that the above conditions are quite difficult to realize in concrete atomic systems.

Besides the interference effects, the state narrowing mechanism we propose also includes a change in the configuration composition of the levels owing to their mixing by the electromagnetic field. In more detail, this approach is presented in sect.2.

The mixing of two states belonging to the autoionization and the discrete spectra under the action of a resonance field has been also examined in a series of studies made by Andryushin, Kazakov and Fedorov (see e.g. ref.[4]). It was found that the maximum possible narrowing of an autoionization state is only half of its initial width. At the same time, as is well known, in order to create inverse population one has to reduce the probability of the autoionization decay by several orders of magnitude.

In contrast to ref.[4], we have considered the mixing of two groups of levels with different parity belonging to the discrete and autoionization spectra. As a rule, the terms of alkaline earth atoms have a multiplet structure owing to the fine–structure splitting induced by spin–orbital interaction. As will be shown in the next section, owing to the field–induced mixing of multiplet terms there appear states with zero dipole moment. However, some of the levels in these multiplets may have an admixture of states of the discrete spectrum whose life time is determined by the radiation decay processes only. As concerns the creation of inverse population, it is the latter levels that are of interest. The fact that the quasi–energies of this sort of states lie in the high–energy autoionization region, while the widths of these levels are mainly determined by the radiation decay processes, gives one a basis for creating laser amplification for short–wave transitions.

In this study, the numerical calculations of the multiplet term mixing and of the autoionization state narrowing in the field of laser radiation have been made with the strong–bond method with account of the super–fine splitting of levels in the field.

Since in the process of excitation of an autoionization resonance there usually appears a whole series of possible decay channels, we have calculated the populations for the whole system of levels (Fig.1) and for ionization, autoionization and spontaneous relaxation channels. The data on the distribution of level population and on its dependence on time and on the strength of the external field enable one to select the most efficient regime of exciting the atom for creating the inversion.

For making concrete calculations of the state mixing and of the population distribution in an atom Ba we have used the MHFD method to obtain the necessary matrix elements of the Coulomb, exchange and dipole interactions.

2. The Theoretical Model of Autoionization State Narrowing

Let us now give a theoretical description to the narrowing of autoionization states of an atom under the action of a laser field. In our description, we have used

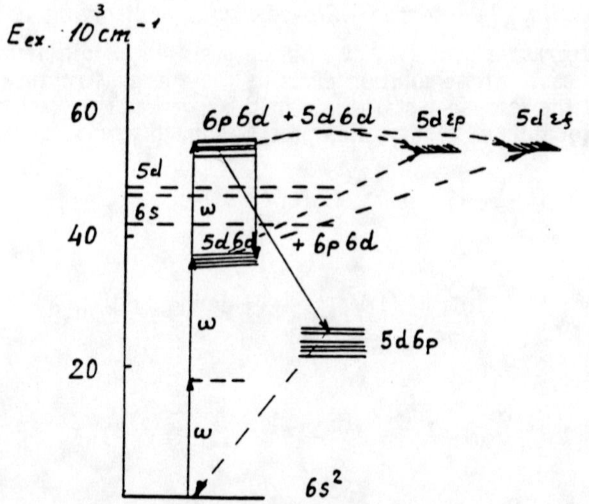

Fig.1. Level diagram of barium atom.

the real system of terms of the Ba atom that has two groups of multiplet terms 5d6d and 6p6d configurations belonging to the discrete and the autoionization spectra, respectively (Fig.1). For efficient narrowing of the autoionization states the excitation scheme should include the so–called multiple resonances [3], which means that (i) the radiation is resonant or quasi–resonant to the levels of both multiplets, and (ii) the levels of the multiplets have opposite parity and are connected by one–photon transitions. These conditions ensure the greatest mixing of levels with substantially different widths, which may lead to stabilization of the short–lived levels of the autoionization spectrum.

In the resonance approximation, the wave function of an atomic state with account of laser radiation reads as

$$\psi^f(\vec{r},t) = \sum_{\alpha=1}^{k} C_\alpha(t)\varphi_\alpha(\vec{r})e^{in_\alpha \omega t} + \int d\varepsilon' b_{\varepsilon'}(t)\varphi_{\varepsilon'}(\vec{r})e^{in_{\varepsilon'}\omega t} +$$

$$\int d\varepsilon'' b_{\varepsilon''}(t)\varphi_{\varepsilon''}(\vec{r})e^{in_{\varepsilon''}\omega t} \qquad (2)$$

Here α sums up the states of the discrete spectrum and the discrete components of the autoionization spectrum; φ_α, $\varphi_{\varepsilon'}$, and $\varphi_{\varepsilon''}$ are the wave functions of an isolated atom; the integral terms in (1) are associated with the wave functions of the continuum spectrum with energies E and E+ω (Fig.1) also taken into consideration; and n_α, $n_{\varepsilon'}$, and $n_{\varepsilon''}$ are the numbers of absorbed photons necessary for excitation of the corresponding state.

By substituting (1) into the wave equation, multiplying it by the basis functions, and integrating over $\vec{r}$ [3], we obtain a system of differential equations for the resonance case, whose solution gives us the energy structure of the atom and the widths of the resonance states in a field, as well as the coefficients of state mixing and the probabilities of ionization processes for different channels:

$$i\frac{dC_0(t)}{dt} = C_0(t)(E_0+3\omega) + \sum_{\alpha} C_{\alpha}(t)V_{\alpha 0}\mathcal{E}^2 +$$

$$\sum_{\beta} C_{\beta}(t)V_{\beta 0}\mathcal{E}^3 + \int d\varepsilon' b_{\varepsilon'}(t)V_{\varepsilon' 0}\mathcal{E}^3 + \int d\varepsilon'' b_{\varepsilon''}(t)V_{\varepsilon'' 0}\mathcal{E}^4$$

$$i\frac{dC_{\alpha}(t)}{dt} = C_0(t)V_{\alpha 0}\mathcal{E}^2 + C_{\alpha}(t)(E_{\alpha}+\omega) + \sum_{\beta} C_{\beta}(t)V_{\beta\alpha}\mathcal{E} +$$

$$\int d\varepsilon' b_{\varepsilon'}(t)V_{\varepsilon'\alpha}\mathcal{E} + \int d\varepsilon'' b_{\varepsilon''}(t)V_{\varepsilon''\alpha}\mathcal{E}^2$$

....

$$i\frac{dC_{\beta}(t)}{dt} = C_0(t)V_{0\beta}\mathcal{E}^3 + \sum_{\alpha} C_{\alpha}(t)V_{\beta\alpha}\mathcal{E} + C_{\beta}(t)E_{\beta} +$$

$$\int d\varepsilon' b_{\varepsilon'}(t)W_{\varepsilon'\beta} + \int d\varepsilon'' b_{\varepsilon''}(t)V_{\varepsilon''\beta}\mathcal{E}$$

.... (3)

$$i\frac{db_E(t)}{dt} = C_0(t)V_{E0}\mathcal{E}^3 + \sum_{\alpha} C_{\alpha}(t)V_{E\alpha}\mathcal{E} +$$

$$\sum_{\beta} C_{\beta}(t)W_{\varepsilon\beta} + b_{\varepsilon}(t)\varepsilon$$

E_0, E_{α}, E_{β}, and E are the energies of the unperturbed states of the main, discrete, autoionization and continuous spectra, respectively; $V_{\alpha\beta}$ and $W_{E\beta}$ are the matrix elements of the dipole $\alpha \longrightarrow \beta$ and autoionization $\beta \longrightarrow$ E transitions calculated according to refs.[1,5].

$$V_{\beta\alpha} = \frac{\mathcal{E}}{\sqrt{2}C}\frac{[j_{\beta}]^{1/2}}{[j_{\alpha}]^{1/2}} C_{10,j_{\beta}1/2}\, C^{j_{\alpha}m_{\alpha}}_{j_{\beta}m_{\beta},1-1} \times$$

$$\sum_{\nu=\pm1} <\Phi_\alpha(\nu)|\Phi_\beta(\nu)>|K_\alpha - K_\beta - \nu|, \tag{4}$$

where $C^{jm}_{j_1m_1,j_2m_2}$ – Klebsh–Gordan coefficient;

$\Phi(\nu) = \begin{cases} P(r), \nu = 1 \\ Q(r), \nu = -1 \end{cases}$; P(r), Q(r) – large and small component of wave function.

$$W_{E\beta} = <ab(\beta)|r_{12}^{-1}|cd(E)> = [j_a, j_b, j_c, j_d]^{1/2} \times$$

$$\sum_\nu \frac{R^\nu(a,b,c,d)}{[\nu]^2} C^{\nu 0}_{j_a - 1/2,\, j_c\, 1/2} C^{\nu 0}_{j_b - 1/2,\, j_d\, 1/2} \times$$

$$\sum_{\substack{m_a, m_b,\\ m_c, m_d}} (-1)^{m_c + m_d + 1} C^{\nu(m_d - m_c)}_{j_a m_a, j_c - m_c} C^{\nu(m_b - m_d)}_{j_b m_b, j_d - m_d}; \tag{5}$$

$$R^\nu(a,b,c,d) = \int_0^\infty dr_1 \int_0^\infty dr_2 \, [P_a(r_1)P_c(r_1) + Q_a(r_1)Q_c(r_1)] \times$$

$$[P_b(r_2)P_d(r_2) + Q_b(r_2)Q_d(r_2)] \frac{r_<^\nu}{r_>^{\nu+1}}.$$

By making a Laplace transformation of (3), we obtain a system of algebraic equations whose coefficient matrix, after diagonalization, gives us the energies and widths of atomic levels in an external field. The dependence of these characteristics upon the field strength, obtained by numerical calculations, is presented in Fig.2. Figure 2a shows the quasi–energy structure of multiplet terms Ba, where the spacings between the levels are actually the resonance detunings. In a field $\mathcal{E}>10^{-3}$ a.u. the levels cross one another, which is accompanied by strong interaction. There are several intersection points in the middle of the multiplets that result in states with unusual properties. Figure 2b shows the widths of the states in the versus logε. The strong interaction stabilizes some of the states, whose widths get nearer to the radiation widths. The configuration composition of this kind of levels at different logε is shown in Fig.3. With increase of the field, the admixture of states with opposite parity may change considerably. As $\mathcal{E}$ approaches 10^{-2} a.u., there appear states that have the admixture only of the 5d6d configurations, whose properties are characterized by the states of the discrete spectrum. It is these levels that may have the property of being stable and thus are of special interest to us.

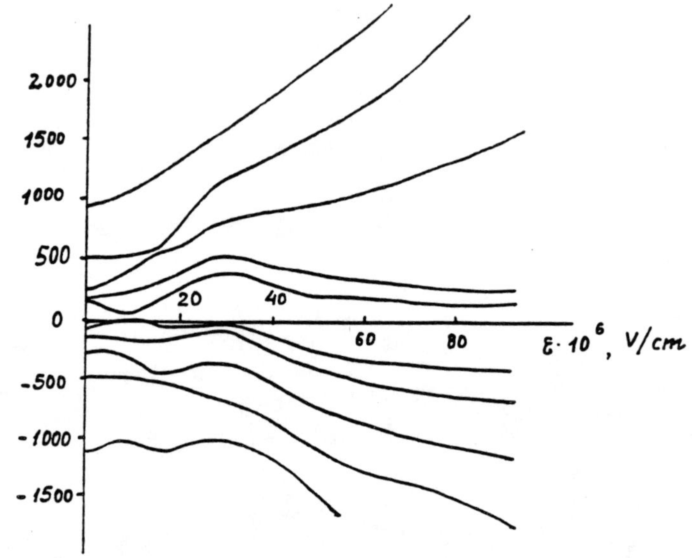

Fig.2a. Quasienergy structure of Ba atom plotted as a function of an external field strength.

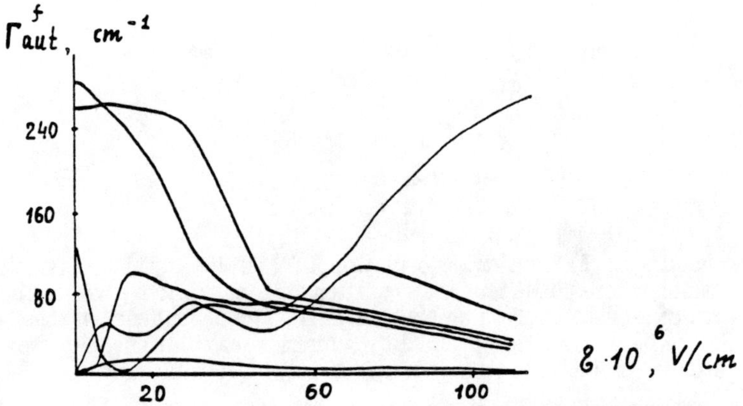

Fig.2b. Widths of levels in field plotted as functions of field strength.

3. Inverse Population of Autoionization States

Let us now examine the population of atomic states Ba, which become highly stable with increase of the laser field strength and for which the probabilities of stontaneous decays into continuum and of decays down to discrete levels become comparable in magnitude. Figure 1b shows the diagram of Ba excitations that we have used when calculating the populations. Since the process of level occupation is nonstationary, let us write down the system of kinetic equations for time–dependent characteristics:

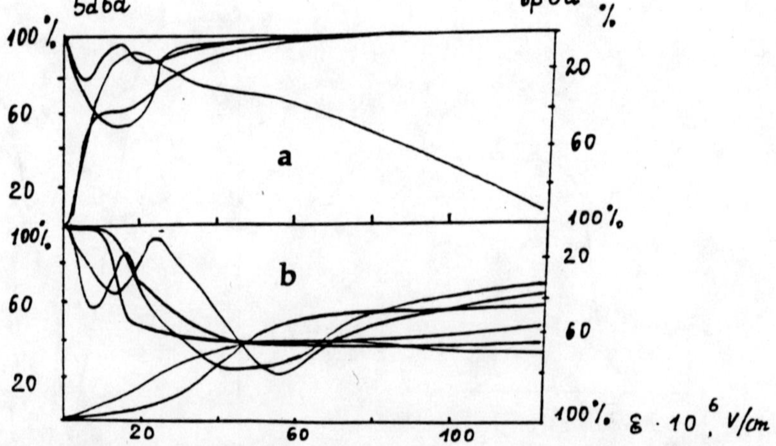

Fig.3. Squares of the coefficients representing of admixture of states of different parity plotted as a function of field strength. a) Strong interaction of levels. b) Weak interaction.

$$\frac{dn_d(t)}{dt} = M_{d0}{}^{(2)}n_0(t)-(M_{0d}{}^{(2)}+\Gamma_{kd})n_d(t)+\Gamma_{kd}n_k(t)$$

$$\frac{dn_0(t)}{dt} = -M_{d0}{}^{(2)}n_0(t)+M_{0d}{}^{(2)}n_d(t)+\Gamma_{0m}n_k(t)$$

$$\frac{dn_k(t)}{dt} = \Gamma_{kd}n_d(t)-(\Gamma_{kd}+2\pi|W_{E1}|^2+P_i+\Gamma_{mk})n_k(t) \qquad (6)$$

$$\frac{dn_{E1}}{dt} = 2\pi|W_{E1}|^2n_k(t)$$

$$\frac{dn_{E2}}{dt} = P_in_k(t)$$

$$\frac{dn_m}{dt} = \Gamma_{mk}n_k(t) - \Gamma_{0m}n_m(t),$$

where $M_{d0}{}^{(2)} = \pi\sqrt{2}|V_{d0}|^2\mathcal{E}^4\Delta\omega^{-1}$;
Γ_{kd}, Γ_{mk} – probabilities of $d \to k$ and $k \to m$ transitions;
P_i – ionization probability of k state.
Γ_{0m} is the probability of the spontaneous $m \to 0$ transition.

The solution of (6) can be presented in the form:

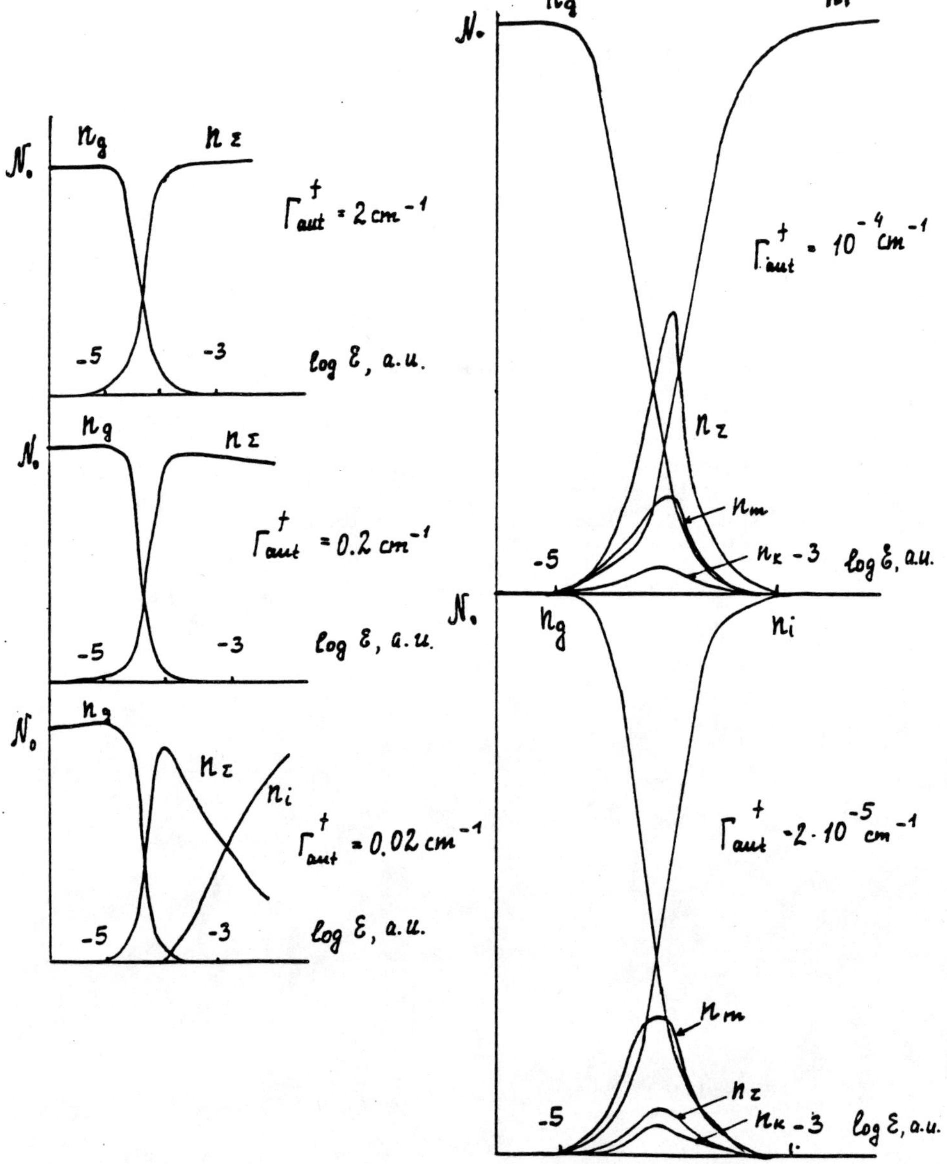

Fig.4. Populations of levels are plotted as function of log ε for different widths of autoionizing level.

$$n_k(t) = \sum_m B_m Z_{km} e^{i\eta_m t} \quad (7)$$

where B_m are obtained from the initial perturbation switching–on conditions and Z_{km}, η_m are the eigenvectors and eigenvalues of the coefficient matrix of system (6). By changing the field strength, we obtain the dependencies $n_k(\mathcal{E})$ for certain moments of recording time. For $t = \tau$ (pulse duration 30 ns) the corresponding diagrams are shown in Figs.4(a,...,f). Each of the latter figures corresponds to a certain width of the autoionization state. We see that for unstable states (Figs.4a–4c), the principal channel along which the population of the ground state is transported is the autoionization decay n_E. In the case 4c ($\Gamma_A = 0.02$ cm^{-1}), there appears an additional channel of ionization upwards to the E'' state of the continuous spectrum. For narrow autoionization resonances (Figs.4e, 4), the low–lying discrete levels n_m and the autoionization state n_k are also substantially occupied. These populations reach their maximum at $\mathcal{E} = 10^{-4}$ a.u. or $5\cdot10^5$ W/cm. The fact that $n_m > n_k$ indicates that the process of spontaneous $k \to m$ decay ends by the time τ. In order to record the population inversion of the a–state in our calculations, let us analyze the time–dependence of the occupation process (Fig.5). The results of calculations show that within the time interval $0 < t \lesssim 4$ ns we can obtain the population inversion in the autoionization state.

4. Conclusion

The present study has shown that it is possible to create the population inversion for autoionization atomic states under the action of a resonance laser field.

It was not our aim to develop a unique scheme for producing short–wave laser radiation. The problem was to make a detailed examination of the state narrowing and population inversion processes in the case of barium atoms widely used in experiments. Calculations have shown that with resonance excitation and mixing of terms with a multiplet structure one may expect to obtain a special class of states with unusual properties. In particular, these states are stable and are located in the continuous region of the spectrum. It is these states that are occupied inversely under the action of a trial field with strength $\mathcal{E} = 5\cdot10^5$ V/cm. Their decay downward, after amplification, may give short–wave laser radiation.

The particular case of inverse occupation of the autoionization states of the Ba atom we considered in this paper is nonetheless of sufficiently general character. The narrowing of these states is not an accidental manifestation of the complicated process of state mixing. The principal condition necessary for this process to occur is that the field should mix not two resonance levels but a whole group of levels to produce intersections and quasi–intersections. As was shown, in the process of strong interactions there are formed states that have the property of stability and contain the admixture of discrete configurations only.

In the case of the barium atom excitation we have considered, there is probably not enough short radiation to be of practical interest. Radiation with shorter wavelength can be produced, for example, by using noble–gas atoms as well as the near–threshold spectra of multicharge ions.

One should also pay attention to the fact that the mixing may compensate

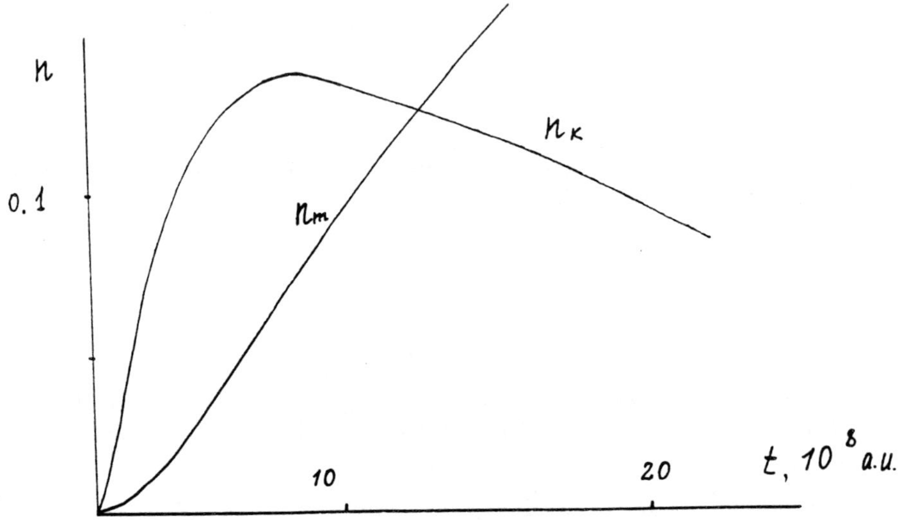

Fig.5. Dependence of autoionizing (k) and discrete (m) states on the time of excitation.

only the autoionization width of the level, whereas the ionization broadening still remains. However, as was shown in ref.[1]. this type of broadening becomes essential only in the fields with $\mathcal{E}>10^7$ V/cm.

Another important problem is the width of the laser radiation. To produce multilevel interaction, the field strength should be of the order of or greater than the resonance detunings: $\mathcal{E}\cdot v \geq \Delta$. In view of the limits on $\mathcal{E}$ from above ($\mathcal{E}<10^7$ V/cm), one can use wide–band radiation to reduce the resonance detunings.

Finally, in our opinion, the optimal scheme for this problem is the one with two lasers. The first laser excites the atoms or ions to sufficiently high–lying states of the discrete spectrum. The intensity of this radiation does not have to be high: around 10^6 W/cm^2. The second laser, in resonance with the transition between the discrete and the autoionization terms, mixes and narrows the states, leading to spontaneous emission.

In conclusion, I express my gratitude to Prof. N.B.Delone for this attention to this work. On his initiative, the effect of state mixing was used in the scheme of creating inverse occupation of autoionization states.

REFERENCES

[1] S.A.Kotochigova, Sov. Phys. JETP 67 (4) 706–713 (1988)
[2] Yu.I.Geller and A.K.Popov, Laser Stimulation of Nonlinear Resonances in Continuous Spectra, Nauka, Novosibirsk (1981)
[3] S.A.Kotochigova, JOSA B (1990)
[4] A.I.Andryushin, A.E.Kazakov and M.V.Fedorov, Sov. Phys. JETP 61, 678 (1985)

X-RAY GAIN AND REABSORPTION OF RESONANCE RADIATION IN LASER-PRODUCED PLASMAS

W. Brunner, R.W. John, and Th. Schlegel
Central Institute of Optics and Spectroscopy, Academy of Sciences, Rudower Chaussee 6, Berlin, 1199, Germany

1. INTRODUCTION

In various schemes which were investigated to demonstrate X-ray amplification in laser-produced plasmas, the reabsorption of resonance radiation creates one of the major limitations on population inversion and gain.
The plasmas considered have not negligible optical thicknesses in the center of the strongest resonance lines; reabsorption of radiation from the p $\longrightarrow$ 1 transition repopulates the level p, by which, according to

$$G=\frac{c^2}{8\pi}\frac{A_{qp}}{\omega_{qp}^2}L_{qp}(\omega)\left(N_q-\frac{g_q}{g_p}N_p\right), \tag{1}$$

the gain for lasing transitions q $\longrightarrow$ p is reduced.
So, in the gain analysis of experiments performed as well as in order to adjust the parameters of an experiment to achieve optimal X-ray gain, a detailed knowledge of the spatial and temporal behaviour of the reabsorption-caused increase in the lower-laser-level population density is needed. To this end, one has simultaneously to solve the coupled rate equations for the ionic population densities and the equation of radiation transport.
In previous papers [1,2], in the case of spatially varying absorption a simple approximate formula describing the effect of radiation reabsorption on the population density of the upper resonance level was derived and used in an analysis of the X-ray gain in laser-produced plasmas. However, because of the Doppler frequency shift of radiation due to the differential expansion of the laser-produced plasma, the bulk motion of the plasma helps to diminish the reabsorption of radiation in comparison to the value it would have for an equivalent

medium globally at rest.
Besides we should like to mention that in a recombination/rapid cooling scheme the trapping of radiation may be further reduced by additional ionizing irradiation diminishing the population density of the ground level (cf. [3,4]).
In the case of a plasma with sufficiently large velocity gradients - across the emitting and reabsorbing plasma the motional Doppler frequency shift is much larger than the local linewidth -, a method to account for the effect of radiation reabsorption on the population densities is the Sobolev approximation ([5], cf. also [6]). In that the effect of reabsorption is considered in the escape factor approximation. Methods resting on the latter approximation are also used in connection with X-ray gain calculations (see.,e.g., [7-10], cf. also [11]).
Now, with regard to the ablative motion of laser-produced plasmas, in studying the reabsorption also the case is of particular interest that the dynamical Doppler frequency shift is of the order of the local linewidth (cf. [12]) which gives rise to a refined treatment.

2. BASIC EQUATIONS

Our starting point is a differentially expanding laser plasma of intermediate density which consists of hydrogen-like ions, as it may have developed from a fully ionized recombining laser-produced plasma. We describe the effect of radiation reabsorption on the populations in the framework of a collisional-radiative (CR) model extended by including the transport of line radiation. In the coefficients for emission and absorption the expansion-caused Doppler frequency shift is considered [13]. We look at the expansion velocity of the laser plasma as a given vector function of position, $\underline{v}\,(\underline{r})$. For the population densities $N_q\,(\underline{r},t)$ of level q (we have dropped the index of the ionic state) coupled with the photon densities $n_{qp}\,(\omega,\underline{r},\underline{\Omega},t)$ of $q \rightleftharpoons p$ radiation, q = 1,2,3..., p = 1,2,3, ..., we write the rate equations in the following form

$$\frac{\partial N_q(\underline{r},t)}{\partial t} = -\frac{N_q(\underline{r},t)}{\tau}\delta_{q1} - N_q(\underline{r},t)W_q^{ci} + N(\underline{r},t)(W_q^{cr} + W_q^{rr}) -$$

$$- N_q(\underline{r},t)\sum_{p\neq q} W_{qp}^{c} + \sum_{p\neq q} N_p(\underline{r},t)W_{pq}^{c} - N_q(\underline{r},t)\sum_{p<q} A_{qp} + \sum_{p>q} N_p(\underline{r},t)A_{pq} +$$

$$\pm\int_{-\infty}^{\infty} d\omega \int d\Omega \sum_{p\neq q} k_{pq}(\omega, \omega_{qp}[1 + \tfrac{1}{c}\underline{v}(\underline{r})\underline{\Omega}], \underline{r}, t)\, n(\omega, \underline{r}, \underline{\Omega}, t), \qquad (2)$$

with the respective collisional and radiative transition rates and the effective absorption coefficient

$$k_{pq} = \frac{1}{c}\left(B_{pq} N_p - B_{qp} N_q\right) \tag{3}$$

Initial conditions are

$$N_q(\underline{r}, t_o) = \tilde{N}_q(\underline{r}).$$

As transport equation for the photon density $n_{qp}(\omega, \underline{r}, \underline{\Omega}, t)$ of line radiation we have

$$\frac{\partial n_{qp}(\omega, \underline{r}, \underline{\Omega}, t)}{\partial t} + c\, \underline{\Omega}\, \mathrm{grad}\, n_{qp}(\omega, \underline{r}, \underline{\Omega}, t) =$$

$$= -c\, k_{pq}\left(\omega, \omega_{qp}\left[1 + \frac{1}{c}\underline{v}(\underline{r})\underline{\Omega}\right], \underline{r}, t\right) n_{qp}(\omega, \underline{r}, \underline{\Omega}, t) +$$

$$+ \frac{A_{qp}}{4\pi} L_{qp}\left(\omega, \omega_{qp}\left[1 + \frac{1}{c}\underline{v}(\underline{r})\underline{\Omega}\right], \underline{r}, t\right) N_q(\underline{r}, t). \tag{4}$$

Allowance is made for different profiles of the coefficients for emission and absorption.

In order to discuss in more detail the effect of radiation reabsorption, diminished by the dynamical Doppler frequency shift, on the population density, we concentrate on the transition with the highest transition probability and, moreover, with the highest population density of the lower level which gives rise to the absorption. This is the 2⟶1 transition (Lyman-α line for hydrogen-like ions). By treating the coupling of the photon density to the population densities in an iterative way, a single approximate rate equation for the population density N_2 results.

The light travel time across the spatial domain of interest being much shorter than the hydrodynamic timescale, the radiation field can be assumed quasistationary (cf. [14]). Furthermore, in the case of sufficiently long laser pulses producing the plasma the hydrodynamic interaction is relatively slow compared to the radiation and collision processes. Consequently, an approximately stationary solution of the rate equations is physically meaningful. Studying the stationary case, our rate equation for N_2 reads [13]

$$N_2(\underline{r}) - \frac{c}{A_{21}} \int d\omega \int d\Omega\, k_{12}\left(\omega, \omega_{21}\left[1 + \frac{1}{c}\underline{v}(\underline{r})\underline{\Omega}\right], \underline{r}\right) n_{21}(\omega, \underline{r}, \underline{\Omega}) =$$

$$= N_2^o(\underline{r}), \tag{5}$$

$N_2^o(\underline{r})$ is a known function acting as an effective pump density.

3. THE INTEGRAL EQUATION FOR THE REABSORPTION-AFFECTED POPULATION DENSITY

Solving formally the equation of radiative transfer and inserting the resulting expression for the photon density $n_{21}(\omega, \underline{r}, \boldsymbol{\Omega})$ in the rate equation (5), one obtains, for general three-dimensional plasma geometry, an integral equation for the population density $N_2(\underline{r})$. It is supposed that $N_1 \gg N_2$, so we can put

$$k_{12} \approx \frac{1}{c} B_{12} N_1 \tag{6}$$

which reduces the integral equation to a linear one. Assuming the $2 \rightleftharpoons 1$ line radiation to diffuse with complete redistribution in frequency over Gaussian emission and absorption profiles, respectively, the integral equation has the form [13]

$$N_2(\underline{r}) - \frac{k(\underline{r})}{4\pi}\int_V d\underline{r}' \frac{N_2(\underline{r}')}{|\underline{r}-\underline{r}'|^2} F(\underline{r},\underline{r}') \otimes$$
$$\otimes \exp\left\{-4\ln 2 \frac{\left[\frac{\omega_{21}}{c}\left(\underline{v}(\underline{r}) - \underline{v}(\underline{r}')\right)\underline{\Omega}\right]}{(\Delta\omega_{12}(\underline{r}))^2 + (\Delta\omega_{21}(\underline{r}'))^2}\right\} = N_2^0(\underline{r}), \tag{7}$$

the kernel containing the frequency integral

$$F(\underline{r},\underline{r}') = \left(\frac{\ln 2}{\pi}\right)^{1/2} \frac{2}{\Delta\omega_{21}(\underline{r}')} \int_{-\infty}^{\infty} d\omega \exp\left\{-4\ln 2\left(\frac{1}{(\Delta\omega_{12}(\underline{r}))^2} + \frac{1}{(\Delta\omega_{21}(\underline{r}'))^2}\right)\omega^2 - \right.$$
$$- |\underline{r}-\underline{r}'| \int_0^1 d\lambda\, k(\underline{r}-\lambda(\underline{r}-\underline{r}')) \exp\left[-\frac{4\ln 2}{(\Delta\omega_{12}(\underline{r}-\lambda(\underline{r}-\underline{r}')))^2}\left(\omega + \right.\right.$$
$$\left.\left.\left. + \frac{\omega_{21}}{c}\left(\frac{(\Delta\omega_{12}(\underline{r}))^2 \underline{v}(\underline{r}') + (\Delta\omega_{21}(\underline{r}'))^2 \underline{v}(\underline{r})}{(\Delta\omega_{12}(\underline{r}))^2 + (\Delta\omega_{21}(\underline{r}'))^2} - \underline{v}(\underline{r}-\lambda(\underline{r}-\underline{r}'))\right)\underline{\Omega}\right)^2\right]\right\}. \tag{8}$$

The $\underline{r}'$-integration is extended over the plasma volume V. In the case that the frequency shift due to the differential expansion of the laser plasma dominates over the local line-width across the plasma,

$$\frac{2(\ln 2)^{1/2}\omega_{21}}{c\,\Delta\omega_{12}(\underline{r}-\lambda(\underline{r}-\underline{r}'))}\left|\left(\underline{v}(\underline{r}) - \underline{v}(\underline{r}')\right)\underline{\Omega}\right| \gg 1, \quad \underline{r} \neq \underline{r}', \tag{9}$$

after asymptotically evaluating the frequency integral (8) one is led to a simplified integral equation exhibiting the reduction of radiation reabsorption by Doppler decoupling [13]:

$$N_2(\underline{r}) - \frac{k(\underline{r})}{2^{1/2}4\pi}\int_V d\underline{r}'\frac{N_2(\underline{r}')}{|\underline{r}-\underline{r}'|^2}\exp\left\{-2\ln 2\left[\frac{\omega_{21}(\underline{v}(\underline{r})-\underline{v}(\underline{r}')\underline{\Omega})}{c\Delta\omega(\underline{r})}\right]^2\right\}=$$

$$= N_2^0(\underline{r}) . \qquad (10)$$

Concerning the case that the expansion-caused frequency shift is of the order of the local linewidth we refer to a future publication.

REFERENCES

[1] W. Brunner, R.W. John, H. Paul, H. Steudel, Laser and Particle Beams 6, (1988) 723.

[2] W. Brunner, Th. Schlegel, Laser and Particle Beams 6, (1988) 277.

[3] D.G. Goodwin, E.E. Fill, J. Appl. Phys. 64, (1988) 1005.

[4] W. Brunner, R.W. John, Th. Schlegel, Appl. Phys. B 50, (1990) 291.

[5] V.V. Sobolev, Astron. Zh. 34, (1957) 694.

[6] V.V. Sobolev, A Course in Theoretical Astrophysics, 3rd edition, Nauka, Moscow, (1985) p. 328 (in Russian).

[7] A.V. Borovsky, V.V. Korobkin, Ch. K. Mukhtarov, Preprint IOFAN No. 21, Moscow, (1985).

[8] C.H. Nam, E. Valeo, S. Suckewer, U. Feldman, J.Opt.Soc. Am. B 3, (1986) 1199.

[9] D.C. Eder, Phys. Fluids B 1, (1989) 2462.

[10] Y.T. Lee, R.A. London, G.B. Zimmerman, P.L. Hagelstein, Preprint UCRL-102219, (1989), prepared for submission to Phys. Fluids B.

[11] V.I. Derzhiev, A.G. Zhidkov, S.I. Yakovlenko, Radiation of Ions in a Non-Equilibrium Dense Plasma, Energoatomizdat, Moscow, (1986) p. 77 (in Russian).

[12] D. Mostacci, L.M. Montierth, J.-P. Dinguirard, R.L. Morse, Phys. Fluids B 1, (1989) 2106.

[13] W. Brunner, R.W. John, submitted to Laser and Particle Beams (1989).

[14] Ya.B. Zel'dovich, Yu.P. Raizer, Physics of Shock Waves and High-Temperature Hydrodynamic Phenomena, Vol. 1, Academic Press, New York and London, (1966) p. 129.

AMPLIFIED SPONTANEOUS EMISSION FORMATION UNDER THE QUANTA REFLECTION BY THE SIDE SURFACE OF AN ACTIVE MEDIA

A.V. Lukin and P.V. Petrov
National Institute of Technical Physics, P.O. Box 245,
Chelyabinsk 454070, USSR

Laser irradiated exploding think foils have been used to create amplified spontaneous emission at 206.3 and 209.6 Å [1,2]. Such experiments induce the formation of elongated plasma region which has high density, small transverse density gradients and nearly constant temperature. Such conditions allow to avoid refraction of the soft X-rays out of the gain region.

In consequence of strong plasma density gradient changers for maximum gain coefficient the refraction of generated radiation is inevitable. To compensate refraction losses it was suggested in [3] to bend the target surface so that X-rays turn back after refraction. The simulation of quants propagation in that system may be received by the definition of region with quanta reflection by the side surface of an active media. This model may be used for consideration of an active media with peripheral material of higher density than region of gain.

In the stationary case for cylindrical gain region of length L and radius R(z) that shown in Fig.1, the intensity of radiation averaged all over transfer section

$$\mathcal{J}(z) = \int_S \mathcal{J}(\vec{r})\, ds$$

satisfied the system of equations

$$\mathcal{J}(z) = \frac{1}{S(z)} \int ds\, d\vec{\Omega}\, dt\, N_2(r_t(\vec{\Omega}))\, \exp\Big(\int_0^t \text{æ}(r_\xi(\vec{\Omega}))\, d\xi \Big/ 4\pi\tau_{21}, \tag{1}$$

$$\begin{gathered} (N_2(\vec{r}) - N_{20})/\tau_{21} + \sigma_{21}\mathcal{J}(\vec{r})(N_2(\vec{r}) - g_2/g_1 N_1(\vec{r})) = (N_1(\vec{r}) - N_{10})/\tau_1; \\ (N_2(\vec{r}) - N_{20})/\tau_2 + \sigma_{21}\mathcal{J}(\vec{r})(N_2(\vec{r}) - g_2/g_1 N_1(\vec{r})) = 0, \end{gathered} \tag{2}$$

where

$\vec{r}_t(\vec{\Omega})$ - radius-vector to the point of quanta birth, which has in $\vec{r}$-point direction $\vec{\Omega}$;

t - distance along quanta trajectory from point $\vec{r}_t(\vec{\Omega})$ to point $\vec{r}$;

$\varkappa(\vec{r})$ - gain coefficient;
$\varkappa_a(\vec{r})$ - absorption coefficient;
σ_{21} - cross section of work transfer;
$N_i(N_{io})$ - populations of work levels (without radiation, that defined by recombination process and pumping) and their statistic weights (i=1,2);
τ_i, τ_{21} - spontaneous life time of i-s level and time of spontaneous transfer 2→1;

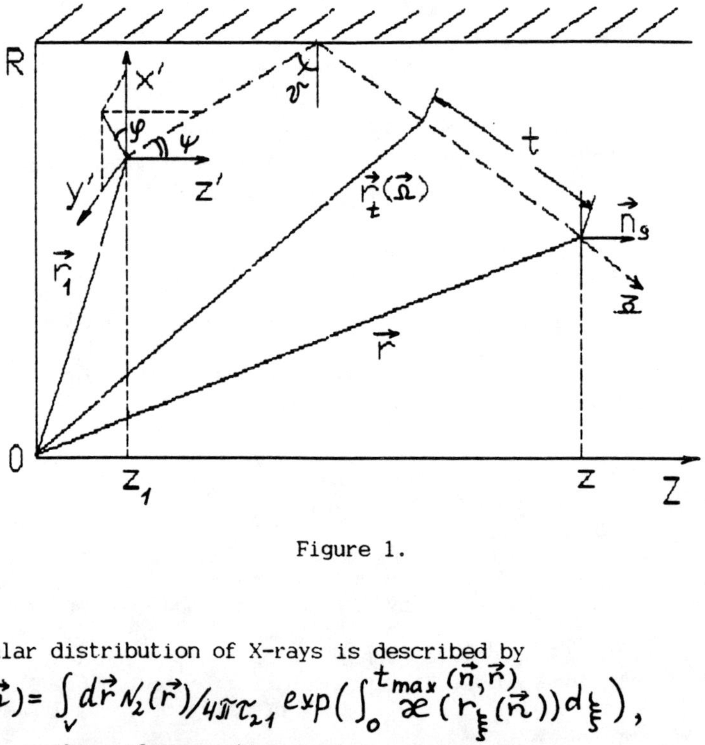

Figure 1.

Angular distribution of X-rays is described by

$$W(\vec{n})=\int_V d\vec{r}\, N_2(\vec{r})/4\pi\tau_{21}\, \exp\Big(\int_0^{t_{max}(\vec{n},\vec{r})} \varkappa(r_\xi(\vec{n}))\,d\xi\Big), \tag{3}$$

where V - volume of generation region;
$t_{max}(\vec{n},\vec{r})$- distance along trajectory of quanta born at point r( ) and escapes active media in direction $\vec{n}$.

Volume element is defined as $d\vec{r}=dz\,ds=t^2 dt\, d\vec{\Omega}$

$$ds/t^2=d\vec{\Omega}/(\vec{n}\vec{\Omega}) \tag{4}$$

and equation (1) goes over into the equation

$$J(z)=\int_V d\vec{r}_1 N_2(\vec{r}_1)/4\pi\tau_{21}\int d\vec{\Omega}/(\vec{\Omega}\vec{n})\, \exp\Big(\int_0^{t(\vec{r},\vec{r}_1)} \varkappa(\vec{r}_\xi(\vec{\Omega}))\,d\xi\Big). \tag{5}$$

It will be assumed that

$$N_i(\vec{r})=N_i(z),\ \varkappa(\vec{r})=\varkappa(z), \tag{6}$$

and using relations:

$$n_i(z)=N_i(z)/N_{i0},\ (i=1,2)\ ;\quad f(z)=J(z)/J_s\ ;\quad l=L\,\varkappa_a\ ;$$
$$r=R\,\varkappa_a\ ;\quad \varkappa^*(z)=\varkappa(z)/\varkappa_a\ ;\quad J_s=1/\sigma_{21}\tau_{21}(1+B)\ ; \tag{7}$$
$$B=(\tau_{21}-\tau_2)/\tau_1\tau_2\tau_{21}\ ;\quad \tau_{21}^*=\tau_{21}\varkappa_a J_s/N_{20}\ ;$$

then system (2,5) may be transformed to dimensionless form

$$f(z)=\int_0^l dz_1 n_2(z_1)\Omega(z,z_1)/4\tau_{21}^*\ \exp\Big[\int_0^{|z-z_1|}\varkappa^*\Big(z-\frac{z-z_1}{|z-z_1|}\xi\Big)d\xi\Big]; \tag{8}$$
$$\Omega(z,z_1)=1/s(z)\int_{S(z_1)} ds\,d\varphi\,d\psi\ \sin\psi/(\vec{\Omega}\vec{n}_s)\ ;$$
$$\varkappa^*=\varkappa_0^*/(1-f(z))-1\ ;\quad \varkappa_0^*=\sigma_{21}(N_{20}-g_2/g_1 N_{10})/\varkappa_a,$$

where φ, ψ - polar and azimuthal angles of quanta born at point $\vec{r}_1$ (fig1).

For R=R(z)=CONST and $0<z<l$

$$\Omega(z,z_1)=-4/r^2\int_0^r \rho\,d\rho\int_0^{\pi}d\varphi\ \log(\cos(\psi_{max}(\varphi))) \tag{9}$$

with $\psi_{max}(\varphi)$- maximum value of azimuthal angle for which photons reach the transfer section with coordinate z.

If photons fall at boundary of active media under angle ϑ, $\vartheta>\vartheta_0=\pi/2-\varepsilon$, ε-slide angle), they reflected and in alternative case escaped gain region then

$$\psi_{max}(\varphi)=\max(\Theta_g(\varphi),\Theta_v(\varphi)), \tag{10}$$

where

$$\Theta_g(\varphi)=\arccos((z-z_1)/((z-z_1)^2+\chi^2(\varphi))^{1/2})\ ; \tag{11}$$
$$\chi^2(\varphi)=r^2-\rho^2\cos^2\varphi-2\rho\cos\varphi(r^2-\rho^2\sin^2\varphi)^{1/2}\ ; \tag{12}$$
$$\Theta_v(\varphi)=\arccos\Big(\Big(\frac{r^2\sin^2\vartheta_0}{r^2-\rho^2}-\frac{\rho^2\sin^2\varphi}{\sin^2\varphi}\Big)^{1/2}\Big). \tag{13}$$

The analytic analysis of amplified spontaneous emission formation with quanta reflection by the side surface of an active media uses results[4]. For little value ε,

$$\varepsilon<\Theta_s=r/l \tag{14}$$

it may be waited that space distribution of $\varkappa(z)$ along z and size of "effective gain" region defined in [4] will invaritable. Then the intensity at foot of active element increases at quantity of photons that escape active media across the side surface under angle when reflection is absent.

$$W_s(\Theta)=W_f(0)(1-l_{eff}/l)\,\Theta/\pi\Theta_s\ ; \tag{15}$$
$$dN_s=2/3\,W_f(0)(1-l_{eff}/l)\,\varepsilon^3/\Theta_s. \tag{16}$$

For $\varepsilon=0$ full quantity cross foot photons is

$$N_{f0}=4/3\,\pi W_f(0)\,\Theta_s^3 \tag{17}$$

and then for case of little ,

$$N_f = N_{f0} + dN_S = 4\pi r^2 f(\ell) \tag{18}$$

and

$$f(\ell,\varepsilon) = f(\ell,\varepsilon=0)\left(1-\left(1-\ell_{eff}/\ell\right)\varepsilon^3/2\pi\Theta_S^3\right). \tag{19}$$

Angular distribution in this approximation may be represented form

$$W(\Theta,\varepsilon) = \begin{cases} W_f(\Theta=0,\varepsilon=0), & 0 \le \Theta \le \varepsilon; \\ W_f(\Theta,\varepsilon=0), & \Theta > \varepsilon. \end{cases} \tag{20}$$

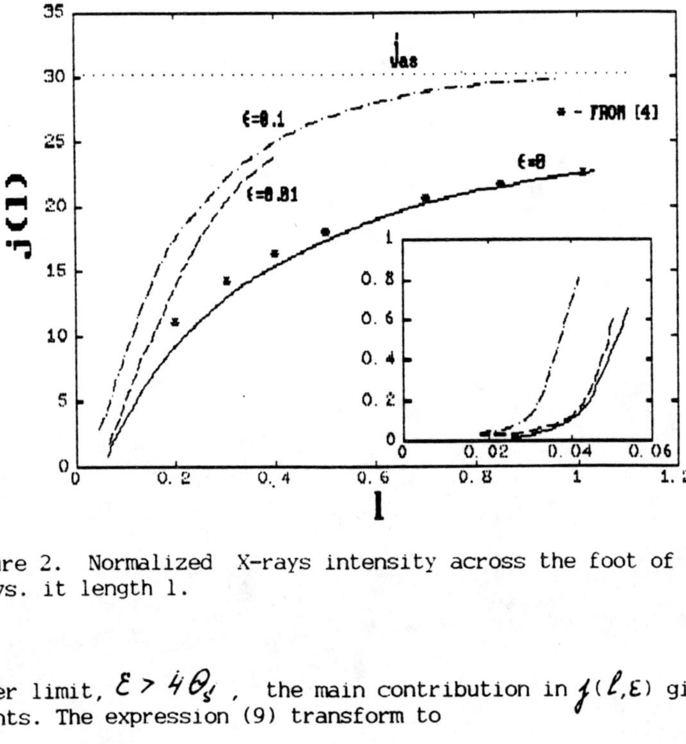

Figure 2. Normalized X-rays intensity across the foot of the active element vs. it length l.

In another limit, $\varepsilon > 4\Theta_S$, the main contribution in $f(\ell,\varepsilon)$ give reflection quants. The expression (9) transform to

$$\Omega(z,z_1) \simeq \varepsilon^2. \tag{21}$$

Inserting this formula in (8), for $z > \ell/2$, we receive differential equation

$$df(z)/dz = \left(x_0^*/(1+f(z)) - 1\right) f(z) \tag{22}$$

with solution for $z > \ell/2$

$$(1 + 1/f_{as}) \ln\left(\frac{f_{as} - f(\ell/2)}{f_{as} - f(z)}\right) = z - \ell/2, \tag{23}$$

where

$$f_{as} = x_0^* - 1.$$

Using correlation between intensity of radiation in the centre and at the end of active element we obtain

$$J(z) \simeq J_{as}(1-\exp(\ell/2)) + \varepsilon^2/2\, J_{sp}\exp(-\ell/2), \tag{24}$$

where $J_{sp} = n_{2s}\ell/2\tau_{21}^*$ - estimation of spontaneous background under saturation.

The space distribution of intensity has form

$$J(z) = J_{as} - (J_{as} - J(\ell))\exp(-(\ell-z)J_{as}/(1+J_{as})) \tag{25}$$

with linear dependence from z if $1-z \ll \ell$.

An angular distribution may be described by

$$W(\varepsilon,\theta) = \begin{cases} J(\ell)\cdot r^2/\varepsilon, & 0 \le \theta \le \varepsilon; \\ J_{sp}V/2\pi\ell, & \theta > \varepsilon. \end{cases} \tag{26}$$

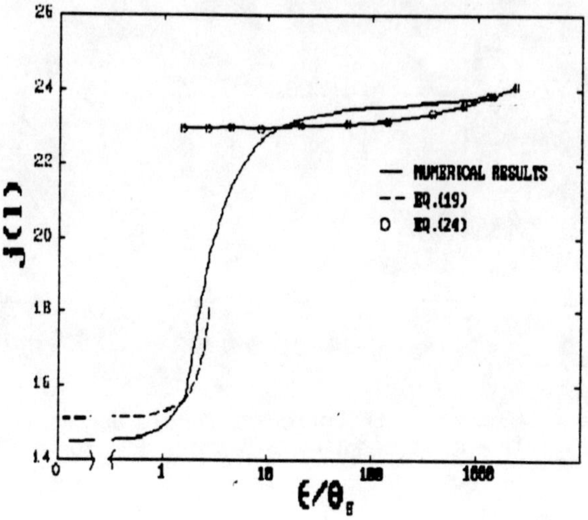

Figure 3. Normalized X-rays intensity at the end active element vs. slide angle

The numerical solution of equation (8) was obtained for next parameters of the active media, which estimated from [6]: $\mathcal{X}_a$ =0.1 sm.$^{-1}$, N_{20}/N_{1s}=8.7, $\mathcal{X}_0^*$=30, B=0.23, τ_{21}^*=0.42.

The dependence of radiation intensity from length of active media with radius r=10^{-3} for different values of sliding angle is shown in Fig.2. For element with fixed length the intensity as function of angle ε is shown in Fig.3, where at $\varepsilon = 3\theta_s$ intensity reaches its asymptotic value. The angular distributions of escape radiation represent in Fig. 4.

From analytic analysis and numerical results it becomes clear that reflection from side surface leads to increasing of X-rays intensity across foot element in case of little flows of radiation ($j(\ell) \ll j_{os}$). The angular distribution keeps geometric divergence if slide angle $\varepsilon < 3\theta_s$.

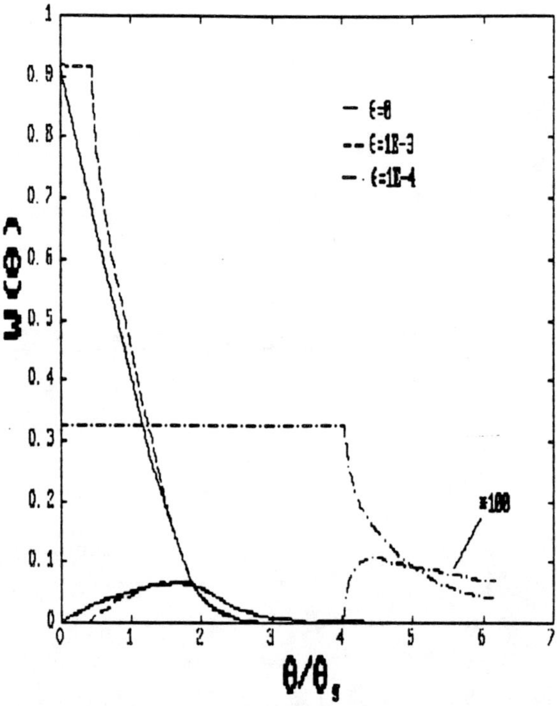

Figure 4. Angular distribution of X-rays across foot and side surface of the active element of length l=3.5 and $r=1^{-3}$ for different slide angle.

REFERENCES

[1] D.L.Matthews, P.L.Hagelstein, M.D.Rosen, M.J.Echart, N.M.Ceglio, A.V.Hazi, H.Medecki, B.J.Mac Gowan, J.E.Trebers, B.L.Whitten, E.M.Campbell, C.W.Hatcker, A.M.Hawruluck, R.L.Kaufman, L.D.Pleasance, G.Rambach, J.H.Scofield, G.Stone and Weaver, Phys. Rev. Lett.54,(1985) 110.

[2] M.D.Rosen, P.L.Hagelstein, D.L.Matthews, E.M.Campbell, A.V.Hazi, B.L. Whitten, B.J.Mac Gowan, R.E.Turner, R.W. Lee, G.Charatis, G.E.Bush, C.L.Shepard, P.D.Rockett and R.R.Johnson, Phys. Rev. Lett.54,(1985) 106.

[3] James G.Lunney, Appl. Phys. Lett.48(14),(1986) 891.

[4] V.B.Kryuchenkov, A.V.Lukin, P.V.Petrov, Yu.I.Chernukhin, Quantum Electronica (Soviet Journal) 19(2),(1990) 183.

[5] J.F.Seely, C.M.Brown, V.Feldman, M.Richardson, B.Yaakobi, W.E.Behring, Optic. Com.54,(1985) 289.

GENERATION OF COHERENT X-RAY RADIATION IN DOUBLE RESONANT PURCELL-SMITH GENERATOR

V.I. Vysotsky, R.N. Kuzmin, and N.V. Maksyuta
Kiev State University 252017, Vladimirskaya 64, Kiev, Ukraine

Let relativistic electron travels in vacuum with the velocity $\vec{v}$ above the corrugated surface $x(z)=b+a\sin^2(\pi z/\Lambda)$ of isotropic non-polar and low-absorbing dielectric (see figure). This electron excites local oscillatious of electron density in the medium, i.e. causes variable polarization of periodic structure of dielectric. Under condetion $a \ll b \ll \Lambda$ of parameters a, b, Λ, wich is called the condetion of quasiplane, this variable polarization can be deseribed with the help of the model of blinking dipole with moment

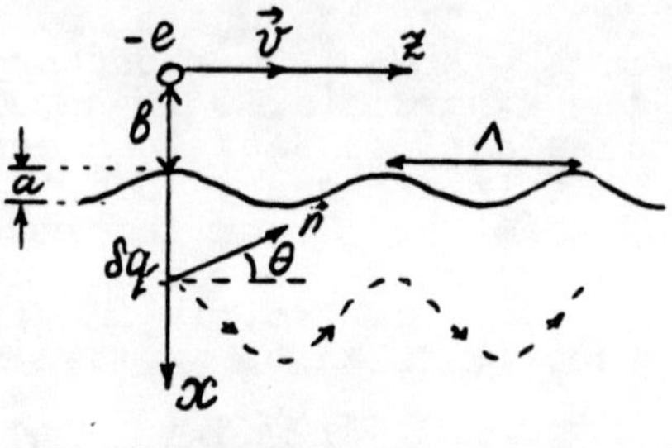

Fig. Schematic representation of radiation by fictitious charge.

$$\vec{p}(t) = 2\vec{e}_x \delta q \cdot (b + \xi a \sin^2(\Omega t/2)), \quad (1)$$

where δq is the value of fictitious charge, ξ - dimensionless parameter. The chosen approach is analogous to the approach used to explain the mechanism of radiation occuring due to uniform motion of an electron near the diffractional metallic lattice. In this case fictitious charge is an mirror-image presentation of an electron, i.e. $\delta q=+e$, $\xi=1$. This problem has been firstly considered by Purcell and Smith [1]. It has been shown, that radiation appears only at interference frequencies

$$\omega_k = \Omega k/(1 - \frac{v}{c}\cos\theta), k = 1,2,... \quad (2)$$

where θ - an angle between unit vector $n = \{\sin\theta\cos\varphi, \sin\theta \cdot \sin\varphi, \cos\theta\}$ in the direction of radiation and velocity vector. In the case of non-absorbing dielectric, which is characterized by the real part $\chi'(\omega)$ of dielectric susceptibility $\chi(\omega)$, one more resonance condition appears, it means that intensity of polarized radiation will be maximum for those interference frequencies (2), which coincide with natural resonance frequencies ω_e of function $\chi(\omega)$. Thus, it is a case of double resonance. The source which helps to implement this is called a double resonance generator of Purcell-Smith. The first work devoted to this problem was [2]. It should be mentioned that in all above-given Purcell-Smith generators dispersion of medium has not been taken into consideration. This condition is decisive for obtaining of supermonochromatic radiation. Let us consider some details.

It is well known (see [3]) that lines of line spectrum (2) have halfwidths

$$\Delta\omega_k \simeq \omega_k(\Lambda/Lk),$$

where L is the length of periodic structure. Thanks to this fact, it can be stated, that for the case, when half-width Γ_e of resonance line satisfies

$$\Omega/(1-\frac{v}{c}\cos\theta) \gg \Gamma_e \gg \Delta\omega_k$$

supermonochromatization is equivalent to selective extraction of a separate line of the line spectrum. Conditionally, it may be called "soft" supermonochromatization. For the second case, when $\Gamma_e \ll \Delta\omega_k$, supermonochromatization occure in the limits of the k-harmonic. It is an example of "rigid" supermonochromatization. Both cases can occur to the some extent.

It is necessary to determine the value of charge δq and parameter ξ and to prove (1). Dipole moment at ω in the frequency range $d\omega$ of an elementary volume dV at the point A(x,y,z) in dielectric, induced by relativistic electron flying above it, can be written in the form

$$d\vec{p}(\omega,\rho) = n\beta(\omega)\vec{E}(\omega,\rho)dVd\omega\ , \qquad (3)$$

where n is a number of dipoles per unit of dielectric volume, $\beta(\omega)$ - polaryzability of a separate dipole at ω, $\vec{E}(\omega,\rho)$ - Fourier component of electric field of an electron; the Cartesian coordinates are

$$E_{x,y}(\omega,\rho) = \frac{2e\omega}{v^2\gamma}K_1\left(\frac{\rho\omega}{v\gamma}\right)\left(\frac{x,y}{\rho}\right),$$
$$E_z(\omega,\rho) = -\frac{2ie\omega}{v^2\gamma^2}K_o\left(\frac{\rho\omega}{v\gamma}\right) \qquad (4)$$

where $K_o(x), K_1(x)$ - Mc Donald function. Let's define Cartesian coordinates of the moment $\vec{p}(t)$ components. It is necessary to substitute (4) in (3) and to integrate along x going from $b + a\sin^2(\pi z/\Lambda)$ to $+\infty$, along y from $-\infty$ to $+\infty$, along z convolation of $f(z(t)) = \frac{b}{a}\cdot\delta(z - z(t))$ should be taken ($\delta(x)$ - delta function, $z(t) = vt$) and at last integration along a contour of resonance line of excitation should be made. It is correct due to comparative contribution to the resulting value gives high enough number of dipoles, and a convalation for z is made because electric field of an electron has noticeable magnitude in $\Delta z \sim b/\gamma$. After integration along coordinates we have

$$dp_x(\omega, z(t)) = \frac{\chi'(\omega)be}{2\omega}\exp\left[-\frac{\omega}{v\gamma}(b + a\sin^2\frac{\Omega z(t)}{2})\right]d\omega,$$

$$dp_y(\omega, z(t)) = 0, \qquad (5)$$

$$dp_z(\omega, z(t)) = -\frac{2i}{\gamma}dp_x(\omega, z(t)).$$

One can see from (5) that at the limit of $\gamma \gg 1$ z-component of induced dipole moment is γ times less, than x-component, i.e. excitation of dipole oscillation along the direction of electron velocity can be neglected. Let us integrate along a line contour

$$p_x(t) = \int \chi'(\omega_e)\Gamma_e \cdot \delta(\omega - \omega_e)dp_x(\omega, z(t)) =$$

$$= \frac{\chi_e'\Gamma_e be}{2\omega_e}\exp\left[-\frac{\omega_e}{v\gamma}(b + a\sin^2\Omega t/2)\right]. \qquad (6)$$

And then, considering $\omega_e a/\gamma v \lesssim 1$, go to the next expression

$$p_x(t) \approx \frac{\chi_e'\Gamma_e e}{2\omega_e}\exp(-\frac{\omega_e b}{v\gamma})(b + \frac{\omega_e ba}{v\gamma}\sin^2(\Omega t/2)), \qquad (7)$$

so having proved (1) and recieved

$$\xi = \frac{\omega_e b}{v\gamma}$$

$$\delta q = \frac{\chi_e'\Gamma_e e}{4\omega_e}\exp(-\omega_e b/v\gamma) \qquad (8)$$

Thus, the idea based on the model of blinking dipole helps to determine spectral-angular and polarization properties of radiation. Using [5], we can also evaluote the intensity of diffraction radiation with the help of this simple model. As far as velocity and acceleration are mutually perpendicular, we can use the ready expression for intensity of 6 and substitute the square of amplitude of the second derivative of for time of the blinking dipole moment (7), we'll recieve that the number of quanta with the energy $\hbar\omega_e$, radiating at $\theta_e \approx \arccos\frac{c}{v}(1 - \Omega/\omega_e)$ in the frequency range $\Delta\omega \approx \Gamma_e$ during the flight of one electron per path unit, equals to:

$$N \approx (\pi\chi_e'^2 e^2 a^2 b^2 \Gamma_e^3 \omega_e / 32\hbar\Lambda v^5 \gamma^2)\exp(-2\omega_e b/v\gamma) \qquad (9)$$

For example, for carbon ($\hbar\omega_e \approx 284$ eV, $\chi_e' \approx 10^{-2}$, $\Gamma_e \approx 0{,}5\cdot 10^{16}$ Hz) at $a \approx 5\cdot 10^{-4}$ cm, $\Lambda \approx 0{,}1$ cm, $b \approx 3\cdot 10^{-3}$ cm, $\gamma \approx 10^4$ we obtain that at angle $\theta_e \approx 3\cdot 10^{-3}$ rad in space angle $\Delta O_e \approx 2\pi\theta_e\Delta\theta_e \approx 10^{-6}$ sterad one electron initiates radiation $N \approx 2\cdot 10^{-7}$ quanta/cm.

In conclusion it should be noted, that this problem lotes has been considered in [8], where a new theory of coherent radiation by local surface plasmon excitated by relativistic electrons has been proposed.

REFERENCES

[1] S.J.Smith,Doctoral thesis.Harward Univ.Cambridge.Mass. USA,(1953)p.27.

[2] V.I.Vysotsky,R.N.Kuzmin,N.V.Maksyuta,Thesis of XIX All-Union conference in physics of interaction of charged particles with crystals,Moscow Univ. Publ.,(1989)40.

[3] D.F.Alferov,Yu.A.Bashmakov,E.G.Bessonov,Undulator radiation,Works of Lebedev Phys.Inst.,M.:Nauka(1975)v.80,p. 100.

[4] J.Jackson,Classical electrodynamics,M.:Mir(1965)p.512.

[5] F.A.Korolev,Theoretical optics,M.:"Vyssh. shkola"(In Russ.),(1966)p.555.

[6] L.D.Landau,The classical Theory of Fields,M.:Nauka (In Russ.),(1987)p.512.

[7] V.A.Bazylev,N.K.Zhevago,The radiation of fast particles in substance and in external fields,M.:Nauka (In Russ.), (1987)p.272.

[8] N.K.Zhevago,ZhETF (In Russ.)97(1990)503.

INNER SHELL VACANCIES PRODUCTION AS A WAY TO OBTAIN INVERSE POPULATIONS

P.A. Golovinskiy and M.A. Preobrazhenskiy
Voronezh Engineering and Building Institute, Voronezh, USSR

The goal of this paper is to determine theoretically gain coefficients in the photoionization laser pumped by broadband soft-X-ray radiation from laser-prodused plasmas.

The scheme of invers population production is the next. The laser irradiation interacts with solid matal surface produses plasma. It is the source of soft-X-ray radiation, which interacts with the gas of neutral atoms. The result of this process is inner-shall ionization of atoms with creation of the vacancies in atomic inner shell or exitation of residual ion. Both of this process can be used for invers population production. But otherwise experimental conditions for using of the first ones may be more realistic [1]. We consider opportunity of practical realisation of this conditions.

The muster equations for the two chennals (ionisation with extation and inner-vacacies production) are different. For the first process it has simple form:

$$\frac{dn_0}{dt} = - W_i n_0$$

$$\frac{dn_2}{dt} = W_{1ex} n_0 - W n_2 \qquad (1)$$

$$\frac{dn_1}{dt} = W_{21} n_2$$

where n_0 is a number of atoms in ground state, n_1, n_2 are numbers of ions in levels |1> and |2> (see Fig.1).

Fig.1

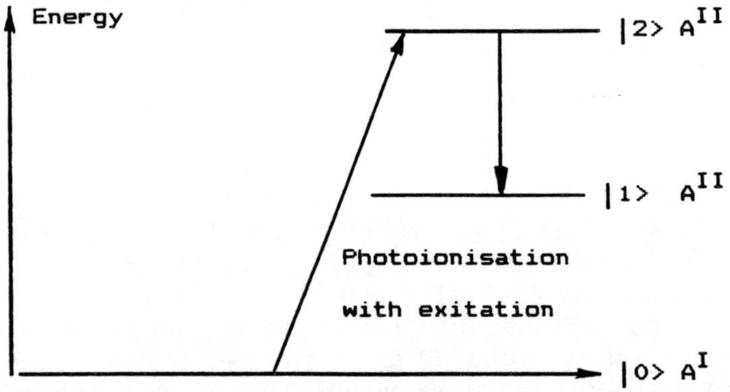

W_1 is total ionisation probability of outer-shell atomic ground state, W_{1ex} is ionisation with |2> state exitation probability, W is total decay pobability of |2> state, W_{21} is transition |2>-|1> probability. The second process can be described by more complicated equations due to necessarity to take into account Auger effect.

The interaction of atoms with X-rays we discribe in RPA (random phase aproximation). The renormalizised by electron correlation interaction operator of interaction with external electromagnetic field Φ_{effqi} satisfies the equation:

$$\Phi_{eff} = \bar{r}\,\bar{e} - \sum_{q,j} \frac{2\,\omega_{qj} \langle j|\ |\bar{r}-\bar{r}^1|^{-1}\ |q\rangle\, \Phi_{eff\ qj}}{\omega_{qj}^2 - \omega^2} \qquad (2)$$

where $\bar{e}$ is unit polarisation vector, q,j are quantum nambers of initial intermadicl states of atomic electrons, which can do wirtual transitions, ω is external field friquency, ω_{qj} is the energy of virtual transition, Φ_{effqj} is the renomalized matrix element for j-q transition. When

$$W_1 = \frac{4\pi\rho F}{\omega\, c} \int |M|^2 d\,\Omega_{\bar{p}} \qquad (3)$$

$M = \langle f|\Phi_{eff}|0\rangle$, $|0\rangle$ and $|f\rangle$ are initial and final states, $\bar{p}$ is electron momentum in final state, F is photon flux, ρ is final stats decity. For ionisation with exitation

$$M = \sum_{k} \frac{\langle m\ f|\,|\bar{r}-\bar{r}^1|^{-1}|i\ k\rangle\, \Phi_{eff\ kj}}{\omega_{kj}-\omega} \qquad (4)$$

For calculation of W_1 and W_{1ex} we have the effective program fr PC with the bank of atomic data. The atomic potential has the GSZ [2] ore local dencity approximation [3] form. It containes the blok for calculation of an angle coefficients of correlation interaction matrix elements for arbitrary electronic configurations. Oscillator strength of bound-bound tranzitons can be calculated by the program too.

The value range of cross-section in maximum is from 0.5 to 40 MB. The lagest cross-sections has d-electrons.

Plasma sources may have different spectra forms. One of the tipical is blakbordy sorce. The average cross-section in this case is:

$$\sigma = \int_0^{\infty} \sigma(\omega) f(\omega) d\,\omega \qquad (5)$$

where

$$f(\omega) = \frac{(X/\pi)^2}{(e^X-1)\ 0.244T}\ , \qquad X = \omega/T\ ,\ \omega_{max} = 2.822T$$

In the table 1 we show our results for cross-section. The temperature has been optimised cross-sections are maximum.

Table 1 alows us to estimate the upper bound of the gain coeffitients. Let us consider the Xe gas medium. The optimum temperature is T=4.88 Ry. It was shown [3], that the rate of Ague process in scheme Fig.1, are lager then radiative ones. The invers population can be produced if next relation is valid:

Atom	Shell	Temperature,Ry	Cross-section, a.u	τ,ps
Cs	4d 5s 5p	5.52	25.27 0.76 3.56	$3.35\ 10^{-2}$ 1.11 0.238
Ba	4d 5s	5.81	21.72 0.57	$3.34\ 10^{-2}$ 1.27
Zn	3d 4s	5.88	20.69 $2.11\ 10^{-3}$	$3.38\ 10^{-2}$ 332
Xe	4d 5s 5p	4.88	23.36 0.7 3.85	$5.24\ 10^{-2}$ 1.75 0.318
Kr	3d 4s 4p	15.14	28.63 $2.63\ 10^{-2}$ 2.1	$1.45\ 10^{-3}$ 1.74 $1.39\ 10^{-2}$
I	4d 5s 5p	3.8	21.89 0.3 1.66	0.118 8.65 1.56

$$\frac{1}{\tau} > \frac{g_1}{g_2} \frac{1}{\tau_{rad}}$$

where τ is vacancies production time, τ_{rad} is radiation time of life, g_1 and g_2 are statistical weightes for upper and doun states. For $5s^0 5p^6\ {}^1S_0 \longrightarrow 5s^1 5p^5\ {}^1P_1$ transition in Xe III

$\tau_{rad} = 4.75 \pm 0.15$ nsec [3]. Our calculations showes, that $\tau \simeq 5$ 10^{-2} psec in optimum. One can conclude, that scheme under discussion can be realised very easy due to $\tau \ll \tau_{rad}$.

REFERENCES

[1] M.Ya.Amusya,M.L.Shmatov,Pisma v Zh.T.F,14,(1988),1249

[2] Green A.E.S.,Sellin D.L.,Zachor A.S.,Phis.Rev.,184,(1969),1

[3] Zangwill A.,Levin Z.H.,Am.J.Phis.,53,(1985),1177

[4] Kapteyn H.C.,Lee R.W.,Folcone R.W.,Phis.Rev.Lett.,57,(1986),2939.

THE PROBLEM OF SMALL SCALE X-RAY LASERS

Yu.V. Afanas'ev and V.N. Shlyaptsev
P.N. Lebedev Institute of Physics, Leninsky Prospect 53, 117924, Moscow, USSR

INTRODUCTION

The systematic financing of X-ray laser programs that was made within the last 5-10 years in different laboratories of the world has given appreciable practical results. They included the demonstration of the laser action in X-ray range, the obtaining of reliable data and their verification by other experiments, the achievement of high gain in one-pass and multi-pass generation regimes with the use of multilayer mirrors etc. Such results, were obtained, as a rule, on large high-power unique laser facilities, such as NOVA, VULCAN, PHEBUS, GEKKO. But high cost and unusual parameters of such lasers make them quite unattainable for the majority of average and small laboratories, universities and companies. As a result the activity of such smaller laboratories is mostly directed to minor,or preparatory, or theoretical investigations of X-ray laser problems, for example, the analysis of the gain coefficients by the relative intensities of spectral lines, all kinds of spectroscopic studies, target design and manufacture of X-ray optics, the development of experimental schemes, the investigation of the influence of various physical processes on the kinetics and dynamics of the laser produced plasma etc.

At present, there appeared a feasibility for the production of laser action in soft X-ray region with the use of comparatively cheap and rather compact lasers of the terawatt power range of pico- and femto-second duration. They are characterized by not so high output energy of ~ 0.1-10 J per pulse. But their development is closely connected with the perspective of the laser action within 1-10 **Å**, when the target flux density will reach 10^{17}-10^{20} W/cm^2. In this paper we investigate a nonstationary collisional (transient inversion - TI) scheme of X-ray laser whose active medium is created by short laser pulses [1]. It was obtained that the X-ray lasers at $\lambda \sim$ 100-1000 Å can be realized, at present, with considerably smaller values of pumping energy, of about 10 J. Such "table-top" laser facilities will essentially extend the field of practical usage of X-ray lasers [2].

RESULTS

As it was shown earlier, during finite time of transient interatomic relaxation processes τ_{ex} it is possible to create higher values of inversion, than those in steady-state (SSI) and quasisteady-state (QSSI) inversion schemes [1,3]. To realize transient inversion (TI) scheme it is necessary to select an appropriate pair of atomic levels with higher velocities of excitation for the upper level, to create temperature of maximum abundance of working ions and then rapidly increase temperature in time scale $t \leqslant \tau_{ex}$. The latter is not so low: $\tau_{ex} \sim 10^{-10} - 10^{-13}$c depending on atomic number, charge, quantum number, electron density and temperature. In this scheme it is possible to obtain inversion even under condition of metastable lower laser level.

To examine the possibility of obtaining laser action in the 200 Å region with using less than 10 J of external laser pumping energy, we investigate primarily full time-dependent level-by-level population kinetics in the case of homogeneous plasma without motion:

$$\frac{d N_k}{d t} = \sum N_i S_{ik} - N_k \sum S_{ki}$$

where N_k is the population of k-th level of a fixed ion (assumed to be Ne-like); $S_{ik} = S_{ik}(z, T_e, n_e)$ is the total probability of collisional excitation (deexcitation) and radiative decay due to i - k transition; z is the ion charge; and T_e and n_e are the electron temperature and density. Initial population of levels corresponds to the temperature of maximum abundance of Ne-like ions. Then we provide approximately twofold temperature jump. Analysis of transient processes relaxation of levels population to the next equilibrium state for different densities ($n_e = 10^{16} - 10^{22}$ cm^{-3}) and spectroscopic numbers ($z_s = 3 - 17$) enable us to obtain inversion in 200-300 Å region with FeXVII at $n_e = 10^{21}$ cm^{-3} within the time interval $t \simeq 10$ ps [1].

The next step is connected with hydrodynamic methods of obtaining a such temperature jumps of necessary magnitude. Using 1-dimensional code RADEX [4] we calculate plasma parameters in cylindrical geometry under conditions of target irradiation with pulse duration $\tau_L \sim 1 - 10$ ps of 1.06μ and 0.53μ Nd-glass, 0.69μ ruby and 30-300 ps, 10.6μ CO_2-laser.

Calculations show that because of not so high power densities $q_L \sim 10^{14}$ W/cm^2, electronic temperatures $T_e \leqslant 1$ keV and validity of inequality $\tau_e \ll \tau_L, \tau_{ex}$, where τ_e is the time of electronic elastic collisions, the electronic distribution function is near maxwellian here. More topical problem, as compared to the long laser pulse situation, is connected with non-Spitzer heat conductivity since the gradient length L_T should be greater than electron mean free path L_{ei}. Fig.1 shows the profiles of plasma parameters n_e, T_e and L_{ei}/L_T, calculated in the Spitzer approximation.

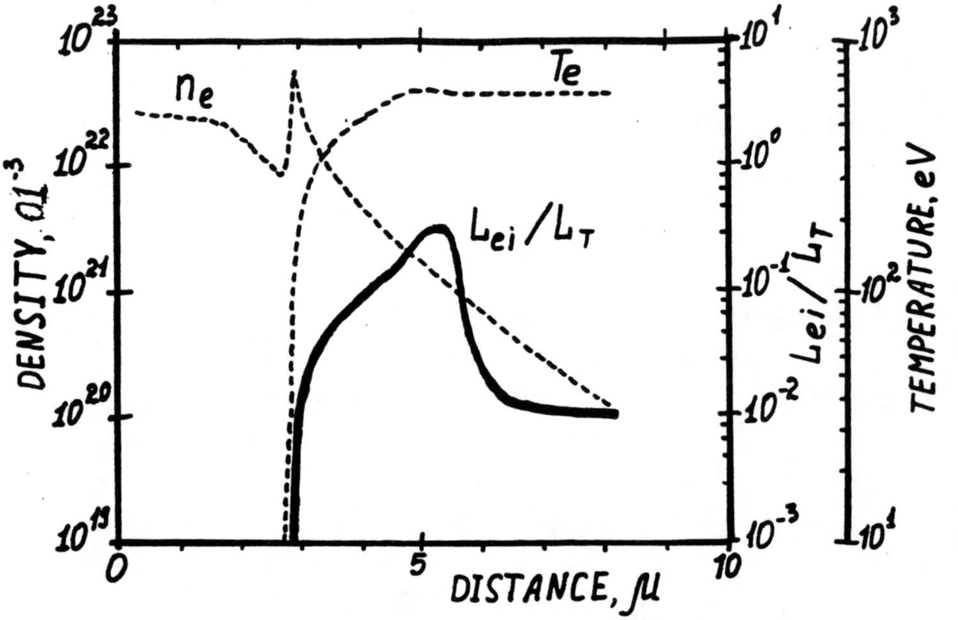

Fig.1. Profiles of n_e, T_e and L_{ei}/L_T for copper target irradiated by 10 ps $5 \cdot 10^{13}$ W/cm^2 ruby laser

We can see that the value of L_{ei}/L_T is 1-2 orders higher, than it is permissible by theclassical theory of conductivity. Therefore we used results of flux-limited heat conductivity theory of ion-sound turbulence [4].

To create the temperature jump for lasing on FeXVII ions we used special profile of a heating laser pulse with a relatively low-power prepulse $q_L \simeq (2 \div 5) \cdot 10^{12}$ W/cm^2 (whose time duration was taken from 0 to 500 ps),and high -power main pulse ($q_L \simeq (0.5 \div 2) \cdot 10^{14}$ W/cm^2 $1 \div 10$ ps). Fig.2 a,b demonstrates the results of RADEX calculations of electron temperature and maximum gain coefficient profiles for 200 ps prepulse and 10 ps 2ω Nd-laser main pulse.

Fig.2 confirms, that it is possible to considerably increase the electron temperature of plasma corona within 10 ps and faster. In spite of prepulse parameters were optimized for QSSI, the gain coefficient in TI scheme is an order higher (cf. Fig.2 b curves with t = -50 ps and t = 0).

To achieve laser action at shorter wavelengths by using transient inversion scheme it can be useful to consider the ions of other isoelectronic sequences (where the gain can also be high), with higher, as compared to Ne-like ions, quantum efficiency, especially taking into consideration the transitions with various main quantum numbers $\Delta n=1,2$.

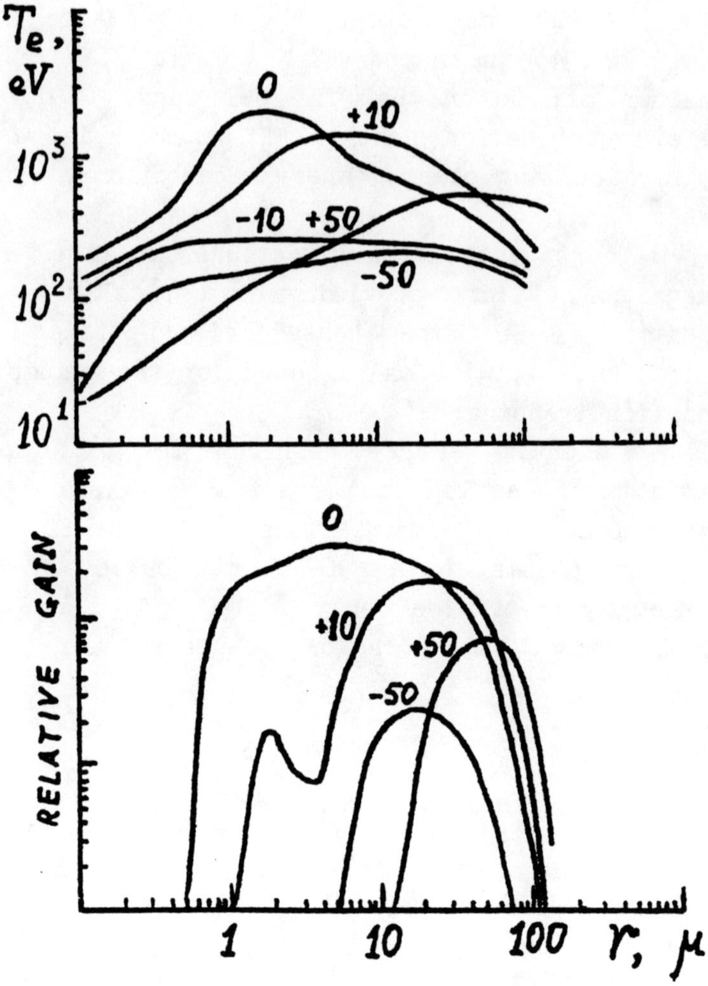

Fig.2 Plasma temperature (a) and maximum gain (b) profiles for 2ω Nd-laser $3\cdot10^{12}$ W/cm^2 prepulse and $1\cdot10^{14}$ W/cm^2 10 ps main pulse iron target irradiation. Different curves labeled by the numbers correspond to different times from the center of the main Gaussian pulse.

CONCLUSIONS

It was found that the TI scheme reveals many attractive properties for X-ray laser design:

| The inversion in TI exists at arbitrary electron density n_e and increases with growing n_e.

| Gain coefficient in TI scheme is several times higher, less sensitive to the details of kinetic model and to the number of atomic levels taken into consideration than that in SSI and QSSI. In contrast tothe recombination scheme, TI scheme is working in the rapidly ionizing plasma where recombination processes are weaker.

| There are no restrictions to the active media dimensions for providing optically transparency conditions. In the case of Ne-like ions the gain coefficient is twice as higher under the assumption of full radiation trapping.

| Pumping energy requirements are impressive: to obtain laser action in spectral region of $\lambda \approx 600$ Å with using 3-3 transitions of Ne-like ions and creating plasma with 30-300 ps pulse of CO_2-laser, it is enough to provide only 0.1-0.3 J and 1 J in 1 ps (or 10 J in 10 ps) pulse of Nd-glass laser (FeXVII, $\lambda \approx$ 200 Å, $q_L \sim 10^{14}$ W/cm^2) .

REFERENCES

[1] Yu.V.Afanas'ev, V.N.Slyaptsev,Sov.J.Quant.Electronics, 19, (1989) 1606. Preprint FIAN N 40 (1985)

[2] P.L.Hagelstein. In Proc. of the Fourth Topical Meeting, Sept. 26-29, 1988, North Falmouth MA

[3] A.V.Vinogradov, V.N.Shlyaptsev, Sov.J.Quant. Electronics, 10, (1980) 754; Sov.J.Quant. Electronics, 13, (1983) 303;

D.L.Matthews et al., Phys.Rev.Lett. 54, (1985) 110.

[4] Yu.V.Afanas'ev et al. J. of Sov. Laser Research, 10, (1989) 1

AXICON PUMPING FOR SHORT WAVELENGTH LASERS

V.V. Korobkin*, L.Ya. Polonsky, and L.N. Pyatnitsky
Institute for High Temperatures, Moscow
**General Physics Institute, Moscow*

The main problem in plasma XUV-lasers creation is pumping efficiency. For increasing this one it is necessary to provide axial symmetry and traveling focus regime. The best way to satisfy these requirements is active plasma medium formation by Besselian (B) beam of electromagnetic radiation [1].

B-beam is formed by conical optic elements (axicons) and its radial intensity distribution is outlined by $J_o^2(kr\gamma)$, where k - wave number, r - radius, γ - focusing angle. The beam characterized by constant value of the diameter d_o of its caustic along whole propagation length and by side energy supply to axial zone.

Some versions of such pumping are shown in fig.1. The first scheme corresponds to gas breakdown produced by laser B-beam and continuous extended laser spark creation [2]. This type of laser spark creates much more homogeneous plasma column in a gas, than by means of a spherical lens. Solid state targets like fiber (2), thin tube (3), foil (4) and slab (5) are possible also. The last arrangement (6) is convenient for realization double-wavelength pumping [3].

Traveling focus regime may be realized in these arranges if pumping pulse front duration τ is much shorter than its time propagation along active medium length l: $\tau \ll l/c$. It means, that suitable pumping pulse duration is $\simeq$ 50 ps for the active medium length $\simeq$ 20 cm [2].

Gas targets require smaller pumping energy than solid ones in the same conditions and active medium may be longer.

Numerical estimations of stimulated radiation gain in H-like ions plasma (CVI and HeII for example) are presented in fig.2 [4]. Optimum plasma parameters for lasing effect are for CVI: $N_e \leq 10^{20}$ cm^{-3}, $T_e \leq 20$ eV; and for HeII:

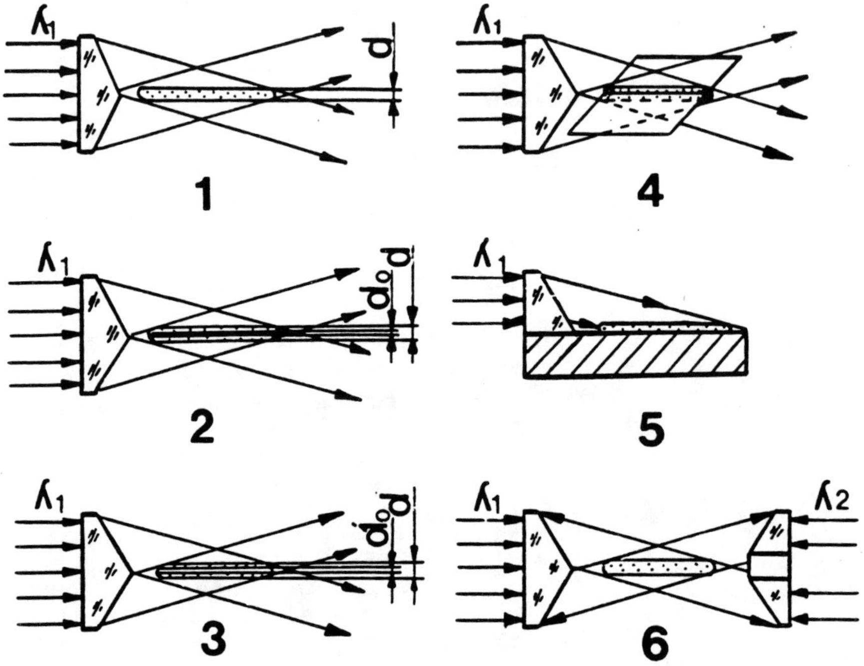

Fig.1

$N_e \le 10^{17}$ cm^{-3}, $T_e \le 5$ eV. Initial values of these parameters must be ten times larger to produce that optimum ones.

In experiments we realized gas breakdown in laser B-beam and showed opportunity of long plasma fiber formation in traveling focus regime. Fibers diameter was about 20 μ and length up to 1 m. In this experiments initial plasma parameters T_e, N_e exceeded 10 eV and 10^{20} cm^{-3} respectively, but unfortunately they were insufficient to produce full ionization of atoms because of low pumping power ($\simeq$ 2 GW). We expect lasing effect in gas target on ions CVI at 100 GW - power pumping and 10 cm of fiber length.

It is interesting to note appearance of regular structure in plasma fiber when B-beam is used for pumping. The structure spasing is $l_1=2\lambda/\gamma^2$. It is caused by radiation intensity amplitude self-modulation along axis B-beam propagation in non-linear medium [5]. The plasma bubbles of the structure help cooling and increase inverse population in active medium.

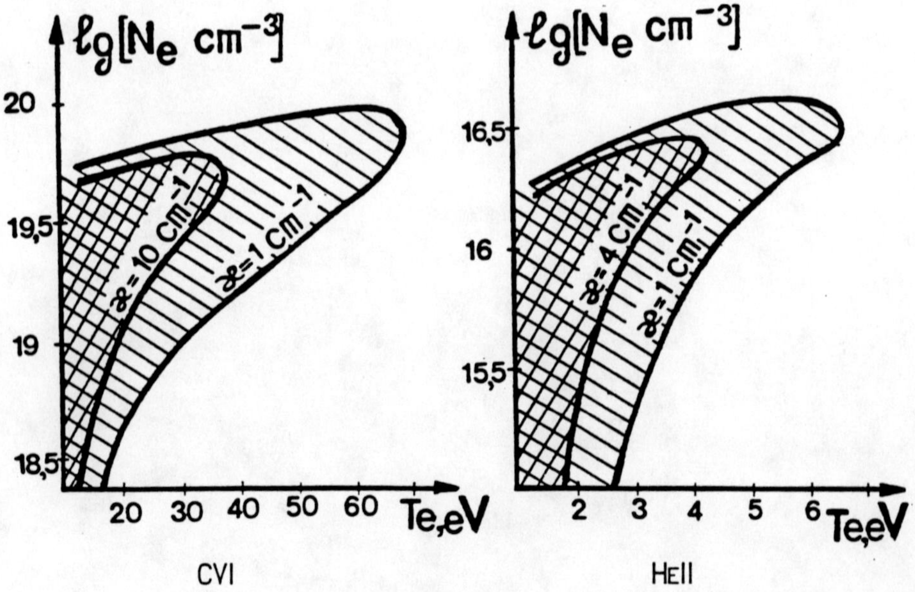

Fig.2

To destroy the periodical structure in plasma fiber it is necessary to prevent radial intensity modulation in the pumping beam. This purpose may be achieved by solid state target arrangement on the beam axis. In this case the beam has conical wave front, but intensity destributions becomes smooth, without zeros. Such pumping beams may be named as conical ones (C-beams).

REFERENCES

[1] V.V.Korobkin, L.Ya.Polonsky, V.P.Poponin, L.N.Pyatnitsky <u>Kvantovaya Electronika</u> (USSR Journ), 13, (1986), 265.

[2] L.N.Pyatnitsky, L.Ya.Polonsky, In Proceed. XIX Intern. Conf. on Phenomena in Ion. Gases, Invited Papers, Belgrade, (1989), p.342.

[3] C.H.Scinner, D.E.Kim, A.Wouters, D.Voorhees, S.Suckewer, In Proceed. Intern. Symp. on X-ray laser microscopy, N.-Y.,(1987), p.36.

[4] A.V.Borovsky, V.V.Korobkin, L.Ya.Polonsky, L.N.Pyatnitsky, M.I.Uvaliev, Kvantovaya Electronika (USSR Journ), 16, (1989), 538.

[5] N.E.Andreev, V.M.Batenin, L.Ya.Margolin, L.Ya.Polonsky, L.N.Pyatnitsky et al. Pisma v JETPh, 15, (1989), 83.

PART II

LASER PLASMA AND APPLICATIONS

INERTIAL CONFINEMENT FUSION BY GEKKO XII MULTI-WAVELENGTH LASERS

C. Yamanaka
Institute for Laser Technology, Institute of Laser Engineering, Osaka University 2-6 Yamada-oka, Suita, Osaka 565, Japan

ABSTRACT

Recent experiments on the GEKKO XII laser (10 kJ, 0.5 μm, 0.3 μm, 2 ns) (Fig. 1) with DT doped plastic shell targets have demonstrated fuel areal densities (ρR) of ~ 0.1 g/cm^2, fuel densities (ρ) of ~ 600 times liquid density (× LD) and fuel temperatures (T_i) of ~ 0.3 keV, which are very near the thermonuclear ignition condition, ρ = 500 - 1000 × LD, ρR ≥ 0.3 g/cm^2, and T_i ≥ 5 keV.(The demonstrated ρ and ρR values refer only to D and T). The high neutron yield 10^{13} had already been demonstrated by the LHART target.[1)] The laser irradiation nonuniformity in these experiments was significantly reduced to a level of 3 % rms by introducing random phase plates, resulting in uniform compression.(Fig. 2) The target irregularity was controlled to a 1 - 2 % level. The fuel ρR was directly measured by the neutron activation of Si which was doped in the plastic targets. The density ρ[2)] was estimated from the measured ρR value by the mass conversion law, where the ablated mass was separately measured from the time dependent x-ray emission of multilayer targets.

Following the investigation of hollow plastic shell implosion, experiments on cryogenic hollow shell targets have been performed. Detailed characteristics of them will be clarified with respect to the plasma interaction and the energy transportation.

Cannonball targets have also been investigated to test the uniform driving pressure on the pellet. Fundamental processes[3)] such as x-ray conversion characteristics from laser, energy confinement in a cavity and x-ray driven burn through of a shell will be thoroughly examined. The ICF research has made a great progress to the ignition.

1. INTRODUCTION

It has long been recognized that one needs to compress thermonuclear fusion fuel up to 500 - 1000 times liquid density (× LD) to achieve thermonuclear ignition and high gain with realistic energy drivers. Previously reported densities were ~100 × LD[4),5)] or 100-200 × LD[6)]. The purpose of the present study is to demonstrate the high density required for ignition and high gain.

In the present experiments, we used deuterium and tritium (DT) exchanged plastic hollow shell targets instead of cryogenically cooled DT solid/liquid targets because (1) the target quality is much better than that of present cryogenic targets so that we can manage the experiments not to be limited by the present technology, (2) the initial density of the fuel part (D and T) in the plastic is approximately equal to the DT liquid density, and (3) increase of the radiation cooling due to the existence of carbons in the plastic is expected not to be a major factor for high density compression. The reason (3) is justified by the calculation that the radiation mean free path is much smaller than the plasma size and the black body radiation loss is much smaller than the plasma internal energy.

2. EXPERIMENTAL CONDITIONS

The experiments were performed on the GEKKO XII Nd:glass laser (λ=0.53μm, 12 beams). The laser energy was 8-10kJ with a 1-ns Gaussian in full width at half maximum (FWHM) or a nominally 1.7-ns (FWHM) flat top pulse having a rise time equivalent to a 1-1.3ns Gaussian pulse. The beam energy imbalance was controlled to be ~1% in standard deviation and the intensity imbalance was estimated to be ~3% in standard deviation. Figure 1 shows the GEKKO XII Glass Laser System.

The targets were DT exchanged plastic targets having a ≈1g/cc initial density and a fuel part of 15-17wt% The target diameters were nominally 500 and 700μm and the thicknesses were 4-13μm. The laser was focused through f/3 lenses behind the target center by 5 target radii. The highest density was observed at the 500μm diameter and 8-10μm thick targets irradiated by the 1.7ns laser pulse.

In order to perform the high density compression experiments we made three major improvements on (1) laser irradiation uniformity, (2) target uniformity, and (3) fuel areal density (<ρR>) measurements.

Fig. 1. GEKKO XII Glass Laser System.

The laser irradiation uniformity was significantly improved by introducing random phase plates (RPP)[7]. Each RPP divided the single large beam into ~20,000 beamlets, each of which had a 2×2mm cross section, producing the focal pattern which has a smooth envelope and small scale spikes due to the interference between the beamlets. The laser intensity distribution on a spherical target was calculated from the measurements on the target plane as a function of spherical harmonic mode number l. This is shown in Fig. 2. The root mean square (rms) nonuniformity was found to be 14% for the nominal 500μm diameter target. Assuming that the distance d between the laser absorption and the ablation surfaces is 5% of the initial target radius, the final nonuniformity including thermal smoothing effect was calculated to be 4.5%. The one-dimensional (1D) hydrodynamic code HIMICO[8] predicted larger d except for the very early time, but we used the above value as a conservative assumption. The laser energy fraction into a circle of 500μm diameter on the target plane was calculated for the RPP case to be 65% of the incident energy.

As for the target fabrication technique we used so-called water/oil/water (W/O/W) emulsion method[9] in which water globules surrounded by oil phase polymer solution are hanged in water. In the past, fabrication of uniform targets was very difficult because of the gravitational effect. However, it was found that the density matching between W/O/W is essential for target

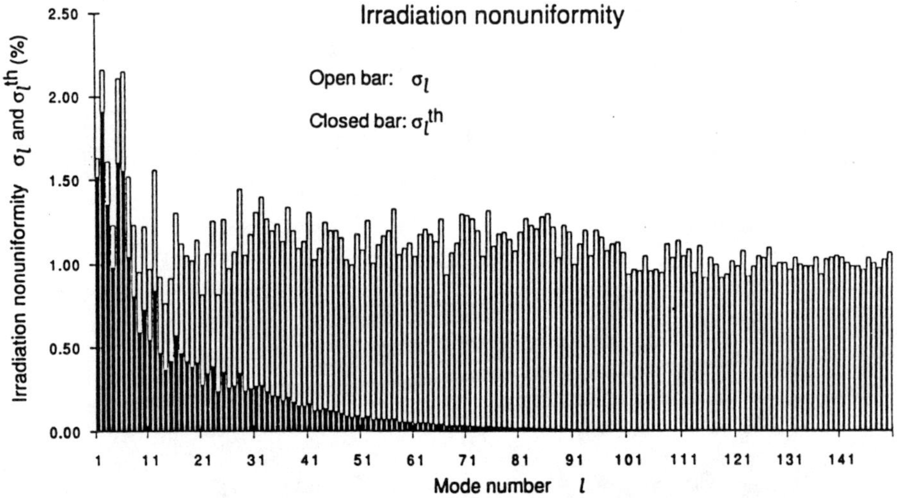

Fig. 2. Irradiation nonuniformity on a spherical surface of a 500μm diameter calculated from measurements on the target plane. Open bars represent the nonuniformity without thermal smoothing, whereas closed bars represent that with thermal smoothing. The absorption-to-ablation distance of 5% of the target radius is assumed.

uniformity[10),11)]. As a result of the density matching, both the deviation from a perfectly spherical shape and nonuniformity of the target thickness were controlled to a 1% level. Typical target performance is shown in Fig. 3.

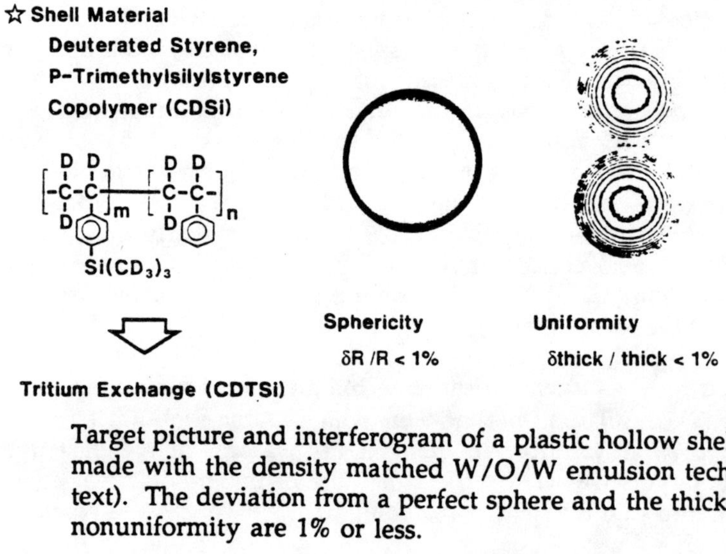

Fig. 3. Target picture and interferogram of a plastic hollow shell target made with the density matched W/O/W emulsion technique (see text). The deviation from a perfect sphere and the thickness nonuniformity are 1% or less.

The target material was a deuterated styrene + deuterated poly-para-trimethylsilylstyrene copolymer $[(C_8D_8)n\text{-}(C_{11}D_{16}Si)_m]$. Si atoms were contained as a chemical composition for fuel <rR> measurements. The Si amount of 5-15wt% was controlled by changing the n/m ratio from 2 to 0. Some of deuterium in the targets was exchanged with tritium, giving tritium-to-deuterium ratios of 3-10%[10].

As for fuel <ρR> measurements, three independent techniques were used: the secondary reaction[12], the knock-on deuteron[13] or proton[14], and the neutron activation[15] techniques. The secondary reaction and knock-on techniques, however, have the maximum measurable <ρR> values above which primary 1-MeV tritons or knock-ons slow-down significantly in the fuel. These maximum <ρR>s are even lowered by the existence of carbon ions in the plastic target to be ~5-20, ~10, ~30mg/cm^2 for the secondary reaction, the knock-on deuteron, and the knock-on proton techniques, respectively. We, therefore, adopted the neutron activation technique for fuel <ρR> measurements as there is practically no <ρR> limitation. The other techniques were used for the cross check.

Si was selected as a tracer atom because the Si activation technique is well developed. The activation reaction is ^{28}Si(n,p)^{28}Al having a cross section of 0.26b for 14-MeV neutrons. The ratio of the number of ^{28}Al to the neutron yield is directly proportional to the fuel <ρR>. The ^{28}Al atom decays back to ^{28}Si, producing β- and γ-ray simultaneously with a half life time of 2.24min and distinct γ-ray energy of 1.78MeV.

The counting system consisted of a gas flow counter for β-ray detection and a NaI scintillation counter for γ-ray detection. The detection efficiency of γ–β coincidence events was absolutely calibrated to be 4%. The γ-ray spectra and the time decay, both of which have a coincidence with β-ray, were measured. The debris collector used in these experiments was a Nb cylinder of 4.4cm diameter and 10cm long. The collector covered a 4π solid angle (actually 85% of 4π) to increase signal counts and to eliminate unwanted effects due to angular distribution of the debris expansion. The debris collection efficiency was calibrated using pre-activated targets containing Na. From the Na activity measurements before and after the laser shots, the collection efficiency was absolutely calibrated to be 28% with shot-to-shot fluctuation of less than ±3%. It was found[11] that the target mass dependence of the collection efficiency is insignificant, whereas the atomic number dependence is significant. In the present experiments it was assumed that the Na collection efficiency is the same as the Al collection efficiency. This assumption was justified by the very good agreement of the fuel <ρR> values measured both with Si activation and knock-on deuteron techniques simultaneously[16].

3. EXPERIMENTAL RESULTS

Typically 40% of the incident laser energy (60% of the laser energy in the 500μm diameter circle) was absorbed by the targets.

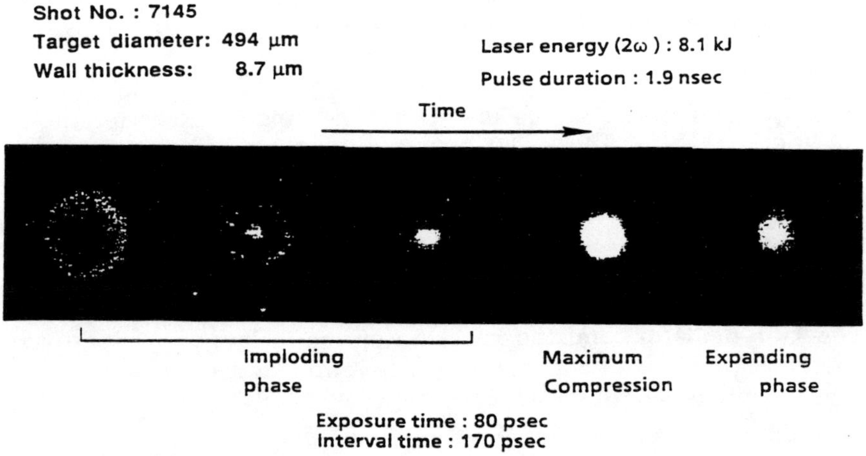

Fig. 4. X-ray framing images of a plastic shell implosion. The central emission appeared in the 2nd, 3rd and 5th frames are due primarily to the insufficient extinction ratio (~100:1) of the camera.

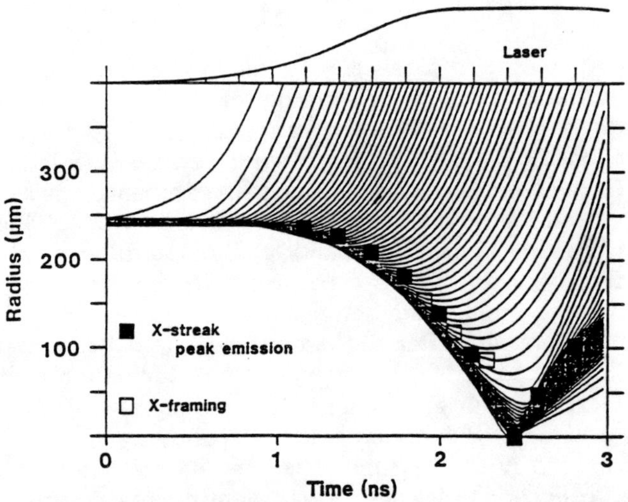

Fig. 5. Comparison of the experimental peak x-ray emission spot with the HIMICO predictions.

Figure 4 shows the space and time resolved x-ray images of a typical high density shot. The data were taken with multi-frame x-ray framing camera having a response to 1-2keV x rays. The maximum compression was observed at the 4th frame. The central emission appeared in the 2nd, 3rd and 5th frames is attributed mainly to the insufficient extinction of the temporal gating of the framing camera. The imploding shell was found to have nearly spherical symmetry. Even in the expanding phase, the sphericity was not too bad. These experimental data and x-ray streak data were compared with predictions from the 1D hydrodynamic simulation (HIMICO) as shown in Fig. 5, suggesting that the implosion was nearly one-dimensional at least in the implosion phase. The implosion velocity was typically (2.5-3)×10^7cm/s.

At the maximum compression, however, no significant information could be derived from the x-ray image because of the high x-ray opacity and the insufficient resolution of the diagnostic instrument. The fuel <ρR> at the maximum compression was measured with Si activation technique. Figure 6 shows the γ-ray decay curve and the γ-ray spectra obtained from several high density shots. As the number of signal counts per shot in these high density shots was typically 2-4 counts/10min, the data were summed up from 8 identical shots. The natural background of these experiments was 0.3/10min. The other backgrounds due to the neutron and plasma bombardments on the collector were much lower than the natural background. Both the time decay of 2.2min. and the distinct spectral peak at 1.8MeV confirmed that the detected events are really signals of ^{28}Al.

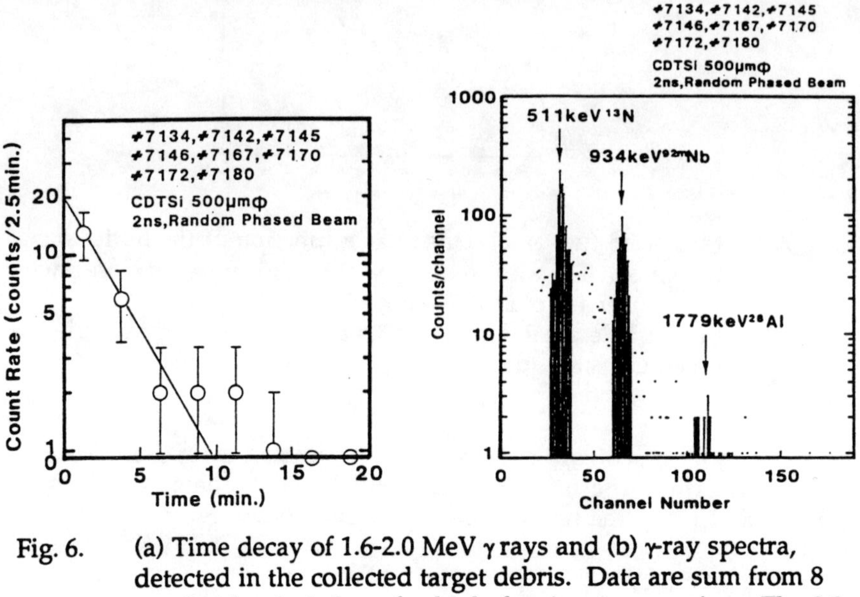

Fig. 6. (a) Time decay of 1.6-2.0 MeV γ rays and (b) γ-ray spectra, detected in the collected target debris. Data are sum from 8 nearly identical shots for high density compression. The 2.2-min half-life time and the γ-ray peak at 1.8MeV endorse the ^{28}Al signals.

Figure 7 shows the Si activation data of fuel <ρR> as a function of the shell thickness. Data taken with the knock-on and the secondary reaction techniques are also shown. Some of these knock-on and secondary reaction data were taken at less uniform conditions of the laser and targets. This may be the reason why these <ρR>s are lower than the activation results. It is clear in Fig. 7 that the fuel <ρR>s are higher by an order of magnitude for the RPP case than for the normal beam case (without RPP). The high fuel areal density for the RPP case is also cross checked with two knock-on data points located at the knock-on limit line, indicating higher <ρR> values than this line. In the above <ρR> determination, we used a hot spark model where neutrons are assumed to be generated at the target center. Any other neutron distributions would result higher <ρR> values.

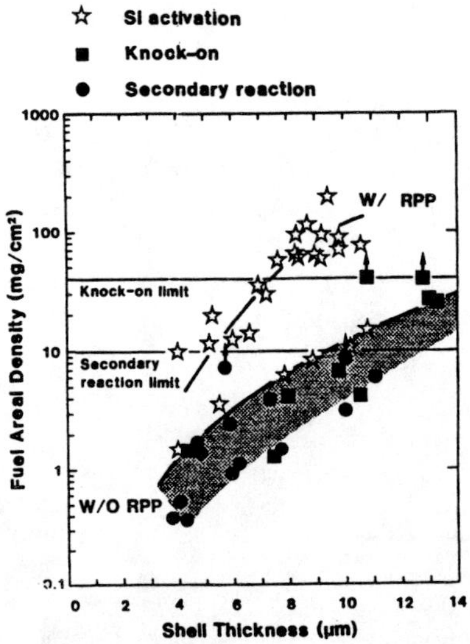

Fig. 7. Measured fuel areal density as a function of the initial shell thickness. The areal density was significantly increased with the random phase plates. Two horizontal lines shows the measurable areal density limits of knock-on and secondary reaction techniques.

The compressed core density can be estimated from the measured <ρR> value by assuming the uniform density distribution having a density ρ and radius R, and using the mass conservation relation,

$$\frac{4\pi}{3}\rho R^3 = M_0 - M_{abl}$$

where M_o is the initial mass which is known, and M_{abl} is the ablated mass up to the maximum compression and was typically $0.65M_o$. The ablated mass was separately measured by using plastic shell targets coated with Si (0.22μm), C_8H_8 (1.2μm), Mg (0.27μm), and C_8H_8 (1.9μm) in order. The target diameter and the mass were designed to be nearly the same as those for high density experiments. The x-ray emission peaks from these tracer layers were measured with a spatially resolved x-ray streak camera. A typical streak image is shown in Fig. 8 (a). Two distinct peaks of x-ray emission (which are indicated by small arrows on Fig. 8 (a)) can be clearly observed, especially on the bottom half of the image. The second x-ray emission can be seen across the entire surface of the target, indicating that the x-ray emission is spatially uniform. Figure 8 (b) shows the x-ray intensity which is spatially integrated over the bottom half of the image. Also shown is a x-ray intensity trace of a plastic shell without any tracer.

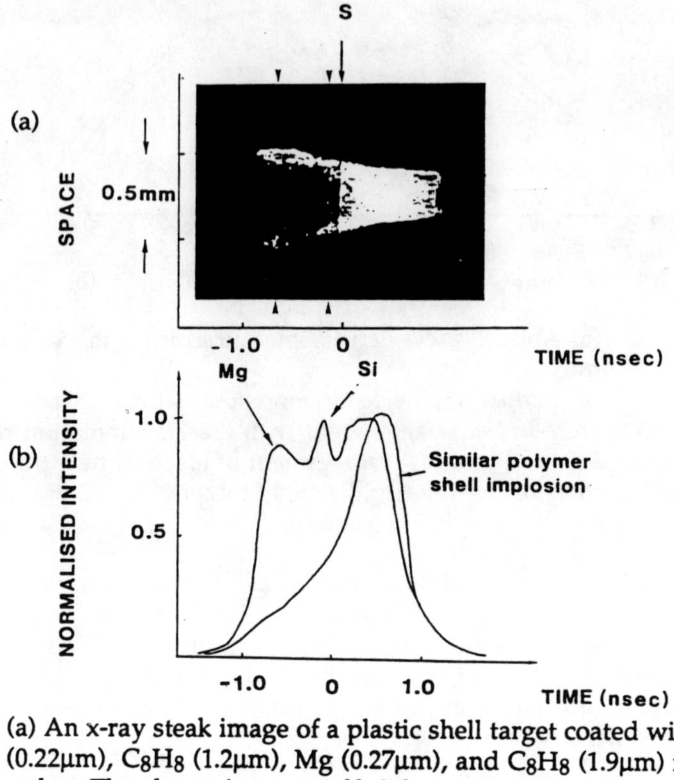

Fig. 8. (a) An x-ray steak image of a plastic shell target coated with Si (0.22μm), C_8H_8 (1.2μm), Mg (0.27μm), and C_8H_8 (1.9μm) in order. The abrupt increase of brightness at "S" is an experimental artifact in the frame memory. The intensity difference between the upper and lower halves of the image is due to a photocathode sensitivity variation. (b) X-ray intensity as a function of time.

Figure 9 (a) shows the areal density ablated as a function of the time of the peak x-ray emission. Also shown are the simulated areal density ablated for the Fokker-Planck (F-P) electron transport as well as those for the Spitzer-Harm (S-H) thermal conductivity with flux limits of f=0.1 and 0.2. Also plotted are the simulated and observed implosion times. (The areal density ablated at the maximum compression is taken to be that of the nearest simulated implosion time.) The F-P calculation shows close agreement with the experiment for both implosion time and the x-ray emission peaks.

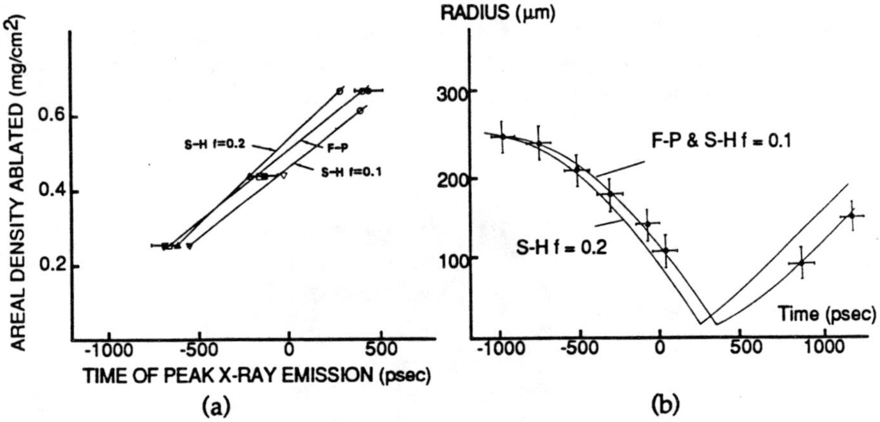

Fig. 9. (a) Ablated areal density measured from the x-ray emission of the tracers vs time.
(b) Radius of peak x-ray emission vs time. Experimental results were in better agreement with the 1D simulation results using the Fokker-Planck (F-P) treatment of electron heat transport and those using the flux limited Spitzer-Harm (S-H) treatment.

Figure 9 (b) shows the radius of peak x-ray emission as a function of time. The x-ray trajectories for the F-P and the S-H f=0.1 simulations are coincident and there is close agreement with the experiment. These agreements are convincing us to use the M_{abl} from the simulation of the F-P electron transport. Uncertainty of the M_{abl} measurements results in only ~10% error of the fuel density.

The estimated core density is plotted in Fig. 10. The vertical axis is the total density in g/cc and the same axis can be read as the fuel density in × LD since 1g/cc density corresponds to the liquid density of the fuel. The average density of the 9 nearly identical shots (500μm dia, 8-11μm thick) was 600g/cc or 600 × LD. The corresponding core radius and the convergence ratio (CR), defined as the core radius divided by the initial target radius, were estimated to be ~8-10μm and 25-30, respectively. The fuel temperature was estimated from the measured neutron yields, areal densities, and fuel masses to be ~0.3keV, being consistent with the measured implosion velocity.

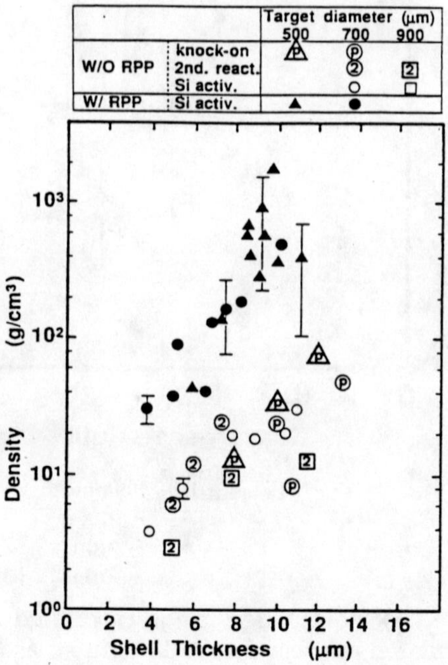

Fig. 10. Inferred density as a function of the initial shell thickness. The vertical axis can also be read as a fuel density in × LD, since 1-g/cc plastic shells contain the liquid density fuel . Average density of 9 identical shots (closed triangle points at 8-11μm thicknesses) was 600g/cc or equivalently 600 × LD.

Although the observed areal densities were 0.6-1.2 of the predicted value from 1D simulations with convergence ratio of 25-30 (Fig. 11), the observed neutron yields were significantly lower than predictions by three orders of magnitude. Figure 12 shows the calculated density and temperature profiles at the maximum compression time. In the simulation most of neutrons is generated in the hot spark appeared at the target center in Fig. 12. The radius of the hot spark was calculated to be ~5μm (convergence ratio of ~50). A possible explanation of the yield discrepancy would be that the residual nonuniformity of the implosion results in the mixing of the cold material into the hot spark region and/or in the insufficient conversion from the kinetic energy to the internal energy.

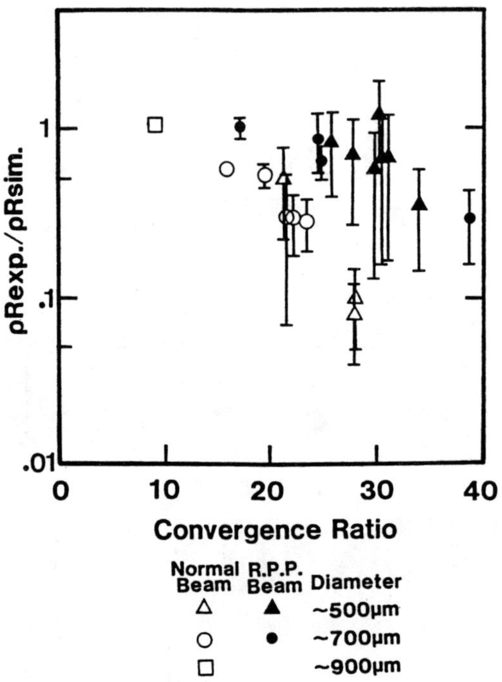

Fig. 11. Measured ρR normalized to the calculated one-dimensional ρR vs the calculated convergence ratio defined as the core radius divided by the initial target radius.

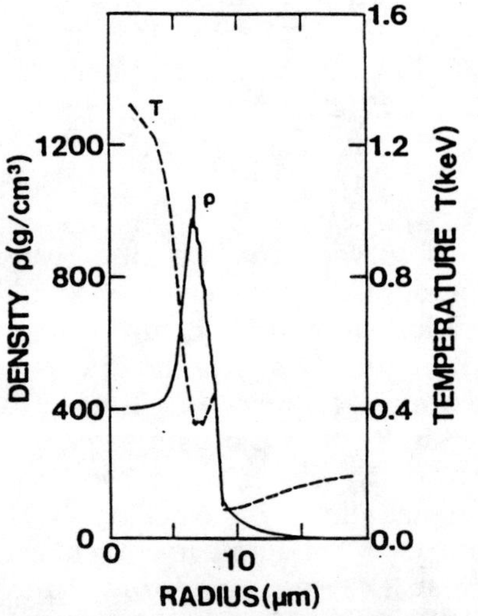

Fig. 12. Simulated density and temperature profiles at the maximum compression time. A hot spot having a radius of ~5μm is created at the center. Most of neutrons is generated at the hot spark.

4. DISCUSSION

We estimated Rayleigh-Taylor instability (RT) growth to find whether the present irradiation uniformity and the target quality are reasonable for the high-density compression (CR~25-30). Two extreme cases were considered: the one is that the target is uniform but the ablation pressure is nonuniform; the second is that the ablation pressure is uniform but there are shell thickness perturbations, which may be caused by the initial laser imprint or initial target imperfection. The initial laser imprint was estimated at the time when the first shock wave traverses through the shell. The target imperfection was found to be significantly smaller than the estimated laser imprint and, therefore, neglected at present.

The basic equation governing the displacement ξ of the fluid interface in the presence of the gravity nonuniformity δg is given by,

$$\frac{d^2\xi}{dt^2} = \gamma^2\xi + \delta g \tag{1}$$

where γ is the RT growth rate. For constant γ and δg the solution is given by,

$$\xi = \frac{\delta g}{2\gamma^2}\left(e^{\frac{\gamma t}{3}} - e^{-\frac{\gamma t}{2}}\right)^2 + \frac{\xi_0}{2}\left(e^{\gamma t} + e^{-\gamma t}\right) \tag{2}$$

where ξ_0 is the initial perturbation.

The first term of the solution represents the perturbation due to the forced RT instability, ξ_{FRT}, whereas the second term is the perturbation due to the normal RT instability ξ_{NRT} starting from the initial perturbation. These two terms were separately estimated. As for the γ value, we used the growth rate reduced by ablation effect[17)]

$$\gamma = 0.9\sqrt{kg} - 3.5kv_a \tag{3}$$

where k is the wave number of the perturbation, g is the gravity and v_a is the mass ablation velocity in the frame of the accelerated shell. The exponent γt in eq. (2) may be estimated by assuming $gt^2 \approx R_0$ (R_0 is the initial target radius), and $v_a t \approx \Delta R\ M_{abl}/M_o$ (ΔR is the in-flight shell thickness) and using the relation $kR_0 = \sqrt{l(l+1)} \equiv l^*$ to be

$$\gamma t = 0.9\sqrt{l^*} - 3.5\frac{M_{abl}}{M_o} l^* / A_{inf} \tag{4}$$

where A_{inf}=$R_0/\Delta R$ is the in-flight-aspect-ratio (A_{inf}=100 is assumed). Putting eq. (4) into the first term of eq. (2) and assuming that $\delta g/g \approx \sigma l^{th}$ (σl^{th} is the irradiation nonuniformity with thermal smoothing effect. (See Fig. 2), we can obtain the final perturbation ξ_{FRT} for the each spherical harmonic mode. The rms perturbation of l =1-150 modes appeared to be 4.2% of the initial target radius. This level of the perturbation may be consistent with the convergence ratio of 25-30 necessary for 600 × LD.

To estimate the initial imprint, we used the experimental mass ablation rate, $\dot{m}=\dot{m}_0\,(I/10^{14}W/cm^2)^\alpha$, where $\dot{m}_0=5\times10^5 g/cm^2 s$ and α=0.45[18]. The mass ablation rate was then integrated for the actual laser pulse shape up to the shock traverse times t_s (calculated from the simulation) to give the initial perturbation,

$$\xi_0=\frac{1}{\rho}\int_0^{ts}\alpha\dot{m}_0\left(I/10^{14}\right)^\alpha\frac{\delta I}{I}dt \qquad (5)$$

where $\rho \approx$4g/cc is the shell density after the first shock compression. The time dependent thermal smoothing effect is taken into account using a relation given by $\delta I/I=\sigma_l \exp(-vt\,l^*/R_0)$, where σ_l is the irradiation nonuniformity without thermal smoothing effect and v is the expansion velocity of the maximum absorption layer. The perturbation grown from this initial perturbation is calculated from the second term of eq. (3). The rms perturbation for l =1-150 modes is calculated to be 9.9%, being somewhat larger than the acceptable level for the convergence ratio of 25-30. However, those modes having the mode number larger than ~100 are expected to be nonlinearly saturated and, therefore, overestimated in the present linear theory calculations. Moreover, we conservatively assumed the nonuniformity of the laser irradiation to be the same as that of the absorbed intensity. If experimentally measured relations between the incident laser intensity and the absorbed intensity is used, a lower nonuniformity would result.

5. SUMMARY

Direct-drive implosion experiments on the GEKKO XII laser (9kJ, 0.5μm, 2ns) with DT exchanged plastic shell targets have demonstrated fuel areal densities of 0.1g/cm^2 and fuel densities of 600 times liquid density (× LD). These densities almost meet one of the ignition criteria of 500-1000 × LD. Although the observed densities and areal densities were in agreement with one dimensional calculation results with convergence ratios of 25-30, the observed neutron yields were significantly lower than the calculations. This suggests that the implosion uniformity is not sufficient to create a hot spark in which most neutrons should be generated. Analysis of Rayleigh-Taylor instabilities suggests that the present implosion nonuniformity is reasonable to achieve the densities of 600 × LD with the convergence ratios of 25-30.

ACKNOWLEDGMENTS

The author acknowledges the efforts of our experimental group as well as the support of the laser group and the target fabrication group. The calibration experiments of the track detector used for the knock-on technique were performed at the AVF cyclotron, Osaka University's Research Center for Nuclear Physics. He also expresses the gratitude to the theoretical staffs of the Institute for the intimate collaboration.

REFERENCES

(1) C. Yamanaka et al., Nature, 319, (1986) 757.

(2) S. Nakai et al., Plasma Physics and Controlled Nuclear Fusion Research, 1988 IAEA, 3, (1989) 3.

(3) K. Kondo et al., Japanese Jour. Appl. Phys., 28, (1989) 1695.

(4) S. Nakai, et al., in Proceedings of the 12th International Conference on Plasma Physics and Controlled Nuclear Fusion Research vol. 3 (IAEA, Vienna), (1989).

(5) F. Ze, in Lawrence Livermore National Laboratory Report No. UCRL-50021-84, 5-35, (1985).

(6) R. L. McCrory, et al., Nature, 335, (1988) 225.

(7) Y. Kato, et al., Phys. Rev. Lett., 53, (1984), 1057

(8) T. Yabe, et al., Nucl. Fusion 21, (1981) 803. Recently, HIMICO was improved to treat non-local heat transport with the Fokker-Planck equation.

(9) U. Kubo, and H. Tsubakihara , Vac. Sci. Technol. A4, (1986) 1134.

(10) M. Takagi, 1990, to be published.

(11) H. Nakaishi, 1990, Osaka University, Doctoral Thesis.

(12) H. Azechi, et al., Appl. Phys. Lett. 49, (1986) 555.

(13) S. Kacenjar, et al., Phys. Rev. Lett. 49, (1982) 463.

(14) H. Nakaishi, et al., Appl. Phys. Lett. 54, (1989a) 1308.

(15) E. M. Campbell, et al., Appl. Phys. Lett. 36, (1980) 965.

(16) Nakaishi, H. et al., Appl. Phys. Lett. 55, (1989b) 2072.

(17) H. Takabe, et al., Phys. Fluids, 28, (1985) 3676.

(18) C. Yamanaka, et al., Laser and Particle Beams, 4, (1986) 43.

REPETITIVE LASER-PLASMA X-RAY SOURCE FOR LITHOGRAPHY AND MICROSCOPY APPLICATIONS

H. Daido, G.M. Zeng, T. Togawa, H. Aritome*, M. Nakatsuka, T. Kanabe†, K. Nishihara, and S. Nakai

Institute of Laser Engineering, Osaka University
2-6 Yamada-oka, Suita, Osaka 565, Japan
**Research Center for Extreme Materials, Osaka University*
1-1 Machikaneyama-cho Toyonaka, Osaka, 560, Japan
†Institute for Laser Technology, Suita, Osaka 565, Japan

Repetitive laser plasma x-ray source for various applications was developed using zigzag glass slab laser and rotating target. As a result of x-ray lithography application, 0.25 µm-wide space mask pattern was replicated with a repetitive x-rays from laser produced molybdenum ions. For x-ray microscopy application, we characterize intense line emissions from laser produced carbon ions with a grazing incidence spectrometer.

I. INTRODUCTION

In general, a laser plasma x-ray source is capable of producing high intensity, short duration (ps-ns), and small source size x-ray pulses. Recent advances of solid and gas laser technology provides a compact (table top) high repetitive systems[1,2]. There are significant advantages of laser-plasma x-ray sources compared with other x-ray sources such as synchrotron radiation machines, discharge machines and so on[3].

In this paper, we present the application of laser plasma x-ray sources for x-ray lithography[1,4] and x-ray microscopy[5,6]. As for the lithography application, the desired properties of x-ray sources are as follows: x-ray spectrum is around 1 nm considering the diffraction and photo-electron migration. The spectral band width should be as narrow as possible, but not so important, thus the near continuum emissions which have relatively high x-ray conversion efficiency from high atomic number materials are favored. The key issue to replicate a lithography pattern is getting high x-ray fluence i.e., intensity times duration times repetition rate of suitable wavelength as high as possible. Finally, the source size should be as small as possible due to minimize the geometrical distortions.

As for the x-ray microscopy application, the desired x-ray wavelength is in the water window region ranging from 2.3 to 4.4 nm which correspond to oxygen and carbon K-edges, respectively for observ-

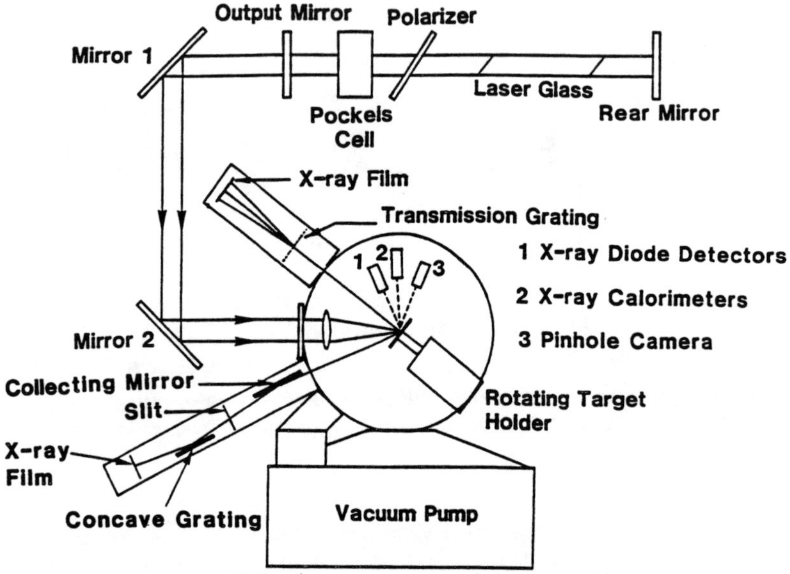

Fig. 1 Experimental setup.

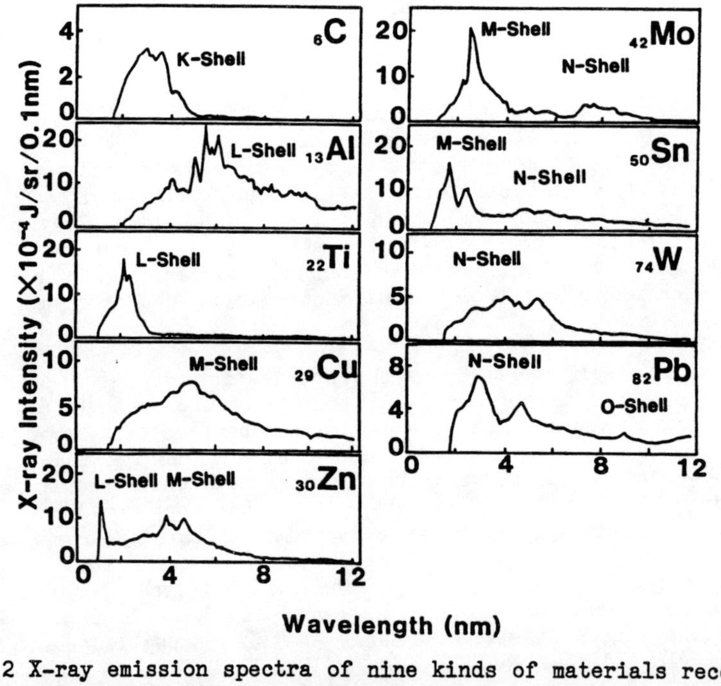

Fig. 2 X-ray emission spectra of nine kinds of materials recorded with the transmission grating spectrometer whose resolution is 0.4 nm in the energy range of 0.1 keV-1.5 keV. The angle of observation was 40 deg. to the normal of the target surface. Note the change of vertical scale for different spectra.

ing a biological specimen. The spectral band width should be as narrow as possible for obtaining high contrast x-ray pictures. Required width is typically W/dW>100 for a Fresnel zone plate, where dW and W are the band width and center wavelength, respectively. Single line emission rather than the quasi-continuum emissions is favored for the microscopy applications. Low atomic number material is suitable for an x-ray source. The most important issue for a x-ray source is that the x-ray intensity should be high enough to take a picture during single x-ray pulse. The source size should be of course as small as possible.

II. EXPERIMENTAL SETUP

Figure 1 shows the schematic diagram of the repetitive table-top laser-plasma x-ray source. A Q-switched zigzag glass slab laser produces 5-10 J in 30-50 ns laser light at 1.05 µm wavelength. The laser light is focused with an f/1.25 focusing lens giving peak irradiance of 2-3 x 10^{12} W/cm^2 on the rotating target which provides fresh target surface at each laser shot with a repetition rate of 0.1 Hz. The vacuum target chamber is evacuated with a turbo molecular pump bellow 10^{-6} Torr. Two spectrometers i.e., transmission grating type and grazing incidence grating type are installed in this chamber giving x-ray spectral pictures.

III. X-RAY LITHOGRAPHY APPLICATION

Various materials from carbon to lead were irradiated by 8 J in 40 ns laser pulses at 1.05 µm wavelength whose peak irradiance was 2.5 x 10^{12} W/cm^2 as shown in Fig. 2. The spectra were taken with the transmission grating spectrometer whose resolution was 0.4 nm. Molybdenum has the highest x-ray conversion efficiency around two to three nm spectral region, leading to a good x-ray source material for contact x-ray lithography. On the other hand, carbon and titanium gave small band width x-rays centered around two to four nm, which were expected to be a good candidate for a source of x-ray microscope.

Figure 3 which was taken with the scanning electron microscope shows the replicated pattern with the laser plasma x-rays using a free-standing gold grating with 0.25 µm-wide space as a mask. The 5.0 J in 40 ns laser pulses irradiated a molybdenum target 60 times. To protect the resist from ultraviolet light came from the laser irradiated target, we used 0.5 µm thick polyimide filter SP-171. The photoresist of 0.32 µm thick FBM-120 was 7.5 cm apart from the target. To avoid the charge up effect, 10-nm thick gold was coated on the resist. The exposed x-ray energy fluence was estimated to be 22 mJ/cm^2. Black parts correspond to the silicon base of the wafer.

IV. X-RAY MICROSCOPY APPLICATION

The other topic using a repetitive laser-plasma x-ray source is described. Figure 4 shows the block diagram of the zone plate x-ray microscope which has just been constructed in our institute. The x-rays in the water window region are collected by the condenser zone plate whose diameter is 1-10 mm. The pinhole selects a quasi monochromatic x-

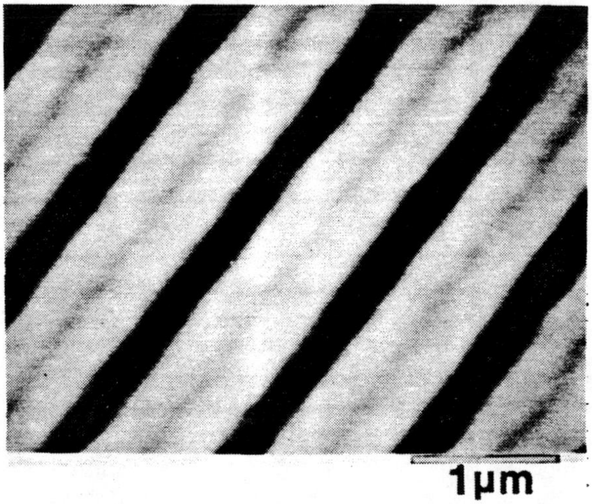

Fig. 3 A 0.25 μm-wide space pattern in a 0.3-μm thick FBM-120 resist by the x-rays from laser-generated molybdenum plasmas. The target-mask distance was 7.5 cm and the laser energy was 5 J/shot. The replication was completed by 60 laser shots. Black parts correspond to the silicon base of the wafer.

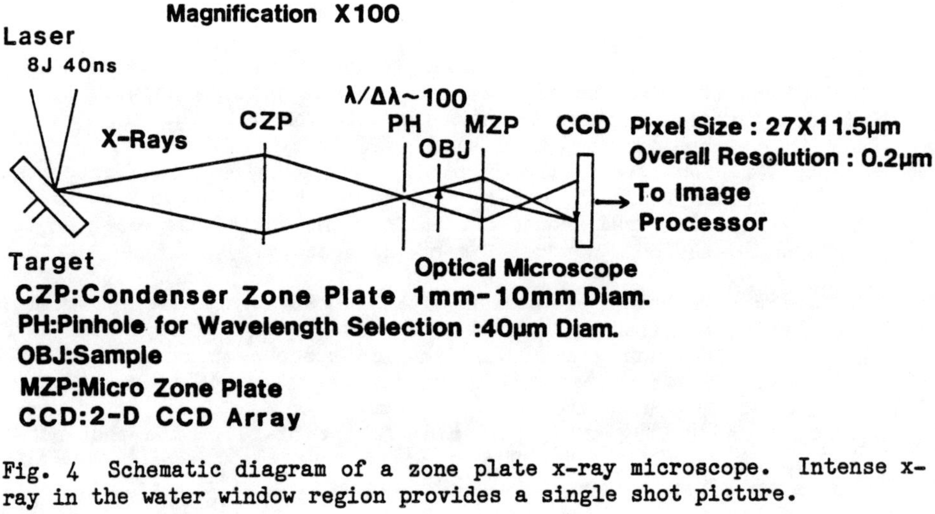

Fig. 4 Schematic diagram of a zone plate x-ray microscope. Intense x-ray in the water window region provides a single shot picture.

rays. Then an object such as a biological specimen is illuminated. The image of the object is taken with the two-dimensional charge coupled device array through the micro-zone plate. We are now preparing such a kind of microscope. The key issues to develop x-ray source for microscopy are following: high intensity in short duration (ps-ns) in narrow spectral width (W/dW100) in small emission diameter.

The spectra of various kinds of materials (carbon, aluminum, titanium, copper, and molybdenum) which emit significant x-rays in the water window region are taken with the grazing incidence spectrometer whose resolution is less than 0.01 nm as shown in Fig.5. Individual line emissions can be seen. The grating efficiency in this spectral region is measured to be almost flat (within 30 deviation). Thus the relative intensity distribution can be easily obtained with the spectral sensitivity of the x-ray film. The spectrographs are taken under the same conditions, therefore, as expected the line emissions from carbon ions are the most intense in water window region in our experimental conditions, although the total conversion efficiency is poor[1].

The transmission grating spectrometer gives the absolute x-ray intensity, but spectral resolution is not sufficient to resolve individual line emissions. The absolute intensity of line emissions from carbon ions is calibrated with a transmission grating spectrometer as shown in Fig. 6. Both integrated photon numbers over spectrum are the same. Then the photon numbers of an individual line emission can be obtained using the result of the gazing incidence spectrometer. Intensity of 7 x 10^{14} photons/sr./0.1 nm/shot is obtained. For x-ray microscopy, x-ray intensity is strong enough to take a picture with single laser shot.

Figure 7 shows the well-known energy level diagram of hydrogenic ions. Basic consideration for increasing the intensity of carbon Lyman alpha line emissions is described. As for the electron density, the well known Griem's boundary[7] of electron density n_e at the temperature of kT_e is given by this formula,

$$n_e = 7 \times 10^{18}(kT_e/Z^2 E_H)^{1/2} Z^7/n^{17/2} \ (cm^{-3}) \quad (1),$$

where Z and E_H are the degree of ionization and ionization energy of hydrogen atom, respectively. n is the principal quantum number. At certain number of n, ten times radiative rate is balanced with collisional rates. If Z=6, and n=2, the boundary electron density is given by,

$$n_e=2.5 \times 10^{20}(T_e)^{1/2}(cm^{-3}) \quad (2).$$

The intensity of Lyman alpha emission I is proportional to $A_{21}hv_{21}N_2$, where A_{21}, hv_{21}, N_2 are the Einstein A coefficient, photon energy of Lyman alpha emission and population density of n=2 level, respectively. N_2 should be as large as possible. Therefore plasma density should be as large as possible but lower than the boundary value. Optimum electron density is expected to be around 10^{21} cm^{-3}. On the other hand, required electron temperature is determined by the ionization balance of the carbon plasmas. Large amount of fully ionized and H-like ions is produced at the electron temperature of around 100 eV. These two are guide line of the plasma conditions.

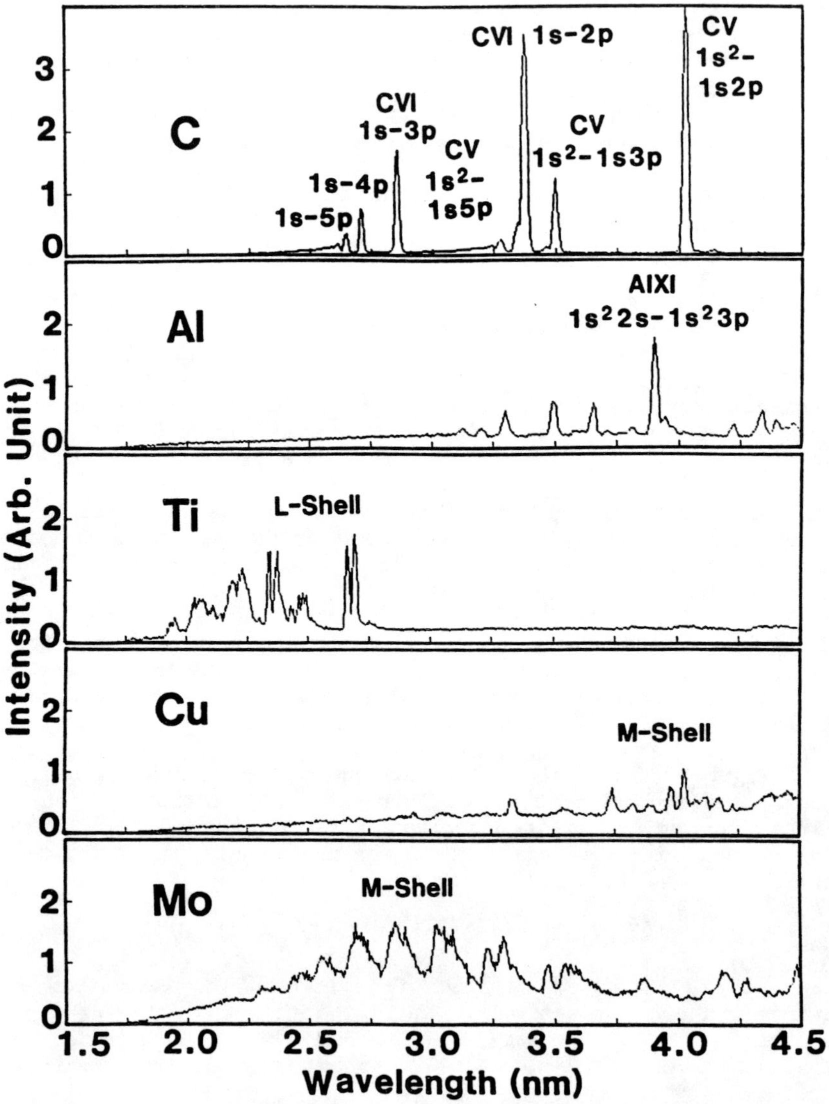

Fig. 5 X-ray emission spectra from five kinds of materials recorded with a grazing-incidence spectrometer. Carbon has the most intense emission in the water window region. The laser intensity was 1.5 x 10^{12} W/cm^2 and the spectra are the integrated over four laser shots.

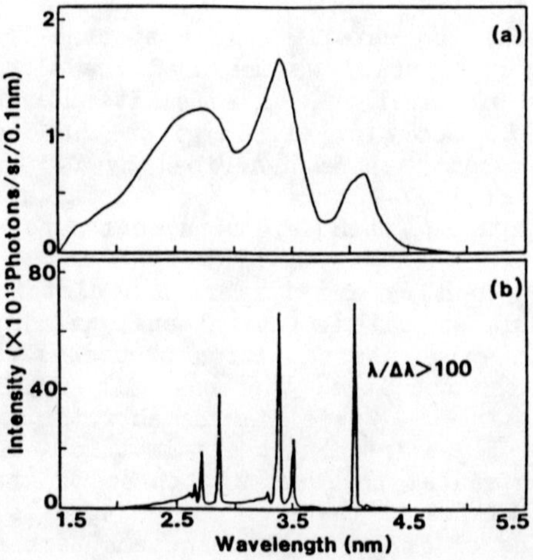

Fig. 6 The absolute intensity of the line emission from carbon plasma was obtained by recording the emission simultaneously with an absolutely calibrated transmission grating spectrometer and a highly resolved grazing-incidence spectrometer. (a) The x-ray spectra recorded with the transmission grating spectrometer. (b) The x-ray spectra recorded with the grazing-incidence spectrometer.

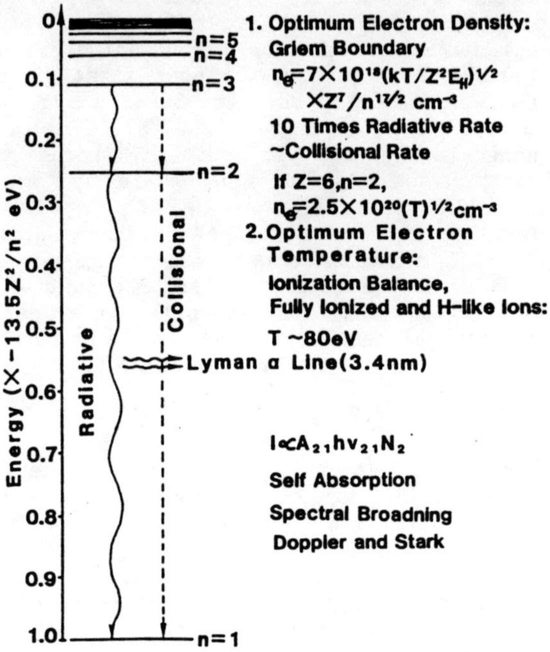

Fig. 7 Grotorian diagram of the hydrogenic ions.

Figure 8 shows the satellite line spectra from the highly ionized carbon. In our experiment, we measured the electron density from the intensity ratios of satellite to satellite lines. The first method described by J. F. Seely et al.[8] gives the density of (5.2-10.0) x $10^{19}cm^{-3}$. The second method described by D. Duston et al.[9] gives electron density of (1.7-9.0) x $10^{19}cm^{-3}$ details of which can be found elsewhere[10]. Both two methods give almost same electron densities.

The electron density and temperature distributions of the carbon plasma are calculated with a one-dimensional hydrodynamic simulation code HISHO[11] in which the radiation processes are taken into account on the basis of the non local thermodynamic equilibrium atomic physics model as shown in Fig. 9. These figures show the simulation results when a 5 J/35 ns laser light irradiates a 1-mm thick carbon plate at the peak laser irradiation and at the decaying phase of the plasma. The carbon line x-rays are emitted in the high temperature low density region which is right hand side of Fig. 9. The electron temperature derived from H-like (1s-3p) line to He-like ($1s^2$-1s3p) line intensity ratio is 80 eV, while the computed temperature in the low density corona region at the peak laser irradiation is around 60 eV. At the decaying phase, the electron temperature of 30-40 eV derived from the slope of the recombination continuum[8] agrees with computed predictions.

The electron density in the high temperature region is around $10^{20}cm^{-3}$ in the simulation result which roughly coincides with the experimental values. The density is about one order of magnitude lower than the Griem's boundary. In order to raise the electron density, shorter pulse width especially for the risetime giving shorter density scale length and shorter wavelength laser light giving higher critical density are needed. Actually when the laser energy was kept 4.5 J and pulse width changed from 45 ns to 35 ns, all lines become nearly twice stronger as shown in Fig. 10. In addition, the large amount of carbon atoms went into higher ionization stages which was measured from the intensity ratio of CVI 2p-1s line to CV 1s2p-$1s^2$ line, which gave the electron temperature of 77 eV and 82 eV in the case of 45 ns and 35 ns, respectively. Higher ionization stage and higher plasma density where the line radiation are emitting give larger amount of line x-rays. The simulation code may be used to optimize the laser parameters as well as experimental investigations for more intense emission.

V. SUMMARY

In summary, x-ray spectrum from various materials were recorded with transmission grating spectrometer. On the basis of the spectroscopic data, repetitive laser plasma x-ray source for various applications was developed using zigzag glass slab laser and rotating target. As a result of x-ray lithography application, 0.25 µm-wide space mask pattern was replicated with a repetitive x-rays from laser produced molybdenum ions. For x-ray microscopy application, we characterize intense line emissions from laser produced carbon ions with a grazing incidence grating spectrometer. A zone plate x-ray microscope for observing a biological specimen was just prepared.

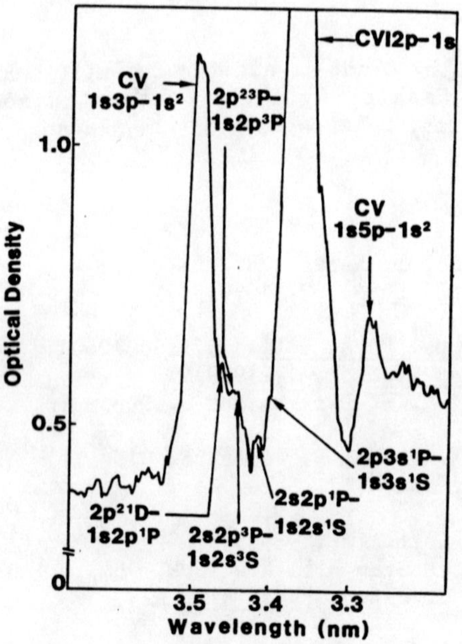

Fig. 8 Satellite line spectra .

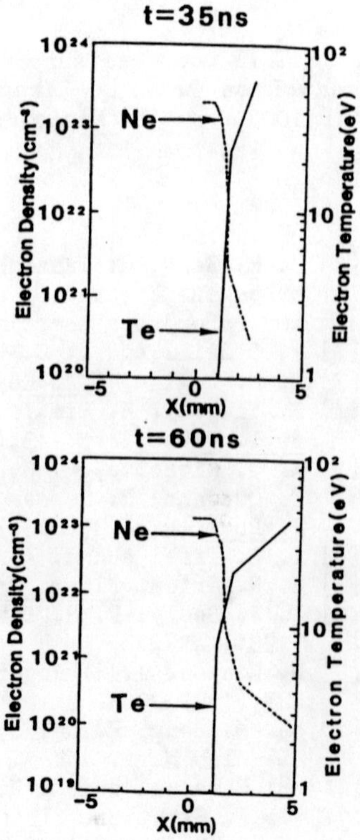

Fig. 9 The electron temperature (solid line) and density (dotted line) distribution .

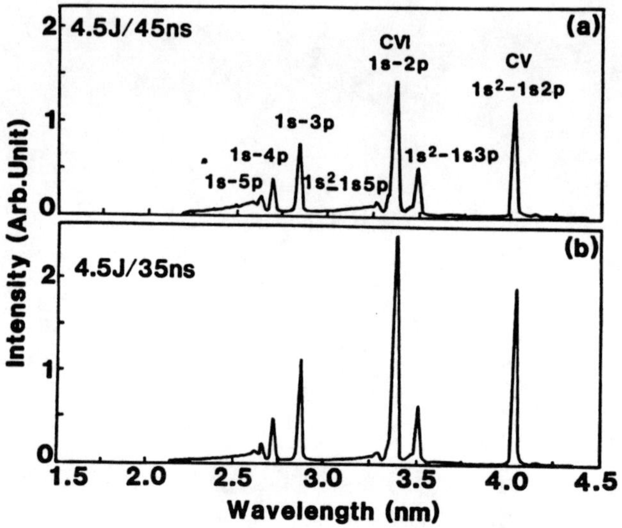

Fig. 10 X-ray line emissions from carbon ions. Intensities of carbon V, VI line emissions become higher when laser pulse duration becomes shorter.

ACKNOWLEDGMENTS

This work was supported in part by Grant-in-aid for Scientific Research on Priority Areas, "X-Ray Imaging Optics", Contract Number 01651003 from the Ministry of Education, Science and Culture, Japan.

REFERENCES

[1] G. M. Zeng, M. Takahashi, H. Daido, T. Kanabe, H. Aritome, M. Nakatsuka, and S. Nakai, J. Appl. Phys. 67,(1990)3597, and references therein.
[2] F. O'Neill and I. C. E. Turcu, Appl. Phys. Lett. 55,(1989)2603.
[3] N. M. Ceglio, J. X-Ray Science and Technol. 1,(1989)7.
[4] B. Yaakobi, H. Kim, J. M. Soures, H. W. Deckman and J. Dunsmuir, Appl. Phys. Lett. 43,(1983)686.
[5] A. G. Michette, P. C. Dheng, R. W. Eason, R. Feder, F. O'Neill, Y. Owadano, R. Rosser, P. Rumsby, and M. J. Shaw, J. Phys. D19,(1986)363.
[6] J. A. Trail and R. L. Byer, Opt. Lett. 14,(1989)539.
[7] H. R. Griem, Plasma Spectroscopy (McGraw-Hill,New York,1964),p.148.
[8] J. F. Seely, R. H. Dixon, and R. C. Elton, Phys. Rev. A23,(1981)1437.
[9] D. Duston, J. E. Rogerson, J. Davis, and M. Blaha, Phys. Rev. A28,(1983)2968.
[10] G. M. Zeng, H. Daido, T. Togawa, H. Aritome, M. Nakatsuka, K. Nishihara, and S. Nakai, submitted for publication.
[11] M. Murakami and K. Nishihara, Jpn. J. Appl. Phys. 26,(1987)1132.
[12] T. F. Stratton, Plasma Diagnostic Techniques, ed. by R. H. Huddlestone and S. L. Leonard (Academic Press, New York, 1965),p.390.

X-UV SPECTROSCOPY OF LASER PRODUCED PLASMAS

R. Benattar and S. Sezen
LULI, Ecole Polytechnique, 91128 Palaiseau Cedex, France

Abstract: Record of photons of energies from several tens to several hundreds of eV, either by imaging or by spectroscopy gives important information on energy transport and radiation heating of targets by powerful focused lasers.

We present a method of probing the ablation layer of laser produced plasmas by time resolved X-UV spectroscopy. Preliminary results are given showing the modification of the spectrum of gold in the range 40-200 Å through a dense plasma of aluminium.

I Introduction

There is a great interest to develop diagnostics in the X-UV range of laser produced plasmas for inertial confinement research. If one observe typical density and temperature profiles of the hydrodynamic flow of a laser produced plasma (figure 1), one can characterize two zones which is possible to diagnose by collecting X-UV photons. First, the thermal transport zone which lies between critical density and ablation front (solid density) radiates from 1 keV to 100 eV . Second, the solid part of the target is preheated or heated by the intense X-ray emission (1 to 3 keV) of the corona and radiates in the low energy part of the X-UV spectrum (10 to 100 eV). This second point has been studied previously by time resolved imaging at 40 and 80 eV of the rear side of thin aluminium foils illuminated by laser. We measured a rear side temperature and showed the effect of radiation on the value of this temperature by comparison with code simulation [1].

In this paper we will focalize our interest on the thermal transport zone (figure 1). This zone exhibits sharp density and temperature gradients particularly for short wavelength laser interaction (4ω). The

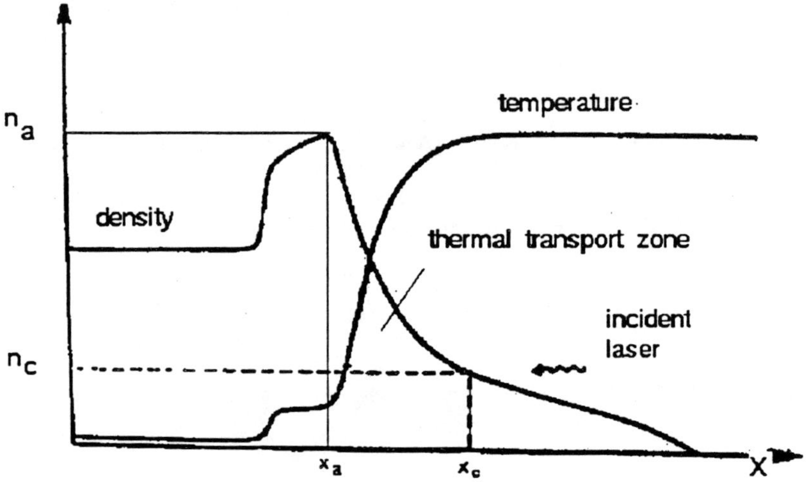

Figure 1

electron temperature varies from more than one keV to several tens of eV. So, for a target material like aluminium, ionic species of different charges follow the temperature profile, beginning at the maximum temperature by totally stripped ions (Al^{13+}). These species radiate, depending on their ionic state, from the keV range (H-like, He-like) to the X-UV range (Li-like, Be-like....). The radiation coming from the thermal transport zone is mainly emitted in the X-UV range (several hundred of eV). But, due to the high density of the material, self absorption effects modify the shape of the line recorded from this region of the plasma and absorption spectroscopy is more adapted to get information from this thermal transport zone.

To probe the thermal transport region of an aluminium laser created plasma, we choose to time resolve the quasi continuous spectrum of a gold plasma in the range 40-120 Å passing through the plasma of aluminium. So, one can deduce a mean mass absorption coefficient $\langle\mu\rangle_\nu$ of the aluminium material at each frequency of the recorded spectrum:

$$\langle\mu\rangle_\nu = 1/L \; . \; \mathrm{Log}(I_{0\nu} / I_\nu)$$

with
$I_{0\nu}$ non perturbed intensity of the gold spectrum at the frequency ν
I_ν intensity of the gold spectrum through the aluminium plasma

L mean path through the aluminium plasma.

Assuming that the radiation, we observe during the laser pulse either in emission or in absorption, comes from the zone of thermal transport, it is possible to deduce by comparison with atomic physics models the temperature and the density of the medium and their evolution during the laser pulse.

II The flat field spectrograph

In order to perform the proposed research, we designed [2] and built a flat field grazing incidence spectrograph. This system is made with an association of a toroidal mirror and a grazing incidence grating. Each element had been designed and manufactured by the Jobin & Yvon french company.

Figure 2 shows the optical arrangement of the spectrograph. Two different gratings are altenatively used. The first one has 1800 lines/mm and gives spectra from 40 to 200Å, the second has 800 lines/mm and works from 100 to 400Å. The gratings are planar and their period per mm are slightly variable in order to record the spectra in a plane.

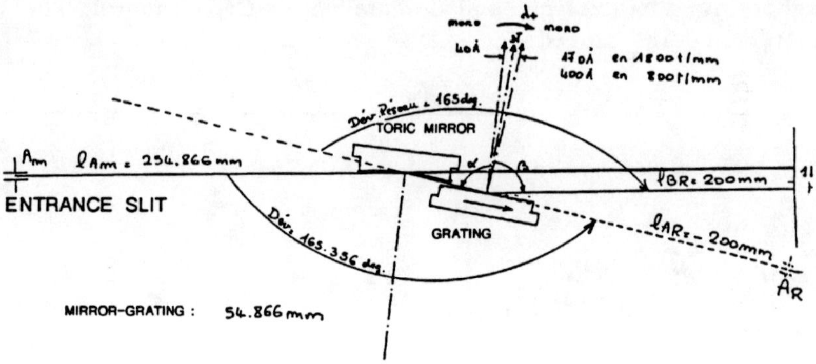

Figure 2.

Figure 3 shows the mechanical arrangement . One can observe the reentrant for the X-ray streak camera with its slit in the detection plane.

The wavelength calibration of the spectrograph with the 1800 lines/mm grating was done using resonnance helium-like and hydrogene-like lines of materials of low Z atomic number as Beryllium and Lithium.

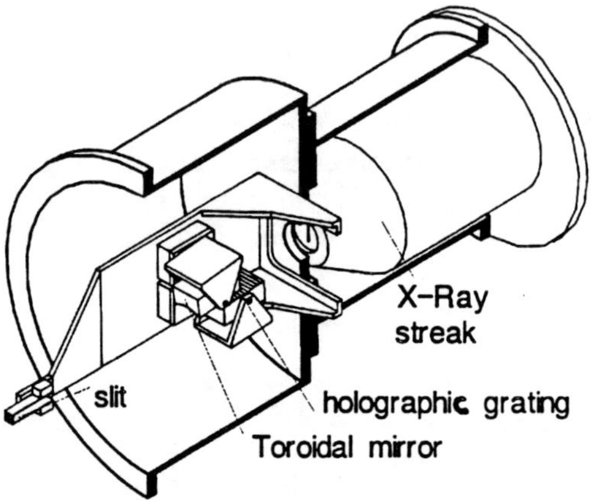

Figure 3.

Figure 4 shows a time integrated spectrum from an aluminium laser created plasma. The laser delivered 1 joule in a 4ns pulse at a wavelength of 0.53 μm. The X-UV radiation is converted through a P22 phosphor into visible light and detected by a CCD camera. The signal is then digitized and stored.

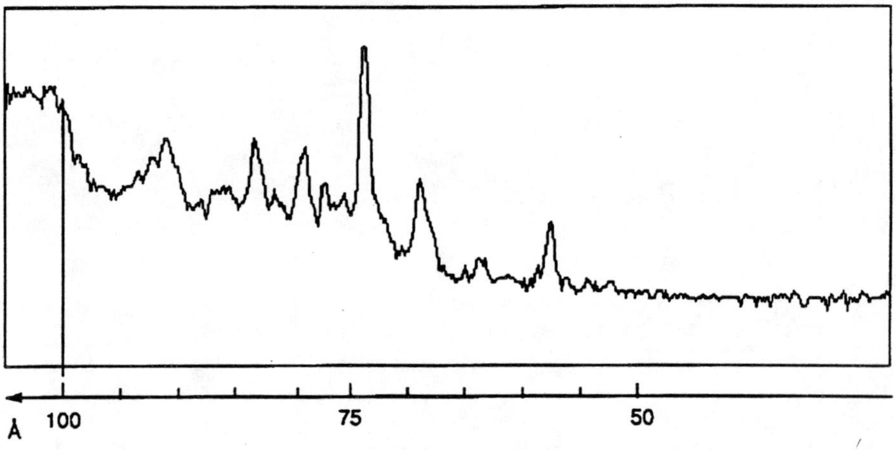

Figure 4.

At this time we have not yet characterized the lines recorded and the ions from which the lines arise. We plan to do this later by running hydrocode simulations including atomic physics.

III Experiment and preliminary results.

The geometry of the experiment is shown on figure 5. The laser running at the wavelength of 0.26 μm is focused on a solid planar target

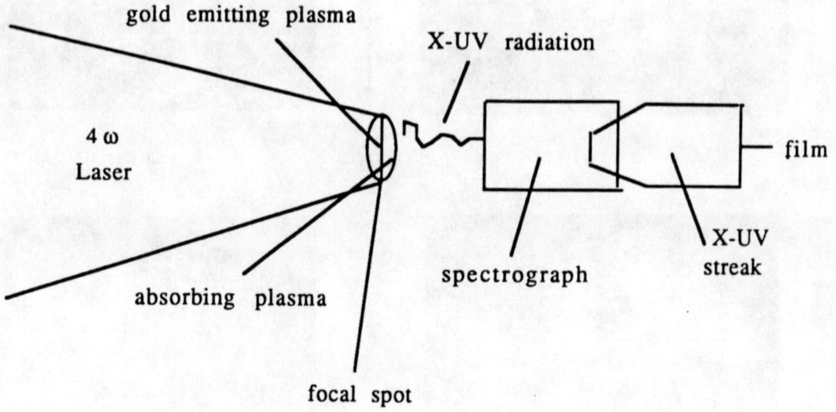

Figure 5.

The target is made of the two materials placed side by side, gold and aluminium. An half part of the focal spot is located on the gold material. The rest is located on the aluminium material. The spectrograph is put at ninety degrees with respect to the laser direction on the side of the material to be probed. So, the spectrum emitted by the more distant part of the target in our case the gold passes through the plasma created on aluminium before entering the spectrograph.

Figure 6 shows typical rough results. Figure 6a is a time resolved spectrum of pure gold between 50 and 120 Å. Figure 6b is a time resolved spectrum of pure aluminium in the same range of wevelength. Figures 6c and 6d show time resolved spectra of gold through a plasma of aluminium. Spectrum of figure 6c is for a weaker emission from gold. This attenuation of the gold spectrum is done by increasing the part of aluminium inside the focal spot by moving laterally the target and consequently the border between the two materials inside the focal spot. These two last spectra exhibit a combination of the spectrum of gold and of that of aluminium. We expect to deduce some absorption after deconvolution of the spectra.

The laser conditions were 0.26 μm wavelength, 15 joules of energy in a

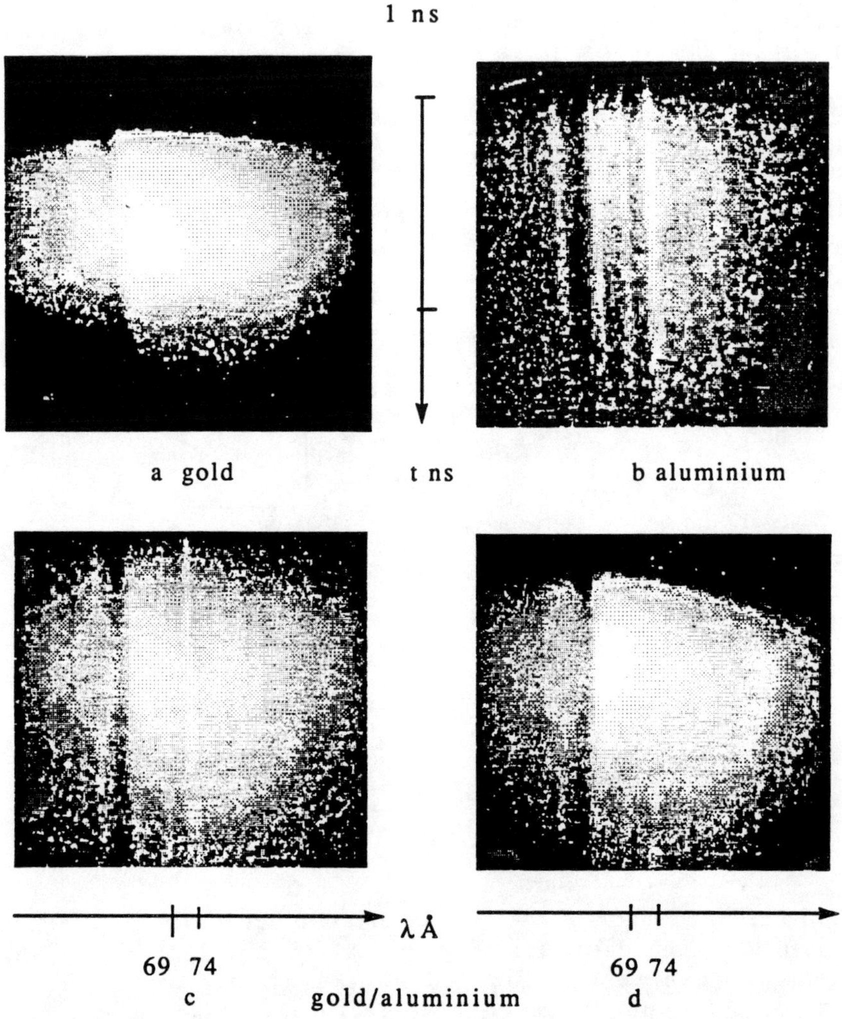

Figure 6.

500 ps pulse duration, focused in a focal spot of about 100 µm diameter, giving a laser intensity of $4\ 10^{14}\ w/cm^2$.

It is necessary to be sure that the recorded spectrum of gold passes through the thermal transport zone of the aluminium plasma and that there is not only addition in intensity of the two spectra on the detector. In order to prevent that, we plan to do other experiments using thin plastic targets on which is deposited several hundred angstroms of gold such as the total thickness of the targets will be less or equal to the ablation depth of the laser for the experimental conditions. So, we will be sure to record the gold spectrum through the thermal transport zone of aluminium.

IV Modelling

In parallel with the experiment a calculation is done in order to simulate the refraction and the absorption of a probe radiation (150Å) of En photon energy through a spherical plasma with radial density and temperature profiles.

The refraction and the absorption are deduced from the absorption mass coefficient μ_a calculated in this step by an LTE average atom model [3], through the Kramers-Kronig dispersion relations [4].

$$f_1 = Z^* + C \int_0^{\infty} \frac{E^2 \mu_a(E)\,dE}{En^2 - E^2}$$

and

$$f_2 = \frac{\Pi}{2} C\, En\, \mu_a(En)$$

The real refractive index is $n = 1 - K f_1$

The absorption coefficient is $\beta = K f_2$

with

$$K = \frac{r_0 \lambda^2 N_a \rho}{2 \Pi A}$$

The refraction angle is calculated step by step inside the plasma by:

$$\Phi = \int_{r\frac{\Pi}{2}}^{1} \frac{dr}{r(r^2 n^2 - p^2 n_0^{2?})^{\frac{1}{2}}}$$

and the absorption by $\mathrm{Exp}(-\beta r)$

Figure 8 shows the result of the calculation for the density and temperature profile of figure 7

This kind of calculation play the role of numerical diagnostic for the description of the medium we are interested with.

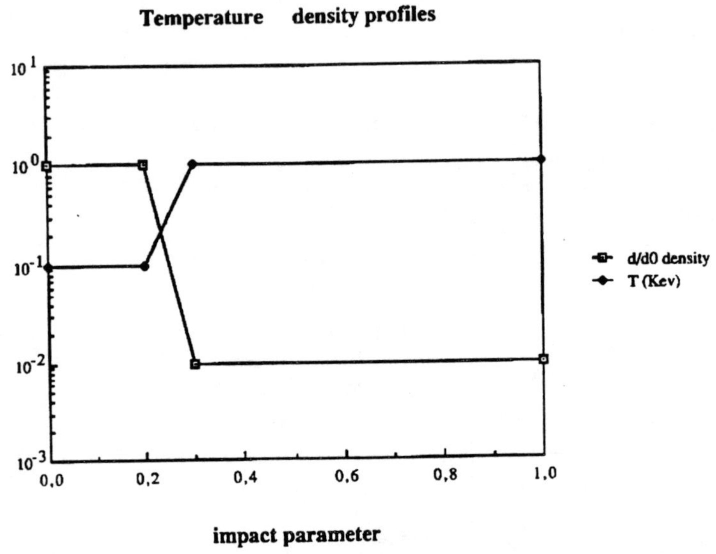

Figure 7.

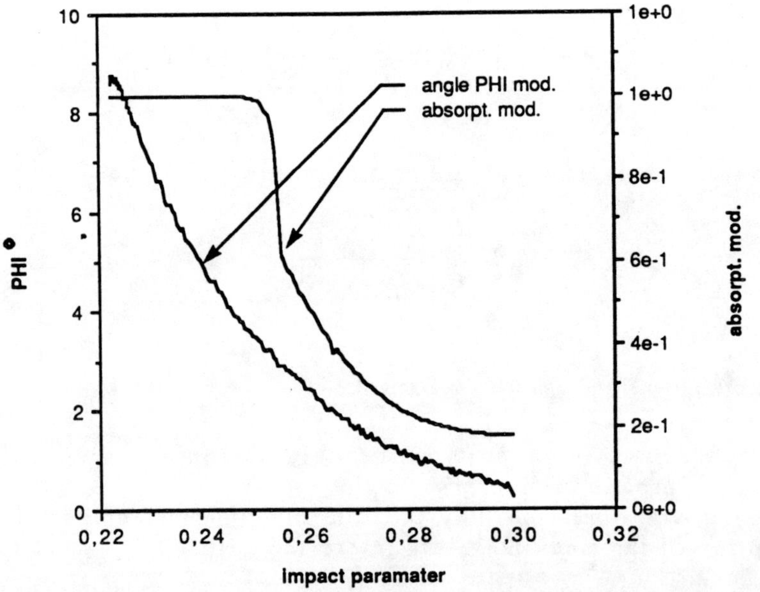

Figure 8.

In order to simulate wholly the experiment, we have to continue the calculation of refraction and absorption for each wavelength of the probe spectrum. On the curves of figure 8, the noticeable effect of absorption happens for impact parameters values where rays are bent by refraction. Consequently it is important to take into account this effect at the densities and temperatures corresponding to the thermal transport zone.

The further step of the model will use non ETL mass absorption coefficients which are more realistic in the studied zone of the plasma and will be applied to density and temperature profiles deduced from hydrocode simulations including atomic physics.

In conclusion, we presented preliminary results exhibiting a modification of a gold spectrum through the dense part of an aluminium laser produced plasma. We have shown that it is very important to time resolve the spectra to be sure that the recorded radiation in the X-UV range comes from the thermal transport zone. A big work in the modelling remain to be done in order to give a good interpretation of the experimental results and to be able to probe the thermal transport region of a laser produced plasma with good accuracy.

REFERENCES

[1] R. Benattar and V. Malka, SPIE **1140**, (1989) 73.

[2] A. Thevenon, V. Millet, J. Flamand and B. Touzet, SPIE **1140**, (1989) 549.

[3] G. D. Tsakiris and K. Eidmann, J. Q. S. R. T. **38**, (1987) 353.

[4] B. L. Henke, P. Lee, T. J. Tanaka, R. L. Shimabukuro and B. K. Fujikawa, At. Data Nucl. Data Tables **27**, (1982) 1.

UV GENERATION IN NE II PUMPED BY THE SHORT WAVELENGTH RADIATION OF LASER PLASMA

V.V. Apollonov, S.I. Derzhavin, D.A. Noraev, and A.A. Sirotkin
General Physics Institute, Vavilov Street 38, SU-11794 Moscow, USSR

The creation of short wave length (SWL) lasers remains the problems so far. Despite of the fact that VUV lasers production was stated by some research groups / 1,2/ to obtain a new effective methods of SW generations is of a certain interest.

In 1967 Duguay and Rentzepis proposed to obtain the inversion in SW region with the help of inner-shell photoionization of atoms by broad band x-ray radiation / 3/. To realize this suggestion the high intensity sources of soft-x-ray radiation were necessary. As the example of such source S.E.Harris and all have used the laser plasma, produced not far from the active medium (AM) to provide the maximum bond "source-AM".

This method has proved to be fruitfull. It allowed to obtain the laser generation in visible and near infra-red region in AM with high gain value /4,5/.

The advantages of photoionization pumping by soft-x-ray radiation from laser produced plasmas (LPP) are following:

1. Short time duration of the radiation pulse from LPP allows to excite effectively a levels with short lifetime which is especially important for VUV and X-ray lasers.
2. The pecularities ofphotoionization cross-sections of atomic shell permits to ionize selectively inner-shells by sufficiently broad band X-ray radiation.
3. The laser radiation transforms to the X-ray flux from LPP with very high efficiency.
4. Photoionization schemes of pumping are defined by relatively narrow laser lines.

Last circumstances show the possibility of using moderate power lasers.

Last time it was shown in Silfvast's works,that X-ray radiation from LPP is also convenient for shake-up processes. Despite the fact that cross-sections of such processes are very smaller than the single-electron ones, the authors with the help of psec pump laser, succeeded to obtain the high gain in spectral region from visible (ArII λ = 4775 A) to VUV (He II,λ = 1640 A) /6,7/.

The comparison of the CO_2- and YAG- lasers applications

to the production of the X-ray flux from LPP was carried out already in the initial works concerned with photopumping /9/.

The value of the electron temperature in LPP can be obtained in the first approximation from the formula:

$T=0,6x(I\quad)^{2/3}$ ev / 10/. I (W/cm^2) and λ (nm) are intensity and wave length of the laser respectively. One can see that long wavelength lasers permits to obtain a higher temperature in equal intensities.

So, the aim of our work is as following:

1. To investigate the possibility of utilizing LPP from CO_2-laser as the X-ray source for the exciting of the two-electron transitions in noble gas atoms.

2. To explore the possibilities of obtaining the generation in shake up schemes with the aid of moderation time-duration laser pulses (τ = 100 nsec).

3. To investigate the different pumping mechanisms formed by CO_2-laser optical breakdown.

The scheme of the experiment is shown in Fig.1. The pulse of CO_2-laser radiation (typical shape is shown in Fig.2) with 100 nsec first peak duration and 0,4 J first peak energy is focused no the target. It is known, the laser radiation focused on a target with high atomic number convers to the shorter wave length radiation flux from LPP, the efficiency of such conversion may achieve 20% and more /11/ . The tantalum strip was used as a target.

For the increasing of AM length the laser radiation is focused by four germanium spheric lenses (F=50 mm). The plasma torches were adjusted in line with 15 mm to one another. It allowed to avoid the intersection of plasma crowns. Especial attention was paid to the equality of the laser intensity in each torch, where it was $2*10^9$ W/cm^2. In proceeding from the given earlier formula the electron temperature of LPP is to be equal 15 eV.

The noble gases were chosen as AM. We chose the argon for the initial experiments, in which Silfvast succeeded to observe the laser generation in visible region. The exciting energy of the upper laser level in ArII (λ= 477 nm) is 45 eV, so it can be easily populated by appeared x-ray flux.

Ar II laser resonator with the 15 cm length was formed by two spherical mirrors with the reflectivety 99,9% and 98% at λ =0,47um.

We obtain the generation on following ArII transitions:

$4p^2P_{3/2}$---$4s^2P_{1/2}$(λ=4765 A); $4p\ ^2P_{3/2}$---$4s\ ^2P_{3/2}$(λ=4545 A) and $4p^2P_{3/2}$---$4s\ ^2D_{5/2}$(λ=4278 A)

All these transition are superlumenescent which could explain the appearance of the generation by small threshold value, but there are many other superlumenescent transitions in ArII without laser generation in our case. One should remark that the electron configuration of the upper laser level was the same in all transitions $^2P_{3/2}$.

On investigating another noble gases we obtained a laser generation on analogous transitions. The most interesting generation was obtained in NeII:

$3p^2P_{3/2}$---$3s^2D_{3/2}$ (λ=3345 A) & $3p^2P_{3/2}$---$3s^2P_{3/2}$ (λ=3324 A)

because this is the shortest wave length generation today with such way of pumping. It is not clear, why the generation may be obtained only between these levels.

The shape of all dependencies in different gases was similar, they differ only for some numerical values, so let us consider the generation in NeII. The oscillograms of the generation pulses from CO_2 and NeII lasers and luminescence of NeII lines without mirrors were shown in Fig.2. The pulse of generation appears on the front of CO_2-laser pulse in 10 - 20 nsec after optical breakdown formation on the target. The generation was observed on the distance 4 : 10 mm from the target, in the sector with angle 100, the top of the angle was placed in the plasma torch (see Fig.3). The dependency of the intensity generation on the distance from the target is shown in Fig.3, on gas pressure in Fig.4, they are not monotonous, there is a maximum of the intensity on ~3 torr. The dependency on the pumping laser energy is not monotonous too, Fig.5. This one was obtained for the case of constant distance from the target.

Silfvast in his work on Cd vapors photopumping stated, that the back front of the generation pulse was connected with arriving plasma. In our case the variation of the distance from the target change neither the shape of the generation pulse nor the time delay of it. Plasma arrive in the cavity caustic only after the end of the generation pulse.

The founders of the photoionization pumping proposed that the inversion results from the direct excitation of upper laser level from the initial atomic state. In the case of Ar and Ne this is two electron processes of "ionization + excitation".

In general one can observe the population processes of $3p\ ^2P_{3/2}$ NeII level by investigating time dependence of the luminescence intensity from this level.

The typical form of the luminescence from NeII line (λ=0,3324 μm) without mirrors is shown in Fig.2. One can see two pronounced peaks. The first one coincide in time with the pulse of generation. The location of the second peak depend upon the pressure of working gas.

The dependence of time delay of second peak relatively to the front of CO_2-laser pulse upon the pressure is shown in Fig.6.

The growth of the delay by the decreasing pressure allows to say that the second peak of the luminescence is owing to electrons relaxation on the level $^2P_{3/2}$ from higher ion levels. The depopulation of this levels is collisional. In our case according to the estimations , the recombination is a slow process because of low pressure and low electron concentration.

Therefore the first peak of the luminescence is connected to the direct excitation of upper laser level. Unfortunately, we failed to understand if this excitation is from the initial atomic state or it is a stage processe. For example, the excitation may result from the initial NeII level pumping and then $^2P_{3/2}$ level may be pumped by more long wave radiation from the initial NeII level. In any case this process may play a role when the population of the initial NeII level will be sufficient.

In consequence of our previous work it was established that the generation were arised from the complex influence of UV radiation and electromagnetic field of CO_2-laser radiation on the gas mixture. UV radiation ionized the AM and released electrons swing in the electro-magnetic field, got the energy which was sufficient for atoms excitation. We observed in the caustic of CO_2-laser radiation an additional concentration of electrons and excited atoms /12/ .

By the intensity of CO_2-laser more than 10^9 W/cm , the atoms of AM may be excited by superthermal electrons flying from LPP. It would be noted that the geometry of their flying is similar to the direction diagram of soft-x-ray from LPP /11/ .

In the case of ArII and NeII by optical break down at low pressure this effects also may participate in the pumping.

To define the role of electron swinging in the excitation process the spatial distributions of the NeII line (λ =3324 A) luminescence were investigated, Fig.7.

It is followed from the geometry of the experimental scheme that the electrons swinging in the EM field on the distance of 5mm from the target in the region of caustic must essentially increase the concentration of excited ions NeII. The smooth shape of the curve and the absence any maximum of the intensity by H=0 mm excludes the possibility of pumping by swinging in the EM field electrons.

Earlier /14/ , it was shown, that the energy of super thermal electrons flew out from LPP depended on the incident laser intensity and for $I=2*10^9$ W/ cm^2 is equal 200 eV.

The role of hot electrons in the formation of inverse population in NeII was explored by using the brake electric field. To obtain the uniform electric field the potential difference of 400 V was applied between the target and the screen (see Fig.1), so the tension between the target and the cavity caustic was 300 V. It allowed to brake the electrons which can take part in the excitation. The experiments show, that the intensity of the generation even in the threshold was affected by no brake neither accelerating field.

So, our experiments proves the photopumping mechanism of NeII lasers. Let us repeat the experimental facts once more:

1. The generation appears on the initial step of optical breakdown formation, the delay of the generation pulse is not affected by the distance between the cavity caustic and the

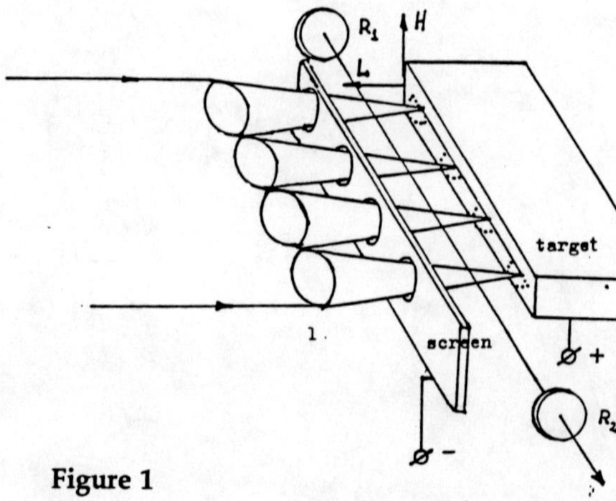

Figure 1

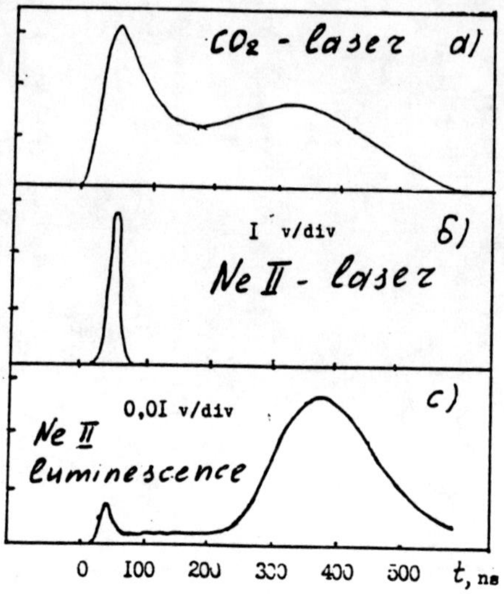

Figure 2

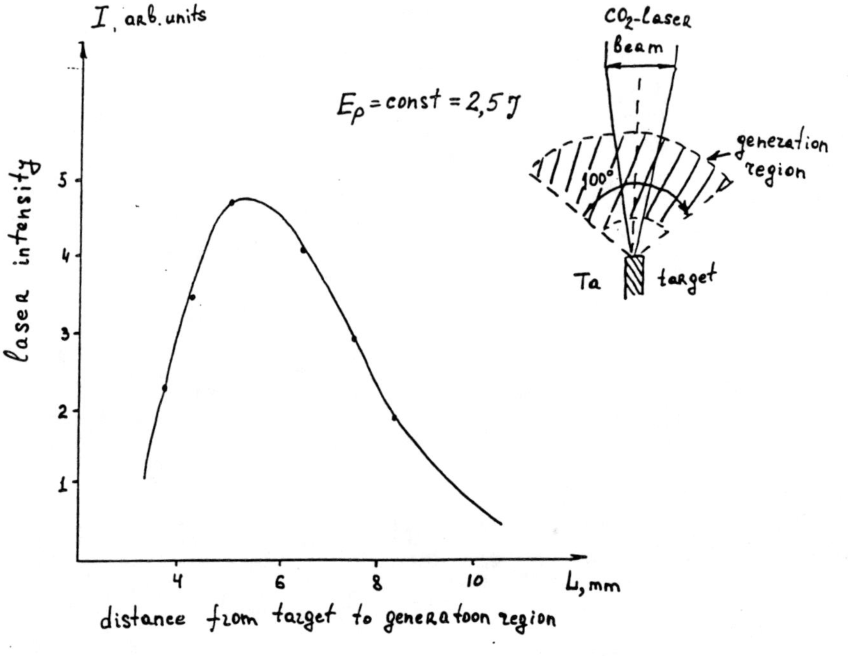

Figure 3

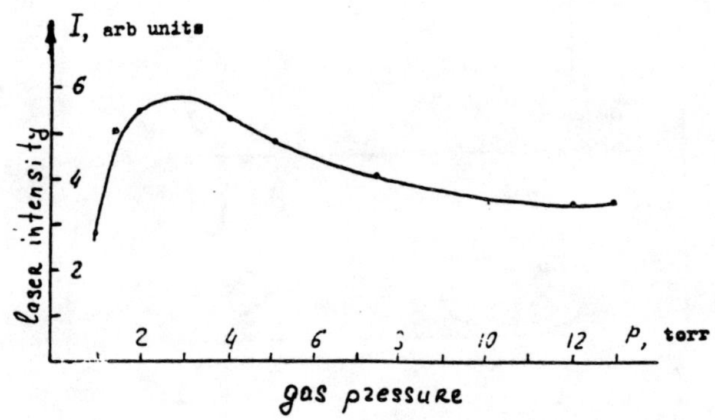

Figure 4

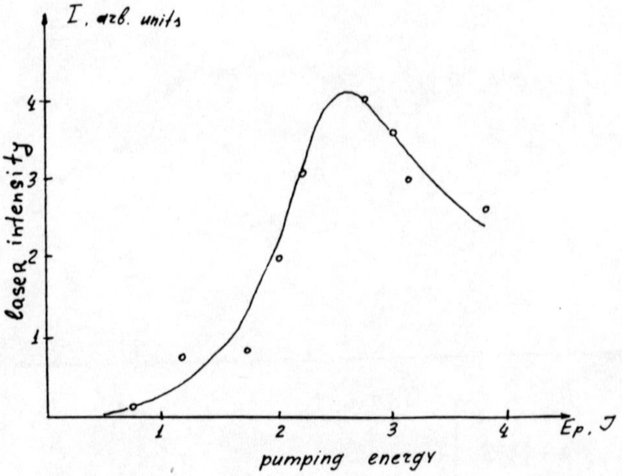

Figure 5

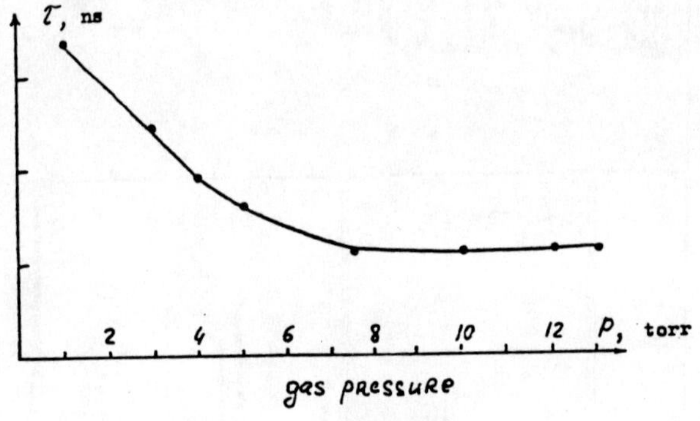

Figure 6

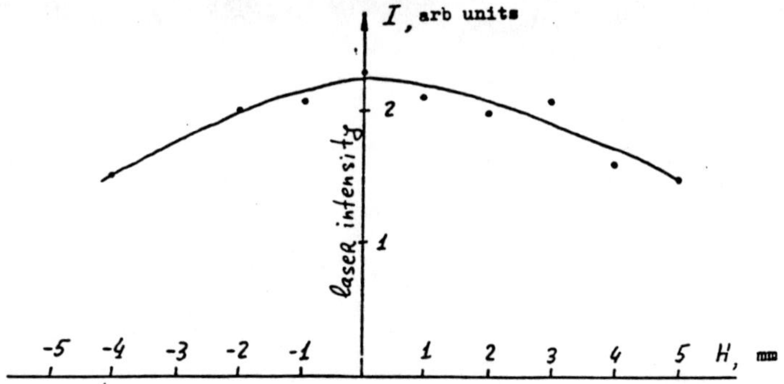

Figure 7

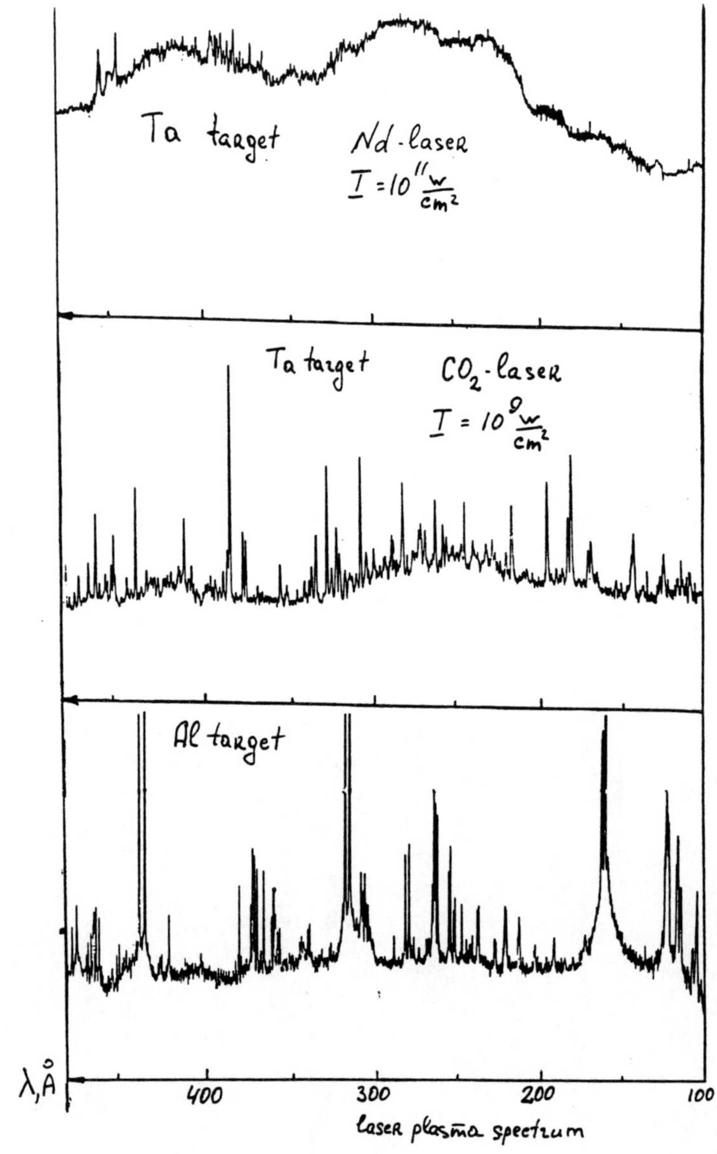
Ta target
Nd-laser
I = 10^11 w/cm^2
Ta target
CO2-laser
I = 10^9 w/cm^2
Al target
λ,Å
400
300
200
100
laser plasma spectrum

Figure 8

target.

2. The spatial distribution of the generation intensity is in a good agreement with the direction diagram of x-ray flux from LPP, which spreads in the angle of 90^{0} /13/ .

3. We failed to discover the nonuniformity of the generation connected with the caustic of CO_2-laser radiation.

4. We failed to discover the influence of the super thermal electrons from LPP on the generation intensity.

5. The inversion is formed only on transitions with upper level $^{2}P_{3/2}$.

In considering the laser plasma as the black body source, the temperature of it was measure we believe that the thermodynamic balance is established in inner region of LPP because of high electron concentration.

To measure the electron temperature we used the lines of buffer gas HeI λ=5876 A & HeII λ=4868 A and the lines of target material CIII λ=2297 A CIУ λ=2530 A.

It was shown, that the using buffer gas lines resulted in lower values of temperature, as the buffer gas was pushed out from hot inner region of LPP. The maximum measured electron temperature was 13 eV which agrees with theoretical one (15eV from the formula (1)) . It is known, that the efficiency of x-ray production in LPP from CO_2-laser is more smaller than from Nd-laser. However the easiness of obtaining the generation in ArII, NeII with the help of CO_2-laser pulse of 100 msec duration state about the perspectives of LPP from CO_2-lasers.

We measured the gain in ArII with the help of "mirrors method ". In the case of CO_2-laser the gain length product GL equals 0,5, which corresponded approximately to the gain 1 cm^{-1}.

The interesting information about the pumping process may be obtained by investigation of x-ray spectrums from LPP. The x-ray spectrums of LPP from CO_2-laser and Nd-laser in the spectral region 100-500 A for Ta target is shown in Fig.8. The intensities of given lasers was so, that the parameter $I\lambda^2$ remained constant in both cases. One can see that the spectra of CO_2-laser produced plasma has a great number of lines. At the same time the tendency of the luminescence decreasing on the edges of this spectral region is observed in both cases.

Thus is was experimentally shown:

1. The x-ray source from CO_2-laser produced plasma may be successfully improved for pumping "shake up" lasers.

2. For the first time, which such way of pumping the laser generation on ArII and NeII transitions

ArII: λ = 4545 A; 4278 A

NeII: λ = 3345 A; 3324 A

in visible and ultra-violet region was obtained.

3. The LPP as the source of x-ray flux was investigated. The given results creates the experimental base for moving in more short wavelength region.

REFERENCES

1. S.Suckewer, C.H.Skinner, H.Milchberg, C.Keane, and D.Voorhess, Phys. Rev.Lett., 55, 1753(1986).

2. DD.L.Matthews, P.L.Hagelstein, M.D.Rosen, M.J.Eckart, N/M/Ceglio, A.U.Hazi, H.Medecki, B.J.MacGowan, J.E.Trebes, B.L.Whitten, E.M.Campbell, C.W.Hatcher, A.M.Hawryluk, R.L.Kauffman, L.D.Pleasance, G.Rambach, J.H.Scofield, G.Stone and T.A.Weaver; Phys. Rev. Lett. 54, 110 (1985).

3. Duguay M.A., Rentsepis P.M. "Some approaches to vacuum UV and X-ray lasers", Appl. Phys. Lett., 1967, 10, p. 1350.

4. W.T.Silfvast, O.R.Wood II, J.J.Macklin and H.Lundberg in Laser Techniques in the Extreme Ultraviolet, S.E.Harris and T.B.Lucatorto, eds. (AIP, N.Y., 1984), p.427.

5. H.C.Kapteyn, R.W.Lee, and R.W.Falcone, Phys.Rev.Lett., 57, 2939 (1986).

6. W.T.Silvast, O.R.Wood II, JOSA, B. v.4, 609 (1987).

7. W.T.Silvast, O.R.Wood II and D.M. Al-Salameh in Short Wavelength Coherent Radiation: Generation and Applications D.T.Attwood and J.Bokor, eds (American Institute of Phusics, New York, 1986), 134.

8. S.E.Harris and R.G.Caro, Opt.Lett.11, 10 (1986).

9. O.R.Wood II, W.T.Silvast, J.J.Macklin and P.J.Maloney, Opt.Lett, 11, 198(1986).

10. R.Ramis and J.R.Sanmartin, Nuclear Fusion 23, 739 (1983).

11. J.M.Bridges, G.L.Cromer and T.J.McIlrath, Appl.Opt., 25 2208 (1986).

12. V.V.Apollonov, S.I.Derzhavin, D.A.Noraev, A.A.Sirotkin Lett.to Sov.Jorn.of Tech. Phys., v.15, n.21, p.12.

13. M.D.Rosen, Phys.Fluids, 22, 2020 (1979).

14. J. de Physique (Paris), 38, 761 (1977).

15. Hans R.Griem "Plasma Spectroscopy", McGraw-Hill Book Company, New York (1963).

LASER-PLASMA SOURCES BASED ON THIN TAPE TARGETS

Michael S. Schulz
Department of Physics, King's College, Strand, London WC2R 2LS, UK

1. — INTRODUCTION

The study of the x–ray emission from laser generated plasmas (LGPs) has usually been carried out using massive solid targets. This leads to three major problems when possible applications of LGP sources are considered. These are: the generation of debris by shock waves in the target; the small number of fresh target sites available; and the difficulty of keeping the target surface at the focus of the laser beam. The first of these can cause damage to delicate optical components and windows, the second can cause problems for applications requiring large number of x–ray pulses (it is difficult to engineer, eg., a spiralling cylinder to give more than a few thousand target sites), while the third becomes important for applications requiring x–rays with energies larger than about 1 keV. All of these problems can be overcome by using thin tapes, of the type used in audio cassette players, as targets. These tapes are typically 10—15 μm thick and so the shock wave punches a hole right through the tape, and little debris is generated in the direction of the focusing optics and diagnostics; standard tapes can be slowed down to provide more than 10^5 fresh target sites; and the tape can be run over a stationary cylinder (with a hole at the target site) to ensure that its surface stays at the laser focus.

In this paper a comparison will be made between the x–ray and debris production from massive solid targets and tapes.

2. — RATIONALE

Conventional targets for LGP applications, such as scanning x–ray microscopy, usually suffer from a number of shortcomings. When a focused laser beam is incident on the target surface it will give rise to the emission of debris particles — a possible mechanism for which will be discussed below. These debris

particles will be deposited on all surfaces facing the target including the focusing optics and the diagnostics. Apart from a decrease in transmission of the optics and diagnostic entrance windows this might not be disastrous; however, if fragile components, such as silicon nitride windows or zone plates, are to be included in the experimental design and if these are facing the target, then it becomes crucial that some form of protection for these components is provided. Silicon nitride windows break after less than 50 shots when placed 10 cm away from the target, and there is a 30% decrease in transmission of the focusing optics after 10^4 shots. An aspect that is connected with the debris problem and which becomes pronounced at repetition rates of $\geq$10 Hz is the presence of a vapour which, due to its refractive index, focuses the laser beam just in front of the target resulting in a greatly reduced irradiance on the target surface and consequently a decrease in the level of excitation.
The second important problem associated with conventional targets is the limited surface area. For example, a typical cylindrical target, with a diameter of 2 cm and a length of 4 cm, coupled to a stepper motor with a 3.5° step angle, would offer about 8000 fresh target sites; in terms of practical imaging applications this would correspond to a maximum image size of 90×90 pixels.
There is a third but less important problem and that is the surface uniformity of the target. The need for this requirement really depends on the application intended. For example, for harder x-rays (>1 keV) the focal position is crucial to within less than 50 μm; however, for softer x-rays it is of the order of 100 μm.
It was decided to attempt to alter the geometry of the target so that there would be (a) a reduction in the amount of debris and (b) prolonged use of the target.

3. — EXPERIMENTAL SET-UP

The basic experimental set-up is shown in figure 1. Details of the types of lasers are given in table 1.

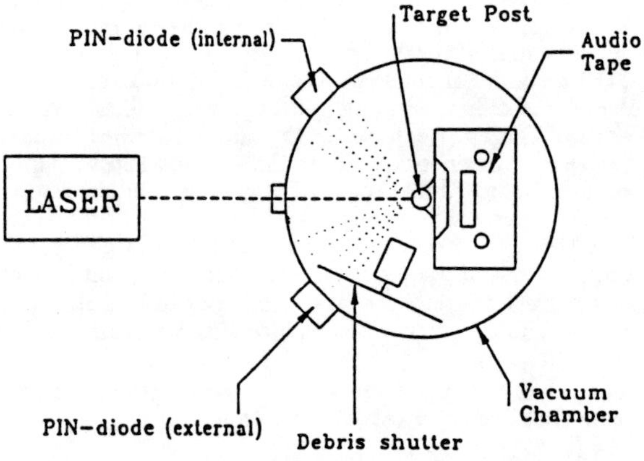

Figure 1. - Basic Experimental Setup

The detection devices are PIN diodes which have an entrance window depth of 0.25 μm and which are used in combination with 2000 Å thick aluminium filters to eliminate the visible and UV radiation. A modified personal hifi system is utilized to drive ordinary audio tapes, which serve as the target, around a stationary post. In figure 1 the component labelled debris shutter is a rotating disc designed to cut out the larger debris particles (>1 μm diameter). The laser is synchronized with the debris shutter so that when it triggers the laser a slot in the disc will allow the radiation through to, in this case, a chamber exit window; however, the larger debris particles, due to their finite velocity, will be cut out by the rotating disc. Details of this debris shutter will be presented in a forthcoming paper.

Table 1. Laser Specifications

Laser	Nd:YAG	KrF
Wavelength (nm)	1064	249
Energy (J)	0.5	0.5
Pulse Length (nsec)	8	25
Repetition Rate (Hz)	10/20	5

4. — DEBRIS PRODUCTION

When a focused laser beam strikes the surface of a massive target it sets up shock waves which propagate through the target (figure 2a). As a result of the reflected shock waves debris particles are ejected from the crater site (figure 2b). A phenomenon that is closely related to the production of debris is the presence of a vapour which establishes itself, after 1 or 2 seconds of continuous firing, between the target and the lens. This medium has a defocusing effect on the beam, hence making it impossible to obtain maximum irradiance on the target. As a consequence of this there will be a considerable reduction (as much as 60% at 10 Hz repetition rate depending on the target material) in the number of x-ray photons produced. If the amount of debris ejected is proportional to the amplitude of the reflected shock wave, then by changing the geometry of the target it should be possible to eliminate or certainly reduce the number of debris particles. The target post is designed so as to offer no resistance to the back of the tape (figure 3). As the laser hits the tape it punches a hole through the tape and the debris particles are ejected in the forward direction of the beam (figure 4).

5. — RESULTS & DISCUSSION

The debris and the defocusing vapour problem is considerably reduced with tape targets, as the life-time of silicon nitride windows is increased by a factor of $\gg$20 when compared with the use with solid massive targets. The remaining debris and vapour particles can be eliminated completely using the debris shutter in conjunction with a low pressure (eg. 30 torr) helium atmosphere.
Typically a C90 audio tape is about 260 metres long and if shots are placed less than 1 mm apart then this provides about 3×10^5 target sites on one tape; this correspond to a maximum image size of 550×550 pixels — a vast improvement when compared to the 90×90 maximum image size obtainable with massive targets.

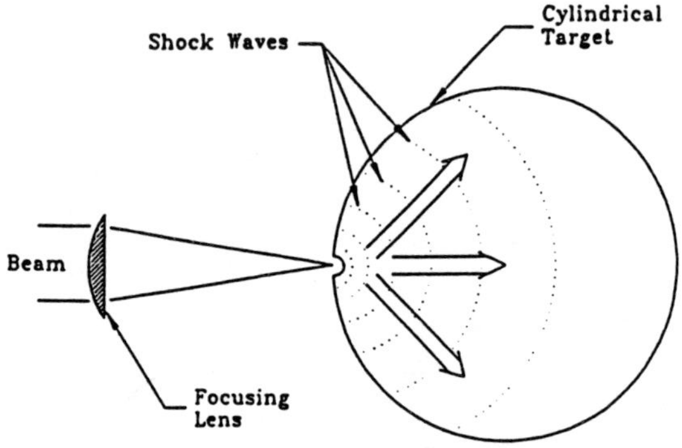

Figure 2a. - Diagram of shock waves propagating within the target.

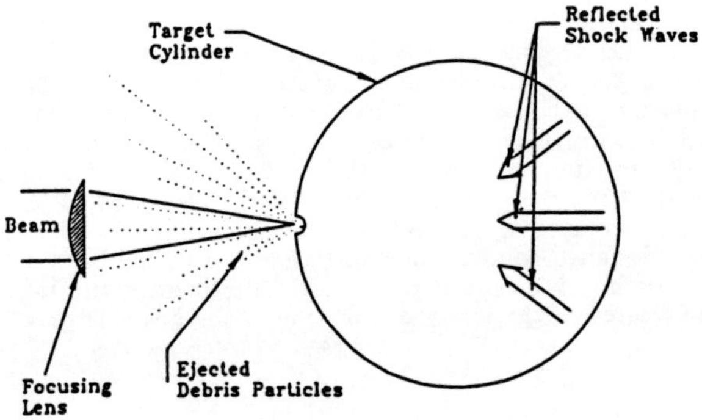

Figure 2b. - Debris particles are ejected as a result of reflected shock waves.

Figure 3. - The Target Post

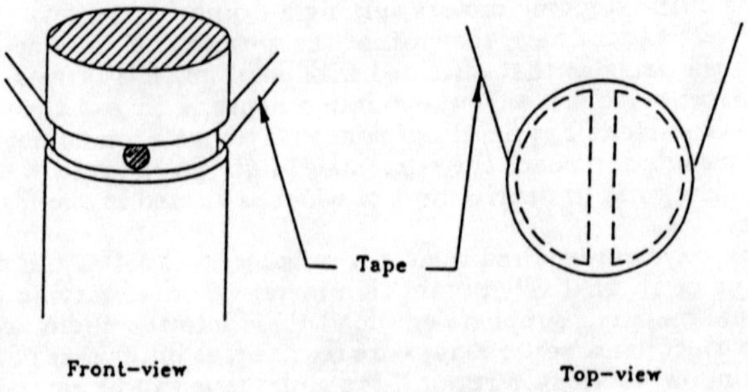

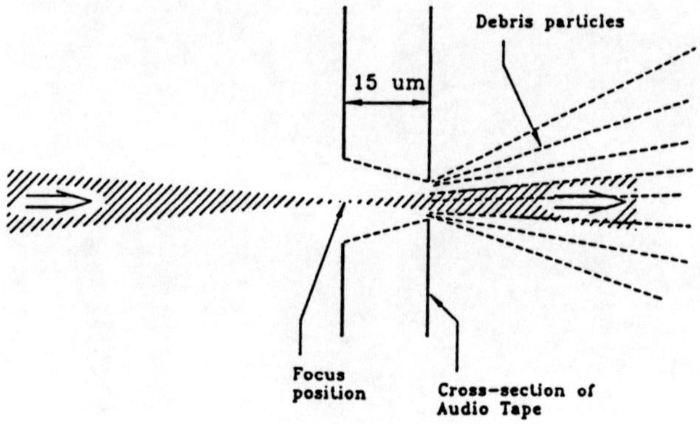

Figure 4. - Schematic diagram of a thin tape LP source

Regarding the surface uniformity of audio tapes, the thickness of a C90 tape is 15±1 μm and since the tape is guided around a stationary post there are no problems with the target moving in and out of focus due to surface irregularities or vibrations of the stepping motors turning a cylindrical target.
Recent investigations have shown that the x–ray signal obtained with audio tape targets is the same as that obtained with solid massive targets. This result is in agreement with the fact that the ablation depth is only of the order of 1 μm. In other words, materials at depths greater than about 1 μm do not contribute to the emitting plasma and hence the remaining 14 μm (in the case of C90 tapes) is bulk material giving rise to debris, most of which is ejected in the forward direction of the beam.
If harder x-rays are required then it is possible to shoot at the metallised surface of the tape or alternatively obtain thin metal tapes (of varying thicknesses) from commercial materials suppliers and load these into the audio tape plastic casing. If, on the other hand, softer x–rays are required, as in the case of x–ray microscopy (water window wavelength region), the audio tape can be reversed and the mylar substrate of the tape will serve as the target. This will result in the production of carbon emission lines in the water window; details of the spectra and more specifically the number of photons in individual lines at different irradiances will be published at a later date.
To summarize, the overwhelming arguments for using tape targets are (a) the very much extended use and (b) an obvious reduction in the amount of debris; and there is also the commercial availability of the audio tapes making custom made target design and accurate machining unnecessary.

X-RAY MICROSCOPY WITH LASER-PLASMA SOURCES

A.G. Michette and M.S. Schulz
Physics Department, King's College, Strand, London WC2R 2LS, UK

1. INTRODUCTION

In previous papers [1,2], the properties required of laser-generated plasmas as sources for x-ray microscopy have been considered in detail. The main conclusion of these deliberations was that pulsed laser-plasma sources can be used for scanning transmission x-ray microscopy, the limiting factor being the repetition rate of the laser. In this paper the practical implementation of such a microscope will be discussed, and areas in need of further development will be identified.

2. THE SOURCE

It has already been demonstrated that interesting information can be obtained about both biological and non-biological material via scanning transmission x-ray microscopy [3,4]. The potential advantages of x-ray microscopy over, for example, electron or optical microscopy are well-known [3] and will not be repeated here. However, the vast majority of experiments in high-resolution (better than about 100 nm) x-ray imaging have been performed using synchrotron sources [4], although some results have been obtained using laser plasmas generated by very large lasers [5]. These facilities are very expensive to build and maintain, and it can be argued that x-ray microscopy cannot become a routine analytical tool to complement electron and optical microscopy unless smaller, less expensive x-ray sources can be used. In addition, the environment at synchrotron sources is far from user-friendly for the microscopist (or the specimen). Plasmas generated by small lasers ($\approx$0.5 J per pulse) can overcome all these problems; for example it is much easier to produce a vertical x-ray beam so that the specimen (potentially hydrated and initially living) can be mounted horizontally.

In order to obtain the monochromaticity requirements (see section 3) of the microscope optical system and the high spectral brilliance needed to allow imaging in short times, a carbon plasma, producing a predominantly line spectrum, must be used; a monochromated spectrum from a high-Z plasma would be insufficiently bright [2]. The $1s^2$-1s2p line (at 4.02 nm) in the C V spectrum is sufficiently narrow to satisfy the monochromaticity requirements; can contain sufficient photons (about

4×10^{12} per steradian) to allow a scanned image to be built up at one pixel per pulse [6]; is far enough from other spectral lines to prevent these degrading the image; and the continuum background adjacent to the line is low enough (line/background > 500 [6]) that image degradation is again minimized.

3. THE MICROSCOPE

All high-resolution images in x-ray microscopy (as distinct from contact microradiography) have so far been obtained using systems with zone plate optics. Although grazing incidence (Wolter) optics [7] or multilayer mirrors [8] can in principle be used, with the advantage of being able to deliver higher fluxes to the specimen, these optics have not to date demonstrated resolutions better than about 1 μm and, in the case of multilayer mirrors, their use has been almost exclusively at wavelengths longer than the most interesting (for biological applications) water-window range (2.3-4.4 nm). Therefore the considerations here will be restricted to microscopes with zone plate optics.

High-resolution zone plates are now made by several groups [9-11] and the technology for their manufacture is advancing towards outermost zone widths (which define the spatial resolution possible) down to 20 nm or less. Currently such zone plates are at best ≈5% efficient in the first diffraction order, but by making better use of phase-modulating effects [12] this could be increased to ≈15% or more. The zone plates must be centrally obstructed and used in conjunction with a downstream aperture to remove the effects of other diffraction orders and of other lines in the plasma spectrum. The zone plates typically have diameters of several tens of micrmetres, focal lengths of about 1 mm, and a few hundred zones (the monochromaticity requirement, due to the inverse dependence of focal length on wavelength, is $\lambda/\Delta\lambda \approx 1.6N$ where N is the number of zones).

With a conversion efficiency of laser radiation into X rays in the $1s^2$-1s2p line of 2.5×10^{-3}, a zone plate diffraction efficiency of 0.15 and a detector efficiency of 0.65 (see section 4), the relationship between the laser irradiance I (W m^{-2}) on target and the number S of x-ray photons in the image is [2]

$$S \approx 6 \times 10^{-5} I \tau_L e^{-\mu x}$$

where τ_L is the laser pulse length (in seconds) and the exponential term represents absorption by the specimen of thickness x and linear absorption coefficient μ. The approxomation sign indicates that some reasonable allowance has been made for transmission of the X rays through vacuum windows, specimen containment membranes and so on. Thus, for example, for an irradiance of 10^{16} W m^{-2} (easily obtainable with a laser with a few hundred millijoules per pulse), a pulse length of 10 ns and $e^{-\mu x} \approx 0.5$, about 3500 x-ray photons per pulse would be detected. This is certainly sufficient for imaging purposes.

4. PRACTICAL CONSIDERATIONS

The feasibility of using a laser-generated plasma as a source for scanning x-ray microscopy is thus determined by the laser properties - primarily, a high repetition rate with a pulse energy of ≈0.5-1 J is required. A repetition rate of 500 Hz would allow a 256×256 image to be obtained in just over two minutes, comparable (or, indeed, better) than with current synchrotron sources. Consideration must also be given to the construction of the target from which the plasma is formed, and as is discussed in another paper in these proceedings [13] tape targets can provide sufficient fresh target sites to allow images with large numbers ($\approx 3 \times 10^5$) pixels to be obtained, while at the same time effectively eliminating the debris problems associated with solid targets.

The major technical problem still to be overcome concerns the detector, which must be capable of peak count rates of several times 10^{11} counts per second (a few thousand photons in a few nanoseconds) or more. To allow the discrimination of grey levels in an image, the detector must also be able to seperate count rates in steps of, say, 10^9 counts per second (corresponding to a few tens of photons), which means that it must be well-protected from noise.

5. CONCLUSIONS

X-ray microscopy is at a very early stage of development compared to optical and electron microscopy. The primary reasons for the slow development have been the lack of good optics and of sufficiently bright sources. Both these problems have to a large extent been overcome, but while it has to rely on large expensive sources such as synchrotrons it is arguable that x-ray microscopy can never become a routine analytical tool like already established techniques. Now that the necessary x-ray optical systems have become well understood and the technology for their manufacture is well advanced, the limiting factor is once again the source. In this paper it has been argued that, for one type of x-ray imaging, scanning microscopy, laser-generated plasmas are extremely promising candidates as laboratory (i.e. small and relatively cheap) sources, and much work is going on to implement such systems.

REFERENCES

1. A G Michette *et al* pp59-62 in *X-Ray Microscopy II, Springer Series in Optical Sciences* **56** (1988)
2. A G Michette *Journal of X-Ray Science and Technology* **2** 1-16 (1990)
3. A G Michette *Rep. Prog. Phys.* **12** 1525-606 (1988)
4. D Sayre *et al* (eds) *X-Ray Microscopy II, Springer Series in Optical Sciences* **56** (1988)
5. A D Stead *et al* pp 433-7 in *X-Ray Microscopy II, Springer Series in Optical Sciences* **56** (1988)
6. M Kühne, private communication (1989)
7. A G Michette, chapter 2 in *Optical Systems for Soft X Rays* (New York: Plenum) (1988)
8. A G Michette, chapter 4 in *Optical Systems for Soft X Rays* (New York: Plenum) (1988)
9. C J Buckley *et al* pp 88-94 in *X-Ray Microscopy II, Springer Series in Optical Sciences* **56** (1988)
10. V Bögli *et al* pp 88-7 in *X-Ray Microscopy II, Springer Series in Optical Sciences* **56** (1988)
11. G Schmahl *et al* pp 63-74 in *X-Ray Microscopy, Springer Series in Optical Sciences* **43** (1984)
12. A G Michette and C J Buckley, pp 477-82 in Proc. SPIE **1140** (1989)
13. M S Schulz, these proceedings

PART III

NONLINEAR X-RAY GENERATION

HIGH-FREQUENCY EMISSION OF A RYDBERG ATOM IN A SUPER-STRONG IR FIELD

N.B. Delone, P.A. Golovinsky, M.Yu. Ivanov, P.B. Lerner, and A.L. Nefedov
General Physics Institute of the USSR Academy of Sciences, 117942 Vavilova, 38, Moscow, USSR

Abstract. We propose and consider a new idea for high-frequency emission production by the use of a Rydberg atom in a strong IR field. Emission of high-frequency quanta $\hbar\omega > 10^2$ eV can be obtained, tunable in the same diapason of about 10^2 eV.

During the last few years the behaviour of atoms in an extremely intense laser fields is a subject of increasing interest. In such fields the atomic systems begin to display new properties. In particular, the phenomenon of ionization suppression (also called as stabilization) in an intense laser field is predicted theoretically in several works [1-6]. If that stabilization just takes plase, we can try to accumulate relatively great energy in atomic system, absorbing the former from the intense laser field without ionization decay of the latter. Than this energy can, in principle, be reirradiated as a single photon, i.e. in the form of high - frequency radiation.

In this paper we show that the stabilization can be obtained best of all for Rydberg states of atoms, and we discuss the possibility of high-frequency emission production using the medium of Rydberg atoms dipped in an intense laser field.

The basic idea of high-frequency emission production is the following. At the first step we prepare the "Rydberg atom " (Fig.1) using, for example, the multiphoton excitation of atoms or ions from their ground state by UV-laser field. There are many experimental works on this subject [7-9]. It would be nice to prepare the inverse population on this step. It's possible theoretically if the time of passing of the Rydberg states through the resonance is greater than the time of Rabi oscillations between Rydberg and ground states (the so-called adiabatic inversion of levels [10]). However, as a matter of fact laser excitation of the Rydberg states from the ground one at any rate prepares population inversion with respect to all low-lying levels except the initial one. Both a single Rydberg level or a Rydberg wave packet can be populated.

Note, that the multiphoton excitation isn't unique process of Rydberg levels population. For example, the cascade photoexcitation [10] is also applicable.

At the second step we prepare the "Stark atom" (Fig. 2), deeping the Rydberg atom in an intense infrared laser field. The laser field frequency ω exceeds the binding energy of Rydberg states $|E_n|$ (n - principle quantum number), but is much less than the binding energy $|E_1|$ of the ground state:

$$|E_n| < \omega << |E_1| \qquad (1)$$

(we use the atomic units $e=\hbar=m=1$). As is well known [10] we obtain in these conditions large positive Stark shift $\Delta E_n = \varepsilon_0^2/4\omega^2$ of Rydberg states (ε_0 - laser field strength) equal to the average quiver energy of a free electron oscillations in a laser field. On the other hand, the Stark shift of the ground state $\Delta E_1 = -\alpha_{st}\varepsilon_0^2/4$ is negative and small (α_{st} - static polarisability, $\alpha_{st} << \omega^{-2}$).

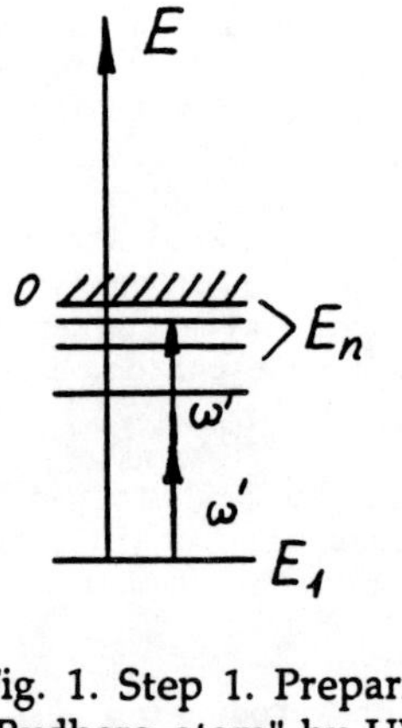

Fig. 1. Step 1. Preparing of the "Rydberg atom" by UV-laser, as a result of the two-photon absorption.

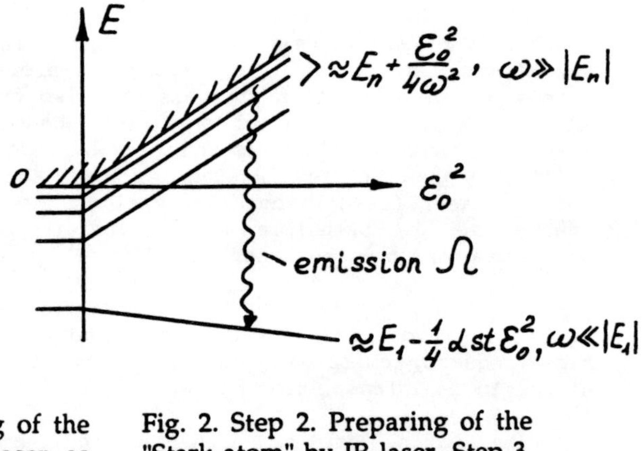

Fig. 2. Step 2. Preparing of the "Stark atom" by IR-laser. Step 3. High-frequency emission from the shifted Rydberg levels.

The distance between the shifted Rydberg and the ground state is, after all, very large, and it can, in principle, be utilised for high frequency emission production (Step 3, see Fig. 2). It should be espesially emphasized that the emission frequency Ω satisfies the equation:

$$\Omega \approx |E_1| + \varepsilon_0^2/4\omega^2 \qquad (2)$$

($|E_n|$ and ΔE_1 are negligibly small), and hence it can be smoothly tuned with changing the field strength ε_0. Thus we can obtain a tunable source of high-frequency emission.

Let us suppose, for example, that we have prepared the Rydberg states of hydrogen with $n \approx 20$ after the Step 1. In this case $|E_1|=0.5$ and $|E_n|=1/2n^2 \approx 0.0012$. Using a modern CO2 laser ($\omega \approx 0.12$ eV = 0.0043 a.u.) and intensity I about 10^{12}- 10^{14} W/cm^2, that corresponds to the field strength $\varepsilon_0 \sim 2.7\ 10^7$- - 2.7 10^8 V/cm (0.0052 - 0.052 a.u.), we are just ensuring the fulfilment of the above conditions (1) and as we see from (2) it is possible to obtain $\Omega \sim 25 - 1000$ eV.

The problem of stability of this "Stark atom" is of great importance, because the intense laser field not only shift the Rydberg levels, but effectively ionize them. That's why we have to answer the question: how much is the photoionization lifetime of the "Stark atom"?

Below is the well known old result, obtained in both isolated level and WKB approximations [11,12]:

$$t_{in} = 2\pi n^3(\varepsilon_c/\varepsilon_0)^2, \quad \varepsilon_0 < \varepsilon_c \approx 2\omega^{5/3}. \qquad (3)$$

The larger is the field strength ε_0, the smaller is the photoionization lifetime t_{in}. The "Stark atom" decays in a Kepler period ($t_{in} = 2\pi n^3$) if the ionization width of the Rydberg state Γ_{in} is equal to the energy distance n^{-3} between the levels, see Fig.3. In this case the field strength $\varepsilon_0 = \varepsilon_c$ is called critical and the isolated level approximation becomes invalid because its applicability region is $\Gamma_{in} << n^{-3}$.

Here we would like to make one passing remark. It appears that increasing of the frequency ω leads to the stabilization of an atom. In the IR-field with $\omega \sim n^{-2} >> n^{-3}$ ($\omega \sim n^{-2}$ provides one-photon ionization) the critical value can be much greater than the atomic field, $\varepsilon_c \sim \omega^{5/3} \sim n^{-10/3} >> n^{-4}$, while in the static field the photoionization occurs at 1 Kepler period when $\varepsilon_0 \sim n^{-4}$.

From our quantitative example we can see that for the CO_2-laser frequency the critical value $\varepsilon_c \sim 10^6$V/cm is really much greater than the atomic field strength $n^{-4} \sim 3\ 10^4$V/cm and it is much less than the laser field strength $\varepsilon_0 \sim 2.7 \cdot 10^7$ -$2.7 \cdot 10^8$ V/cm, which is necessary for large AC-Stark shift (that very field we shall indicate ε_d below, d means "desired").

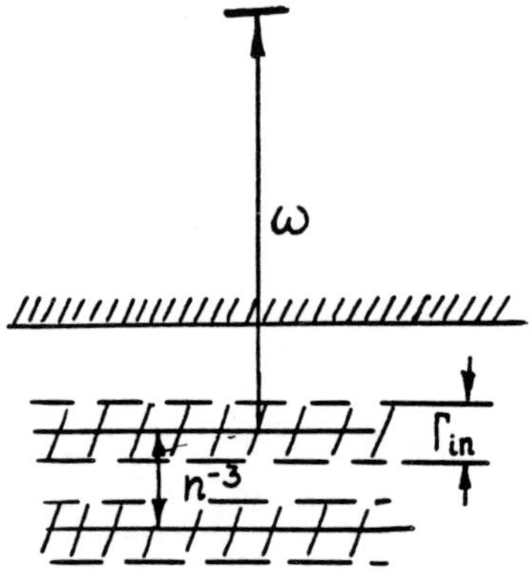

Fig. 3. The scheme of the ionization when $\Gamma_{in} << n^{-3}$.

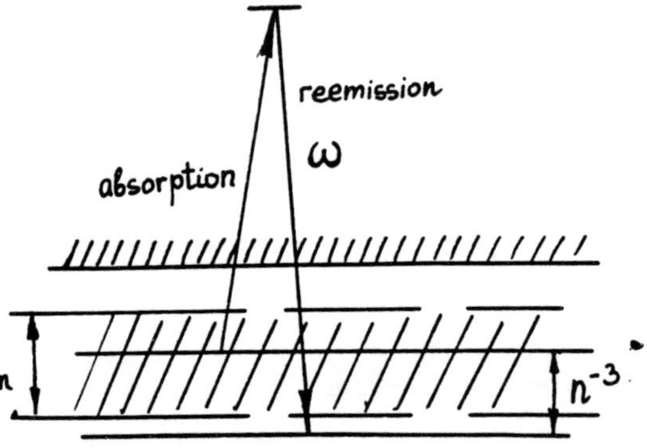

Fig. 4. The scheme of the ionization when $\Gamma_{in} \sim n^{-3}$.

But in the field ε_d the lifetime t_{in} (see Eq. (3)) is much less than the Kepler period. However, in this conditions the isolated level approximation isn't valid and we have a question: what if the stabilization occurs at the fields $\varepsilon_0 > \varepsilon_c$? It is wonder, but results of recent works [3-6] show, that it is just so.

Below is the new result obtained in the resonance approximation [3, 4], which is applicable for not very strong field:

$$t_{in} \sim 2\pi n^3 (\varepsilon_0/\varepsilon_c), \quad \varepsilon_c < \varepsilon_0 < n\varepsilon_c. \qquad (4)$$

We can see the increasing of the lifetime t_{in} with increasing of the field strength ε_0. This phenomenon have been also observed in the numerical calculations by Parker and Stroud [6]. It is an example of the "population trapping" and is connected with the strong interaction of Rydberg states through the continuum (Fig. 4), which provides effective reemission after absorption.

For our quantitative example the laser field ε_d is larger then the upper limit of the resonance approximation $n\varepsilon_c = 2.10^7$ V/cm, and this approximation isn't valid. Nevertheless, the new quantum results look promising because they predict the stabilization (Fig. 5).

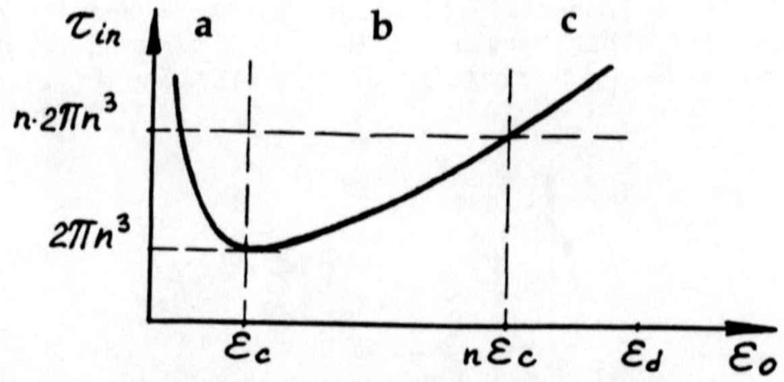

Fig. 5. Photoionization lifetime as a function of the field strength: (a) is the applicability region of the "old" quantum results (Eq. (3)), (b) is the applicability region of the "new" quantum results (Eq. (4)), it is the region of the quantum stabilization, (c) is the region of our classical results (classical stabilization).

To control whether the stabilization exists in stronger fields we have performed the classical calculations.

The classical method was based on the numerical integration of the Newton's equation of motion:

$$d^2\bar{r}/dt^2 = - Z\,\bar{r}/r^3 - \varepsilon_0 \cos(\omega t+f) \qquad (5)$$

where r is the radius-vector of the electron, Z - nuclear charge, f is the initial phase of the field. The initial conditions (velocity v and radius-vector r) were chosen from the classical ensemble using the Monte-Karlo procedure [13]. Thus we have investigated the behaviour of the atom on the grounds of the analysis of the classical trajectories ensemble. This approach allows to obtaine the probability characteristics of the photoionization process [14].

Our classical results are as follows. At first, as in the case of the "old" quantum results, we observed the high-frequency stabilization. The motion of an electron in the field with $\omega > n^{-3}$ and $\varepsilon_0 \sim n^{-4}$ looks like irregular oscillations near the nucleus, the life-time t_{in} being much greater, than $2\pi n^3$.

Secondly, we have found a principally new type of long-living classical trajectories in extremely strong fields, when the amplitude of electron oscillations ε_0/ω^2 exceeds the size of Kepler orbit n^2 and the field frequency ω is much greater than the Kepler frequency n^{-3}:

$$\varepsilon_0 > n^2\omega^2 >> n^{-4}. \qquad (6)$$

An example of such a trajectory is shown in Fig. 6, and its envelope looks like a treble clef. There are about 20 - 30 % of such trajectories in all classical ensemble which photoionization lifetime is larger than 10 Kepler periods.

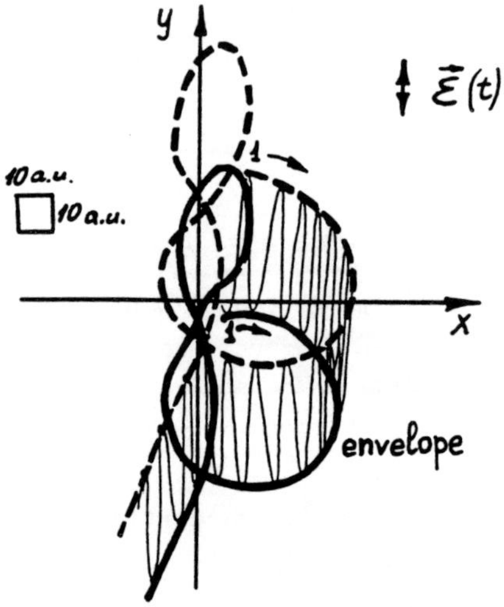

Fig. 6. An example of long-living trajectory in the case n = 3, ε - 0.4, ω = 0.15. Photoionization lifetime in ~10 Kepler periods ~41 periods of the laser field.

In our example the with CO_2-laser the field, which is necessary for high-frequency emission, is in the region of classical stabilization: $\varepsilon_0=2.7\cdot 10^7-2.7\cdot 10^8$V/cm > $n^2\omega^2=3.7\ 10^7$V/cm and ω =0.12 eV >> n^{-3}=.0035 eV

Now we'll discuss some other interesting phenomena attending photoionization in a strong field. There are the bremsstrahlung in the scattering of oscillating electrons on the nucleus and the ionization and excitation of atomic innershell by outer-shell electrons. The last process have been discussed by Boyer and Rhodes [15]. Both of this phenomena have to become apparent just in the same conditions when the stabilization exists. The thing is that in this case the outer electron can receive a large portion of energy from the laser field before the interacting with the atomic core. If the stabilization is absent, the electron go away from the core, which evidently suppress both inner-shell excitation and bremsstrahlung on the parent ion (e.g. harmonic generation).

The above statements serve as theoretical substantiation for the possibility of the Stark atom creation in the modern experimental conditions. Nevertheless, the real utilization of this Stark atom for high-frequency emission production is connected with some difficulties.

First of all, we have supposed above that the IR- field is switched on instantly. However, a real switching has a smooth character. As the field increase its strength pass through a critical point $\varepsilon_0=\varepsilon_c$, when the photoionization time is equal to 1 Kepler period. Hence, as it seems at first sight the creation of the Stark atom is impossible, because it will be ionized at the moment when $\varepsilon_0=\varepsilon_c$, i.e. earlier than the field will achieve the large values $\varepsilon_0>>n^2\omega^2$, which are necessary for stabilization. Nevertheless, we can indicate at least two ways to solve this problem. For example one can use a laser pulse with sufficiently short front. In this case we can pass through a critical point at the time less than 1 Kepler period, and photoionization will not occur. This time is about 1 ps for n~20, and it is in the region of technical resources of modern lasers. On the other hand, if we create the Rydberg wavepacket, it is necessary to pass the critical point at the moment, when the Rydberg electron is at the greatest distance from the nucleus, when the absorption of the energy from the laser field is impossible.

Attention is to be also called to the fact, that organization of stimulated transitions from the Stark states exactly to the low-lying states is a matter of some difficulty. The thing is that the probability of stimulated transition to the other Rydberg states is sufficiently high due to great value of the dipole matrix elements. In addition, the size of the Stark atom is relatively great and it leads to limitation for maximum density of atoms in active medium. That's why it can be difficult to achieve sufficiently high value of the gain.

However, it is worthy to overcome the above difficulties because the scheme proposed gives us a unique possibility to obtaine a source of high frequency emission, tunable in a very broad range. It should be also noted that the method we have proposed to obtaine high energy transitions is principally new. In traditional schemes a population of working levels is combined with transmission of the energy to a quantum system and both these processes are realized by the same source. In the scheme that we have proposed the pumping is devided into two stages. First of all the standard process of population of high-excited states of atomic system is carried out. Than the change of the energy spectrum of the quantum system "atom+field" takes place. In other words, we create a "dressed atom", and a bulk of the energy is brought into the atomic system just at the second stage and not at the stage of population of the upper working levels. It is important to underline that ~100 % population of Rydberg levels at the first step is not necessary. Being high frequency for the Rydberg states, as a matter of fact strong IR-field is low frequency for the ground one. Hence, effective depopulation of the ground state due to tunneling or even to above-barrier decay during the IR-field half-period takes place in sufficiently strong field with intensity greater than about 10^{14}W/cm^2.

Experimental observation of long living Stark atom would be very important on the first stage of experiments on high frequency emission production. The observationof the tunable spontaneous emission at frequencies ~ 10^2eV from Rydberg states in extremely intense IR-field could be a direct evidence of existing of these Stark atoms.

REFERENCES

[1] J.Gersten, M.H.Mittleman, J.Phys.B 9,(1976),2561.
[2] M.Gavrila, J.Z.Kaminski, Phys.Rev.Lett. 52,(1984),613.
[3] M.Fedorov, A.Movsesian, J.Phys.B 21,(1988),L155.
[4] M.Fedorov, M.Ivanov, J.Opt.Soc.Am. B 7,(1990)
[5] Q.Su, J.H.Eberly, J.Javanainen, Phys.Rev.Lett. 64,(1990),862.
[6] J.Parker, C.Stroud, Phys.Rev. A 41,(1990),1602.
[7] J.A.Yeazell, M.Mallaieu, J.Parker, C.R.Stroud, Phys.Rev. A, 40, (1989),5040.
[8] A. ten Wolde, L.D.Noordam, H.G.Muller, A.Lagendijk, H.B. van Linden van den Heuvell, Phys.Rev.Lett. 61,(1988),2099.
[9] D.Normand, L.Lompre, A.L'Huiller, J.Morellec, M.Ferray, J .Lavancier, G.Mainfray, C.Manus, J.Opt.Soc.Am.B 6,(1989),1513.
[10] N.B.Delone, V.P.Krainov, Atom in the Strong Field,.Springer-Verlag, Berlin, Heidelberg,N.Y.,Tokio, (1985).
[11] N.B.Delone, S.P.Goreslavsky, V.P.Krainov, J.Phys.B,15,(1982)L421.
[12] I.Ya.Berson, Sov.Phys. JETP, 56,(1982),737.
[13] G.Bandarage, A.Maquet, J.Cooper, Phys.Rev. A,41,(1990),1744.
[14] J.H.Leopold, I.C.Percival, J.Phys. B 12,(1979),709.
[15] K.Boyer, C.K.Rhodes, Phys.Rev.Lett. 54,(1985),1490.

NONLINEAR OPTICAL FREQUENCY CONVERSION OF $YAlO_3:Nd^{3+}$ LASER RADIATION TO THE UV AND VUV IN Ba VAPOR

M.K. Kodirov, S.A. Myslivets, A.K. Popov, and V.V. Slabko
Institute of Physics and Krasnoyarsk University,
660036 Krasnoyarsk, USSR

As is known, one of the techniques of the UV and VUV generation is frequency conversion of lasers operating in other spectral ranges. Two experimental approaches are possible: 1) resonant and quasi-resonant frequency conversion, i.e. by high-order nonlinear susceptibilities, at low pump power densities; 2) nonresonant frequency conversion at high pump power densities. It should be noted that resonant and nonresonant frequency conversion, along with the increase of the nonlinear susceptibility $\chi^{(n)}$, is accompanied by the competing processes, leading to saturation of the conversion efficiency. Therefore the optimum choice depends on the experimental conditions and the pump sources available [1-3].

Atomic metal vapors are the most popular nonlinear medium used in the experiments. Their major advantage, compared to condensed media, is the broad range of transparency in the VUV and the feasibility of phase matching by adding a buffer gas.

In this paper we report the experimental results on quasi-resonant and nonresonant third- and seventh-harmonics generation of $YAlO_3:Nd^{3+}$ in Ba vapor. Our choice of the system $YAlO_3:Nd^{3+}$ laser+Ba vapor was dictated by the fact that the second-harmonic of this laser is two-photon-resonant with the Ba atomic transition $6s^2\,{}^1S - 8s\,{}^1S$. A relatively small two-photon resonance detuning $\Delta\tilde{\nu}$ = 10 cm^{-1} and high pump power (compared to that of frequency-tunable lasers) encourage us to expect high conversion efficiencies (CE) both in the UV and VUV by using various combinations of frequencies. It should be noted that recently Ba atoms have been extensively used in resonant nonlinear optical experiments due to their convenient energy levels structure and large electric dipole moments of transitions between low lying atomic states [4-6].

A schematics of energy levels of Ba, giving major contribution to the parametric optical wave mixing is shown in Fig.1. We studied the process of third (THG) $2\omega+2\omega-\omega$ (b), fifth- $2\omega+2\omega+\omega$ (d), sixth- $2\omega+2\omega+2\omega$ (f) and seventh- $2\omega+2\omega+\omega+\omega+\omega$ (g) harmonic generation (ω and 2ω are the fundamental and second-harmonics frequencies, respectively). We also studied nonresonant processes of third- $\omega+\omega+\omega$ (a), fourth- $2\omega+\omega+\omega$ (c) and fifth- $\omega+\omega+\omega+\omega+\omega$ (e) harmonics generation by fifth-order processes.

The main experimental results on 3-6-harmonics generation of YAlO :Nd^{3+} laser ($\tau \simeq 10$ nsec, $P_{\omega} \simeq 2\cdot10^{6}$ W, $P_{2\omega} \simeq 3\cdot10^{5}$ W) in Ba atoms as well as the experimental set-up and the

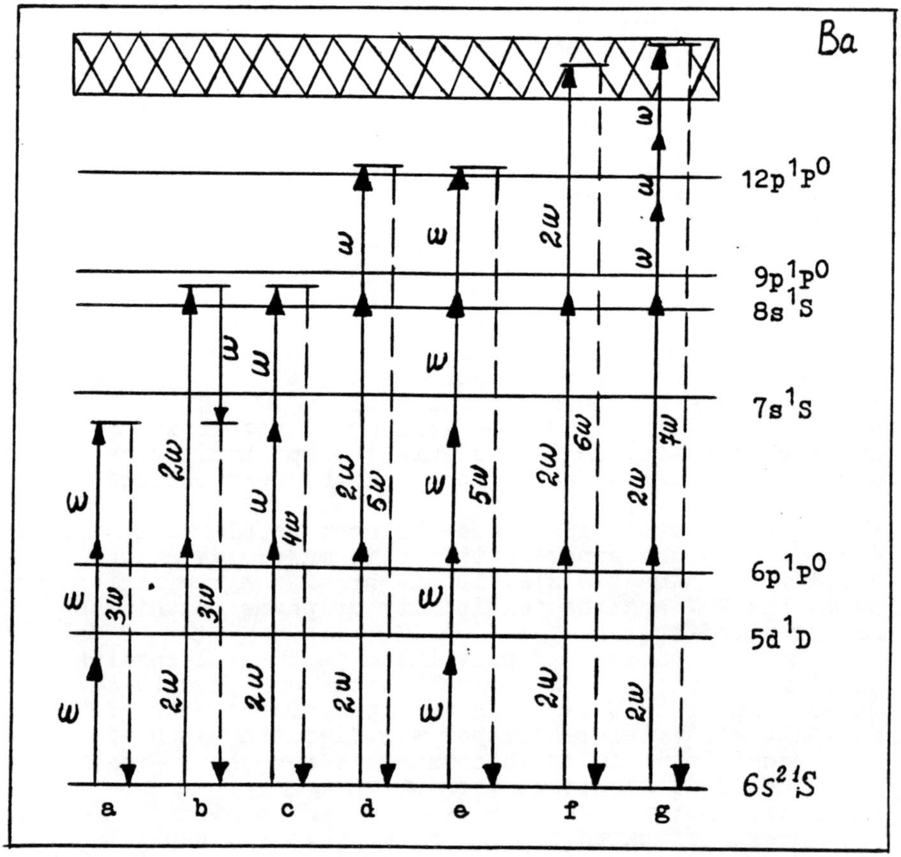

Fig.1. Schematics of Ba atomic energy levels involved in the conversion process

the design of the atomic vapor cell were described in detail in our previous papers [7-9].

Here we state that comparison of the experimental values of CE η_{exp} and nonlinear susceptibilities $\chi^{(n)}_{exp}$ for quasi-resonant (third- $2\omega+2\omega-\omega$, fifth- $2\omega+2\omega+\omega$ and sixth- $2\omega+2\omega+2\omega$ harmonics generation) and nonresonant (third- $\omega+\omega+\omega$, fourth- $2\omega+\omega+\omega$ harmonics generation) processes pumped by low power nanosecond pulses shows that in the former case the η_{exp} and $\chi^{(n)}_{exp}$ values under equal pumps are more than an order of magnitude higher than the nonresonant ones. This is associated with the two-photon transition giving the main contribution to the quasi-resonant conversion process. However the attained conversion efficiency did not exceed $\sim 5\cdot10^{-5}$, e.g. for the sixth-harmonic generation $2\omega+2\omega+2\omega$. Under the second-harmonic pump $P_{2\omega} \simeq 3\cdot10^5$ W the absolute value of the sixth-harmonic ($\lambda \simeq 179{,}9$ nm) power was only some watts.

Since in the third-order nonlinear processes the conversion efficiency η is proportional to the square of the pump power, in order to improve the conversion efficiency (as well as the absolute values of the output power P_{gen}) in parametric processes it is more advantageous to use as a pump high-power phase-matched picosecond laser pulses. Therefore in this paper we shall consider frequency conversion of picosecond YAlO :Nd^{3+} laser pulses in Ba vapor in sixth- and seventh-harmonics VUV generation. The pump beam ($\omega, 2\omega$) was focused by a quartz lens $f \simeq 27{,}5$ cm into a cell containing a mixture of Ba and He, vacuum-sealed to a vacuum monochromator. The output was detected by a solar-blind photomultiplier.

Concentration (temperature) dependences of the generated sixth- and seventh-harmonics power in Ba vapor have been measured, typical of the processes under tight pump focusing. The curves appear to have one maximum at different Ba concentrations, depending on the process studied. Therefore Fig.2 illustrates only the temperature profile of the sixth-harmonic power, for the seventh-harmonic power displays a similar one. As seen from Fig.2, maximum of η for the sixth-harmonic generation is achieved at $T \simeq 920°C$ which corresponds to $N \simeq 5\cdot10^{16}$ cm^{-3}. The maximum value of η_{max} is 0,7% which corresponds to the VUV ($\lambda \simeq 179{,}9$ nm) power $P_{6\omega} \simeq 2\cdot10^6$ W. We note that this value of η is one of the highest experimental values obtained so far.

For the seventh-harmonic ($\lambda \simeq 154{,}2$ nm) generation by fifth-order processes ($2\omega+2\omega+\omega+\omega+\omega$) the maximum of η , compared to the second-harmonic power, was $\sim$ 0,03%, corresponding to a power of $P_{7\omega} \simeq 8{,}3\cdot10^4$ W.

The η estimates for the sixth-harmonic generation under picosecond pulsed pump show that the experimental value $\sim$ 0,7% is much less than the theoretical one, the major limiting effect being the high-frequency Stark effect. Under typical pump intensities in picosecond pulsed regimes the Stark shift of the two-photon transition is considerably larger than the resonance detuning. This results in that in nonresonant conditions the bulk of the pulse is converted

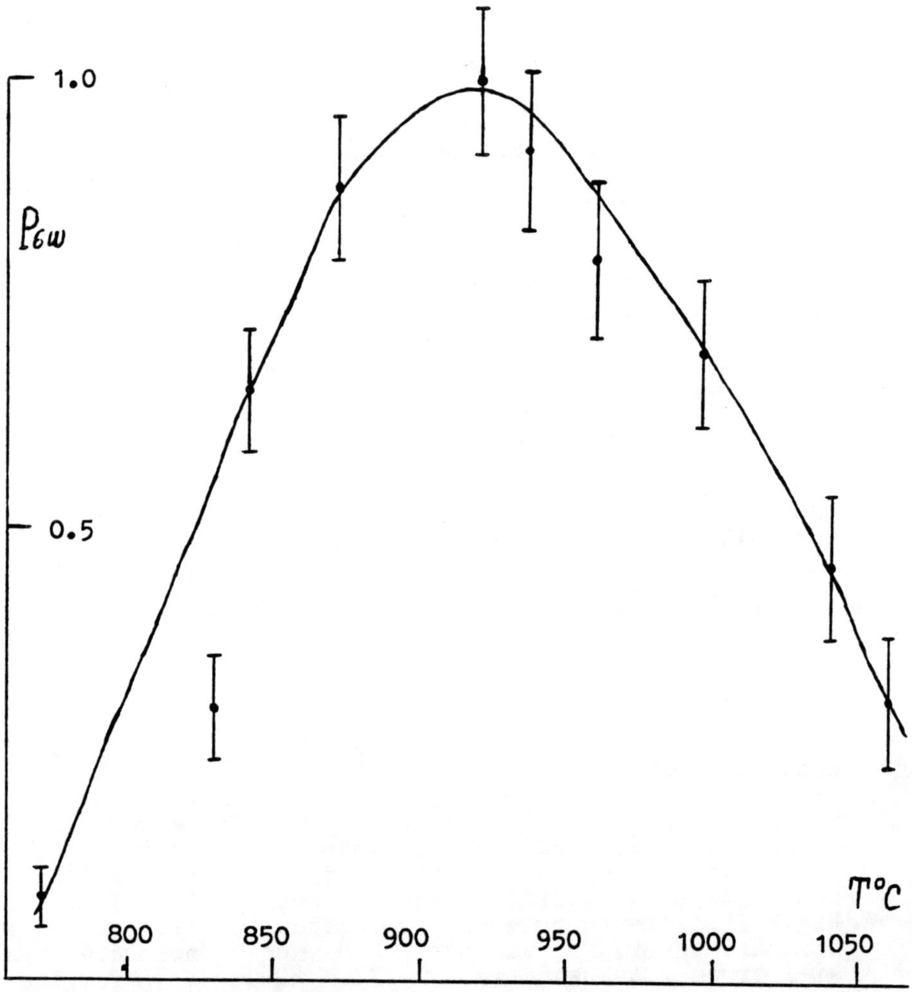

Fig.2. The output power at the sixth-harmonic frequency ($2\omega + 2\omega + 2\omega$) versus the cell temperature (arbitrary units)

as a function of the time and transverse coordinate, and the generated power (or η) linearly grows with the pump instead of the third-order one.

Besides we should note that in the process under discussion ($2\omega+2\omega+2\omega=6\omega$ and $2\omega+2\omega+\omega+\omega+\omega=7\omega$) the dispersion is positive ($\Delta k>0$) and under tight focusing into a homogeneous medium such processes are forbidden [3,10]. It is obvious that the violation of this ban is associated with the medium inhomogeneity. In the interference experiments the homogeneous part of the Ba vapor column was measured to be $L\simeq 7{,}2$ cm which is commensurable with the confocal parameter value $b\simeq 1{,}7$ cm and the medium is not homogeneous [7,10].

Thus, in the situation under consideration where $\Delta k>0$, the tight focusing approximation is not adequate and effective generation is feasible.

Acknowledgements. The authors are thankful to S.V.Karpov for technical assistance with the picosecond experiments.

REFERENCES

1 V.G.Arkhipkin, A.K.Popov, Nonlinear Light Conversion in Gases, Nauka, Novosibirsk (1987) 147 p.

2 S.A.Bakhramov, G.Kh.Tartakovsky, P.K.Khabibullaev, Nonlinear Resonant Processes and Frequency Conversion in Gases, FAN, Tashkent (1981) 160 p.

3 J.Reintjes, Nonlinear Optical Parametric Processes in Liquids and Gases, Mir, Moscow (1987) 512 p.

4 S.O.Sapondzhyan, D.G.Sarkisian, G.A.Torosian, Kvantovaya Elektronika 12, (1985) p.963.

5 A.A.Isaev, G.Yu.Lemmerman, S.V.Markova, G.G.Petrash, Trudy FIAN 181, (1987) p.3.

6 R.Stringat, G.Fabre, and M.Tabel, J.Phys.B:At.Mol. Opt.Phys. 21 (1988) p.4107.

7 A.N.Zhdanov, M.K.Kodirov, V.F.Lukinykh, S.A.Myslivets, A.K.Popov, V.V.Slabko, Parametric Frequency-Mixing Processes of $YAlO:Nd^{3+}$ laser radiation in Ba Vapor. Preprint N 584 F, Institute of Physics, Krasnoyarsk (1989) 39 p.

8 M.K.Kodirov, S.A.Myslivets, A.K.Popov, V.V.Slabko, In Proc. 6-th Conf. "Optics of Lasers", Leningrad (1990), p.304.

9 A.N.Zhdanov, M.K.Kodirov, A.K.Popov, V.V.Slabko, Optika i Spektroskopiya, V.67 (1989) p.197.

10 A.Lago, C.Hilber, and R.Wallenstein, Phys.Rev.A, V.36 (1987) p.3827.

THE RELATIVISTIC DYNAMICS IN PLASMA OF PICOSECOND EXIMER LASER PULSES' FOCAL SPOTS AS OBJECTS FOR X-RADIATION AMPLIFICATION

A.B. Borisov†, A.V. Borovskiy*, V.V. Korobkin*, C.K. Rhodes‡, and O.B. Shiryaev†

†Research Computer Center, Moscow State University, Moscow 119899, USSR
**Coherent and Non-linear Optics Department, The General Physics Institute of the Academy of Science, Moscow 117942, Vavilov Street, 38, USSR*
‡Department of Physics, The University of Illinois at Chicago, Box 4348, Chicago, Illinois 60680, USA

ABSTRACT

The relativistic propagation is studied for intense subpicosecond ultraviolet pulses in homogeneous plasmas and plasma channels which arise from the non-uniform ionization caused by the leading edge of the pulses. Three different modes of such propagation are found. It is established that the defocusing on bell-shaped electronic density profiles can stabilize pulses against radial oscillations in homogeneous plasmas. The dependence of the quasi-stabilized, the pulsing waveguide and the single focuse regimes on the initial transverse amplitude distributions is considered. ,The locus of the first focus for pulses in homogeneous plasmas is calculated.

1. THE BASIC EQUATIONS

The decreasing of the electronic concentration in focal spots formed in the process of propagation of intense short laser pulses in plasma makes it possible to realize amplification of X-radiation due to the suppression of the distributed losses in them. This work presents a theoretical investigation into one of the non-inertial mechanisms of forming focal spots by powerful ($I_m \cong 10^{18}$-10^{21} W/cm^2) supershort ($\tau < 10^{-12}$ s) eximer laser pulses (λ=0.248μm, $r_0 \cong 1$-3μm in fully stripped plasma - the relativistic self-focusing. A part of these results is published in [1,2].

A consecutive multi-photonic ionization of 6-10 atomic electrons occurs during $\Delta t \lesssim 10^{-15}$s (i.e. almost instantly) on the front of a Gaussian or a hyper-Gaussian beam propagating through a gas with the parameters given above at the intensities greater than $I^{*}\cong 10^{16}W/cm^2$ [3]. Thus a bell-shaped plasma channel is formed and the central temporal part of the pulse is transferred along it. The spatial and temporal dynamics of a laser pulse propagation into a plasma channel is determined by the competitive diffraction, defocusing due to refraction on bell-shaped electronic profiles, and self-focusing processes. Defocusing prevails while the plasma channel aperture is falling down to that of the beam.

The heat and strictional ion-displacement self-focusing do not determine the process of the propagation of powerful ultrashort laser pulses with the studied parameters into a gas because those self-focusing mechanisms are inertial [1]. If a laser pulse ionizes a gas fully, the non-inertial Kerr self-focusing [4] is not significant either. In this case the relativistic increase in the electron's mass due to its oscillations in an intense light field [5] and the charge-displacement in which the expulsion of free electrons from the spatial domain occupied by the powerful field by the ponderomotive force [6,7] (while the ions do not move) results can be responsible for the self-focusing of laser pulses.

Considering the self-focusing of the central temporal part of a laser pulse in a plasma channel, we neglect the energy losses through further ionization which in particular causes the broadening of the plasma channel.

Under these conditions the spatial and temporal dynamics of a laser pulse in a plasma channel, including diffraction, defocusing, relativistic and charge-displacement self-focusing, is governed by the non-linear Shrødinger equation [2,6]

$$\left(\frac{1}{v_g}\cdot\frac{\partial}{\partial t}+\frac{\partial}{\partial z}\right)\cdot E+\frac{i}{2k_0}\Delta_\perp E+\frac{ik_0}{2\varepsilon_{RL_0}}\cdot\delta\varepsilon_R\cdot E+\frac{k_0}{2\varepsilon_{RL_0}}\varepsilon_I\cdot E=0, \qquad (1)$$

where $E(t,z,r)$ is the slowly alternating complex amplitude of the light wave, t is the time, z is the coordinate along the axis of propagation, r is the transverse coordinate, $\Delta_\perp=\frac{1}{r}\cdot\frac{\partial}{\partial r}\cdot r\cdot\frac{\partial}{\partial r}$ is the transverse Laplacian, $k_0=k(\omega_0)=\frac{2\pi}{\lambda}$ - the wavenumber, $k(\omega)=(\omega/C)(\varepsilon_{RL}(\omega))^{1/2}$, $v_g=\left(\frac{\partial k}{\partial\omega}(\omega_0)\right)^{-1}$ - the group velocity of the laser pulse in plasma in the linear approximation, $\varepsilon_{RL}(\omega)$-the real linear part of the dielectric swelling of plasma for the frequency ω. $\varepsilon_{RL_0}=\varepsilon_{RL}(\omega_0)$, ω_0 being the central frequency of the pulse.

It is supposed that the light wave is circularly polarized, i.e. the electric field $\mathbf{E}(t,z,r)$ and the slowly alternating complex amplitude $E(t,z,r)$ in (1) are related as follows

$$\mathbf{E}(t,z,r)=(1/2)\left[(\mathbf{e}_x+i\mathbf{e}_y)E(t,z,r)exp[i(\omega_0t-k_0z)]+\text{c.c.}\right].$$

It should be noted that in the case of the circular polarization of the light wave the non-linear term in the studied equation (1) has no high-frequency dependence.

The real nonlinear term $\delta\varepsilon_R(r,\omega,|E|^2)$ describing relativistic, charge-displacement self-focusing and defocusing of the pulse is expressed as

$$\delta\varepsilon_R=\varepsilon_R-\varepsilon_{RL_0},\quad \varepsilon_R=1-\omega_p^2/(\omega_0^2+\nu^2),\quad \varepsilon_{RL_0}=1-\omega_{p_0}^2/(\omega_0^2+\nu_0^2),$$

$$\omega_p^2=4\pi e^2 N_e[I](r)/m_e(I),\quad \omega_{p_0}^2=4\pi e^2 N_{e,0}/m_{e,0},$$

$$I=(C/4\pi)|E|^2.$$

Here ω_p is the electronic plasma frequency, ν-the frequency of collisions between electrons and ions, $m_{e,0}$-the electron's rest mass, and $m_e(I)$ is its relativistic mass, depending on the intensity as [5]

$$m_e(I)=m_{e,0}(1+3I/I_r)^{1/2}.$$

In the last formula

$$I_r=3m_{e,0}^2\omega_0^2C^3/(4\pi e^2)$$

is the parameter called the relativistic intensity [5].

$N_e[I](r)$ is the concentration of the electronic component due to the quasistatic balance of the ponderomotive and the electrostatic forces. It is assumed that $N_i(r,z)=N_{i,0}\cdot f(r,z)$ is the static concentration of ions, where $N_{i,0}\equiv const$ is the maximal ion concentration value, and $f(r,z)$ gives the bell-shaped spatial profile of the concentration of ions, and so $f(r,z)\geq 0$, $\max_{r,z} f(r)=1$ ($f(r,z)\equiv 1$ in the case of homogeneous plasma). Then $N_{e,0}\cdot f(r,z)$ is the initial electronic concentration in the case of quasineutrality, where $N_{e,0}=N_{i,0}\cdot Z$, Z being the ion charge number. The relativistic expression for $N_e[I](r)$, describing the perturbation of local neutrality of plasma by the ponderomotive force, is as follows [2]

$$N_e[I](r)=N_{e,0}\max\{0,f(r,z)+(C^2/\omega_{p_0}^2)\Delta_\perp(1+3I/I_r)^{1/2}]\}.$$

The electronic pressure is small in the case of the beam and plasma parameters studied in this paper and this fact is used in the derivation of the formula for $N_e[I](r)$. The expression $\max\{0,\cdot\}$ is to provide for $N_e[I](r)\geq 0$.

The collisional absorption of light in plasma is given by $\varepsilon_I=-(\omega_p^2/\omega_0^2)\nu=-\mu^-/k_0$, where μ^- is the absorption coefficient. The estimation of μ^- expressed in [5] shows that in the case of the prescribed parameters the collisional absorption is extremely small and does not influence the solution we obtain numerically. Henceforward we assume that $\mu^-=0$.

If $\mu^-=0$, then $\nu=\nu_0=0$, $\varepsilon_{RL}(\omega)=1-\omega_{p_0}^2/\omega^2$ and $v_g=C\cdot\varepsilon_{RL_0}^{1/2}$.

Introducing the frame travelling with the pulse: (t,z,r) (s,z,r), where $s=t-z/v_g$ plays the role of the problem parameter and normalizing the values in (1) as follows

$$r_1=r/r_0,\ z_1=z/(2k_0r_0^2),\ u(r_1,z_1)=E_0^{-1}\cdot E(r,z),$$

assuming E_0 to be the maximal value of $|E(r,0)|$ corresponding to the current value of $t(s)$, we get the problem ($f_1(r_1)=f(r)$, henceforward the subscript "1" is omitted for r_1, z_1)

$$\frac{\partial u}{\partial z} + i\Delta_\perp u + iF[f_1(r),|u|^2]u = 0,\ z > 0, \tag{2}$$

$$u(r,0) = u_0(r),\ \max_r|u_0| = 1, \tag{3}$$

$$\frac{\partial u}{\partial r}(0,z) = 0,\ u(\infty,z) = 0. \tag{4}$$

The non-linear term F describing the relativistic and charge-displacement self-focusing is a real operator [2]

$$F[f_1,\xi]=a_1\left[1-(1+a_2\xi)^{-1/2}\max\left\{0,f_1(r)+a_1^{-1}\Delta_\perp(1+a_2\xi)\right\}\right]. \tag{5}$$

The parameters a_1, a_2 make the π-complex of the problem (2)-(4)

$$a_1=((k_0r_0)^2/\varepsilon_{RL_0})\cdot(\omega_{p_0}/\omega_0)^2,\ a_2=3I_0/I_r,\ I_0=(C/4\pi)|E_0|^2. \tag{6}$$

This work reports on the modes of the relativistic propagation of laser pulses in homogeneous plasmas and in plasma channels. The non-linear term in (2) describing the relativistic self-focusing is

$$F(r,\xi)=a_1\left[1-f_1(r)(1+a_2\xi)^{-1/2}\right]. \tag{7}$$

The relativistic self-focusing is the prevailing mechanism determining the features of the non-linear propagation of laser pulses in plasma channels in the case

$$a_1>>1,\ a_2\approx1. \tag{8}$$

Exactly this case is studied in the present work (see (18)).

The investigation of the temporal dynamics of the relativistic self-focusing of the pulse is reduced to solving a number of problems (2)-(4),(7) for a sequence of the parameter values $a_2=3\cdot I_0(t)/I_r$, where $I_0(t)$ is the entrance pulse intensity on the axis of the beam.

The problem (2)-(4),(7) has at least two first integrals

$$I_1=\int_0^\infty |u(r)|^2 r dr, \tag{9}$$

$$I_2=\int_0^\infty \left\{|\frac{\partial u}{\partial r}|^2-\Phi(r,|u|^2)\right\} r dr, \tag{10}$$

its solution depending essentially on their values [8]. Here

$$\Phi(r,\xi)=\int_0^\xi F(r,\eta)d\eta=a_1\left\{\xi-(2/a_2)f_1(r)[(1+a_2\xi)^{1/2}-1]\right\}. \tag{11}$$

We simulate the self-focusing dynamics of laser pulses for the spatial and temporal Gaussian or hyper-Gaussian initial signals

$$I\Big|_{z=0}=I_0(r,t)=I_m exp\left\{-(t/\tau)^{N_1}-(r/r_0)^{N_2}\right\},\ N_1\geq 2,\ N_2\geq 2 \tag{12}$$

for $I_m \approx I_r \gg I^* = 10^{16} W/cm^2$, r and t being dimensional.

Besides, we concentrate on the central temporal part of the pulse, $|t|\leq t_0$, where t_0 satisfies the

$$I_0(t_0)=I_0(0,t_0)=I_m exp\left\{-(t_0/\tau)^{N_1}\right\} \gg I^*. \tag{13}$$

The initial amplitude distribution of the entrance signal defined by (12) for a flat initial wave front is expressed as

$$u_0(r)=exp\left(-r^{N_2}/2\right),\ N_2\geq 2. \tag{14}$$

The transverse profile of the plasma channel created by the front of the pulse is simulated by a hyper-Gaussian function

$$f(r)=exp\left(-(r/r_*)^{N_3}\right),\ N_3\geq 2, \tag{15}$$

r being the dimensional value. The aperture of the plasma channel r_* is defined by

$$I_0(r_*,t_0)\equiv I_0(t_0)exp\left(-(r_*/r_0)^{N_2}\right)=I^*. \tag{16}$$

In the case of a Gaussian transverse intensity distribution $N_2=2$, the aperture of a plasma channel at $I^{*}=10^{16}W/cm^{2}$, $I_0(t_0)=0.1\cdot I_r$, $I_r=1.34\cdot 10^{20}W/cm^{2}$, makes $r_{*}\cong 2.68r_0$. Then the homogeneous plasma approximation $f(r)\equiv 1$ is valid.

2. THE RELATIVISTIC SELF-FOCUSING OF LASER PULSES IN HOMOGENEOUS PLASMA

Fig. 1 shows the relativistic propagation of a wave front corresponding to the value of t, for which $I_0(t)=(2/9)\cdot I_r$, into a homogeneous plasma along the Oz-axis for the Gaussian initial transverse intensity distribution $N_2=2$ in (14), the flat initial wave front and the following beam and plasma parameters

$$\lambda=0.248\mu m,\ I_r=1.34\cdot 10^{20}W/cm^{2},\ I_0=(2/9)I_r\cong 2.98\cdot 10^{19}W/cm^{2},$$

$$r_0=3\mu m,\ N_0=7.5\cdot 10^{19},\ Z=10. \qquad (17)$$

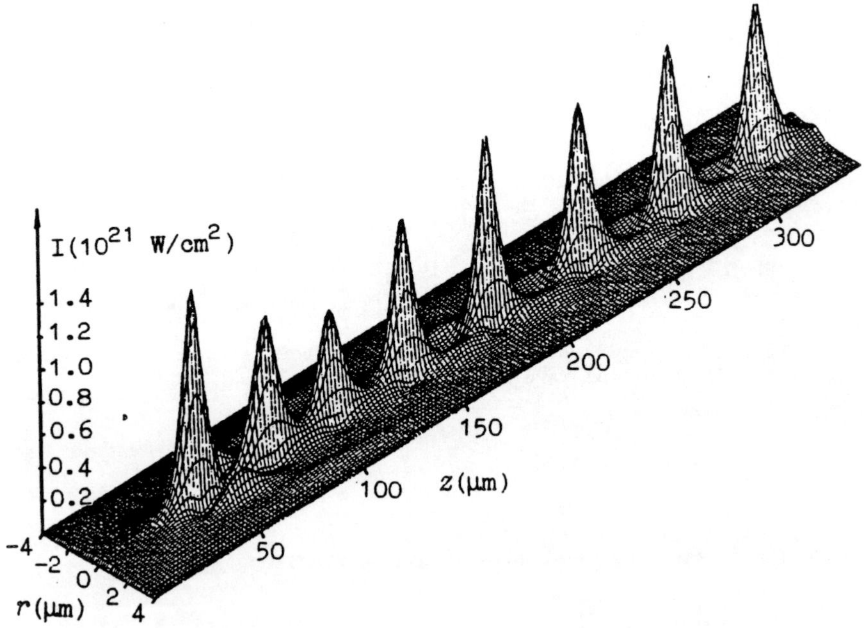

Fig. 1. The intensity distribution $I(r,z)=I_0\cdot|u(r/r_0,z)|^2$ in the case of $u(r_1,0)=exp(-0.5r_1^{2})$, $f(r)\equiv 1$.

For the dimensionless parameters of the problem (2), (3) we have

$$a_1 \cong 2.486 \cdot 10^2, \quad a_2 = 2/3. \tag{18}$$

Fig.2 presents the analogous graph for the plateau-like initial transverse intensity distribution with $N_2=8$ in (14). One may draw a conclusion that the solution of the problem depends essentially on the initial transverse intensity distribution.

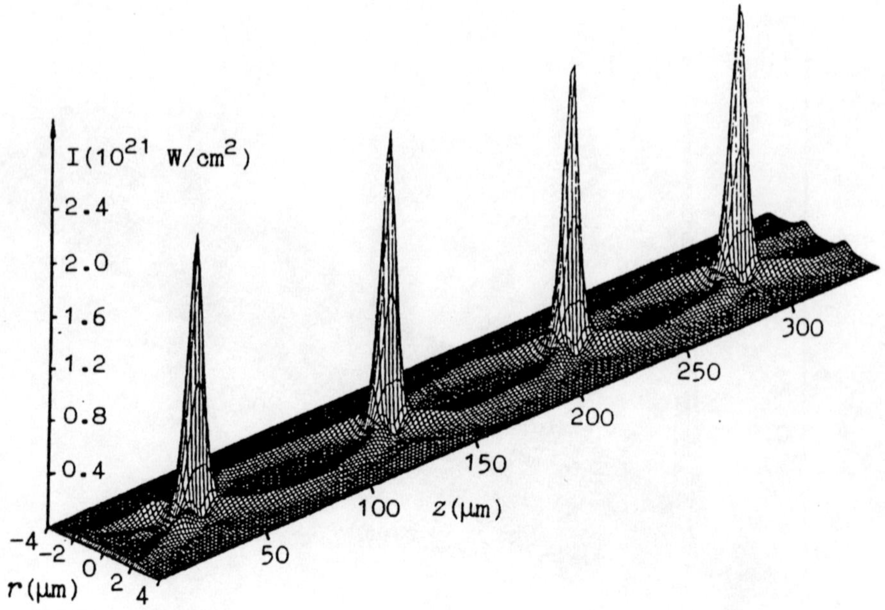

Fig.2. The intensity distribution $I(r,z)=I_0 \cdot |u(r/r_0,z)|^2$ in the case of $u(r_1,0)=exp(-0.5r_1^{\,8})$, $f(r)\equiv 1$.

Note that for the values of the parameters a_1, a_2 defined by (18) and for the initial condition (14) $I_2<0$ in (10). It can be proved in the same way as it is done in [8] that in homogeneous plasma, $f_1(r)\equiv 1$, for $I_2<0$

$$\max_{r,z} |u(r,z)|^2 > (4/a_1 a_2)|I_2|/I_1 \tag{19}$$

i.e. the extreme beam intensity with respect to r has a positive infimum independent of z. Therefore, the powerful relativistic beam self-focusing in homogeneous unabsorbing ($\bar{\mu}=0$) plasma results in a pulsing waveguide regime in case $I_2<0$. Fig.1,2 illustrate the formation of such a regime.

The pulsing waveguide shown in Fig.2 is an alternation of ring structures and focal spots on the axis of the beam. The power trapped in it makes approximately 90% of the total initial power of the beam, unlike the case of the focusing of a beam with a Gaussian initial transverse intensity distribution, in which about 50% of the total initial power is trapped.

In the case $I_2 \geq 0$ an estimation like (19) cannot be obtained. Following [8] we simulate the case $I_2 \geq 0$ by considering a beam initially focused (or defocused) by a lens at the entrance of the medium. Let the focusing length of the lens be $R = k_0 r_0^2 R_f$. Then the initial condition is as follows

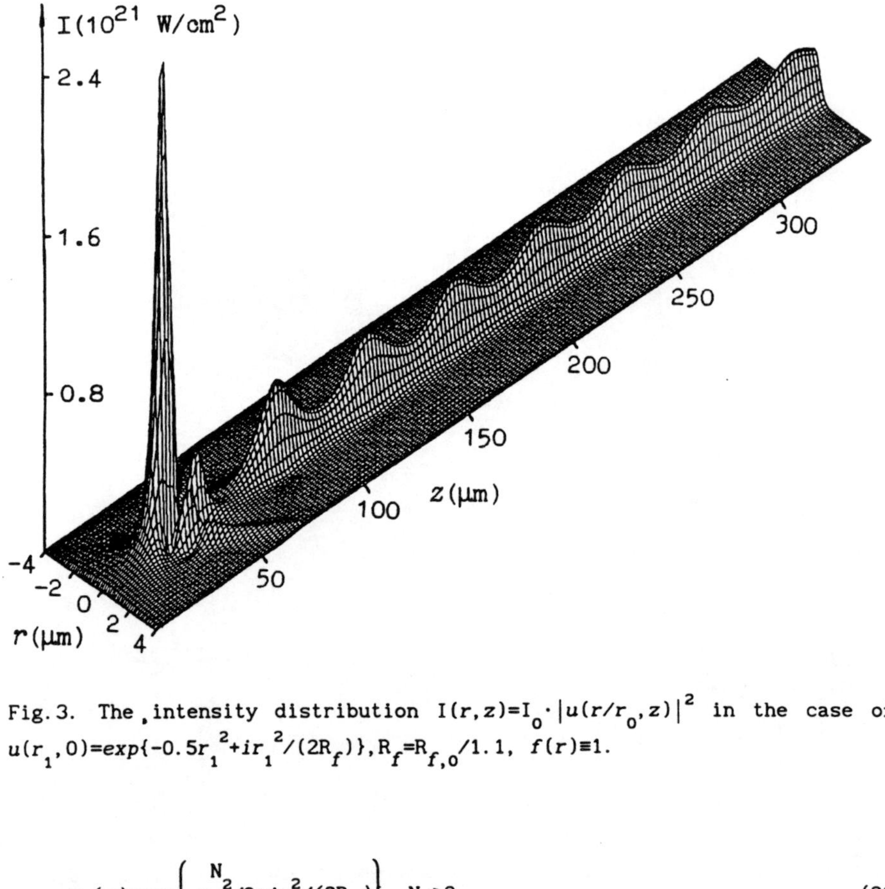

Fig. 3. The intensity distribution $I(r,z)=I_0 \cdot |u(r/r_0,z)|^2$ in the case of $u(r_1,0)=exp\{-0.5r_1^2+ir_1^2/(2R_f)\}, R_f=R_{f,0}/1.1,\ f(r)\equiv 1$.

$$u_0(r)=exp\left(-r^{N_2}/2+ir^2/(2R_f)\right),\ N_2\geq 2. \tag{20}$$

The case $R_f>0$ corresponds to the focusing lens, and the case $R_f<0$ - to the defocusing one.

Present $u(r,z)$ as $\rho \cdot exp(iS)$ and consider the case in which $I_2 \equiv I_2^0 < 0$ when $S(r,0)=0$. Then, for $u_0(r)$ defined by (20) $I_2 \geq 0$ for

$$R_f^2 \leq \left(\int_0^\infty \rho_0^2 r^3 dr \right) / \int_0^\infty \left\{ \Phi(\rho_0^2) - \left(\frac{\partial \rho_0}{\partial r} \right)^2 \right\} \cdot r dr = \left(\int_0^\infty \rho_0^2 r^3 dr \right) / |I_2^0| . \qquad (21)$$

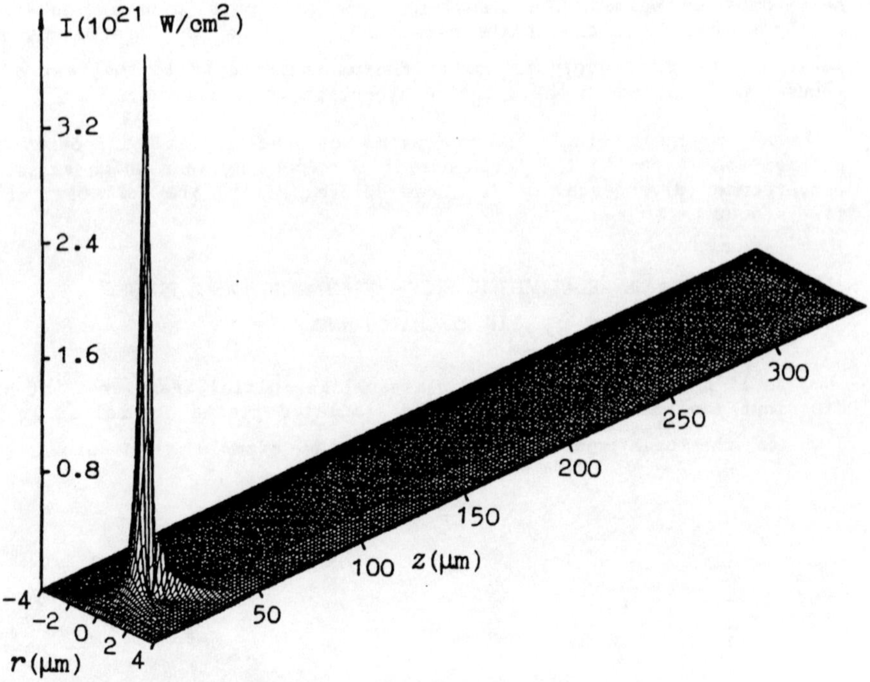

Fig. 4. The intensity distribution $I(r,z)=I_0 \cdot |u(r/r_0,z)|^2$ in the case of $u(r_1,0)=exp\{-0.5r_1^2+ir_1^2/(2R_f)\}, R_f=R_{f,0}/2, f(r)\equiv 1$.

In the case of a Gaussian initial transverse intensity distribution $N_2=2$ the estimation (22) leads to $|R_f| \leq R_{f,0}=(2 \cdot |I_2^0|)^{-1/2}$.

If $|R_f|=R_{f,0}$, $I_2=0$. The case $I_2>0$ corresponds to a higher degree of the beams initial focusing (or defocusing): $0<|R_f|<R_{f,0}$. For $R_{f,0}<|R_f| \leq \infty$ I_2 is negative.

Fig. 1, 3, 4 show the intensity distributions for the relativistic self-focusing in homogeneous plasma of initially focused beams with Gaussian initial transverse intensity distributions, the parameters of the problem being defined by (17), (18), and the values of R_f respectively: $R_f=+\infty$, $R_{f,0}/1.1$, $R_{f,0}/2$. In these figures one may see the dynamics of the transition from the "pulsing waveguide" regime ($R_f=\infty$, Fig. 1) to the regime with a single powerful focus on the axis of the beam ($R_f=R_{f,0}/2$, Fig.4).

The intensity distribution corresponding to $R_f=R_{f,0}/1.1$, $I_2>0$, presented in Fig.3 shows that $I_2=0$ is not the threshold separating the two self-focusing regimes. The transition from the "pulsing waveguide" regime to the single focus one in the case $R_f>0$ occurs as the first focus gains power and is shifted closer to the medium entrance while the rest of the focuses are shifted in the opposite direction and diffused.

The powerful single focus regime of the initially focused beam propagation is due to the influence of the focusing lens which raises the convergence (divergence) of the beam before (after) the formation of the first focus.

3. THE RELATIVISTIC SELF-FOCUSING OF LASER PULSES IN PLASMA CHANNELS

As it follows from (16) for plateau-like initial transverse intensity distributions the aperture r_* of the simulated plasma channel is falling down to the beam aperture r_0. So, in the example given above, when

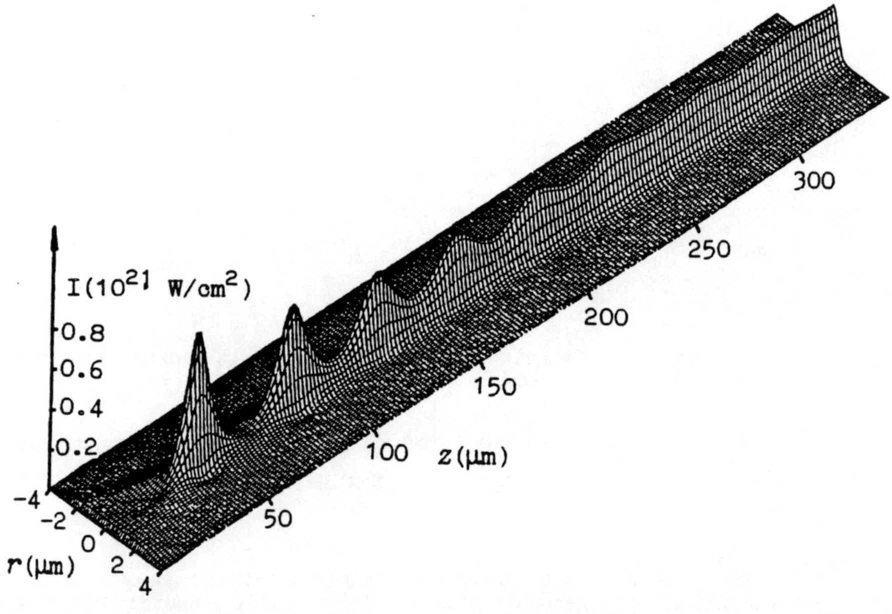

Fig.5 The intensity distribution $I(r,z)=I_0\cdot|u(r/r_0,z)|^2$ in the case of $u(r_1,0)=exp(-0.5r_1^{\ 8})$, $f(r)=exp\{-(r/r_0)^8\}$.

$I_0(t_0)=0.1\cdot I_r$, $I_r=1.34\cdot 10^{20}$W/cm^2, $I^*=10^{16}$W/cm^2 for $N_2=8$ from (16) we have $r_* \cong 1.28r_0$. Thus the beam defocusing which is significant because of the coincidence of the laser beam and the plasma channel apertures should be taken into account, when the evolution of a beam with a plateau-like initial transverse intensity distribution is studied.

Fig. 5 shows the propagation of a wave front into a plasma channel along the *Oz* axis for $N_2=8$ and the values of the parameters given by (17), (18). Here the transverse profile of the plasma channel is simulated by $f(r)$ defined by (15) where $N_3=8$ and $r_*=r_0$. The comparison between Fig. 2,5 establishes that the defocusing of a hyper-Gaussian beam in a plasma channel with an aperture close to that of the beam influences the spatial dynamics of the latter essentially. Defocusing makes a part of the beam spread away from the channel. The rest of it aims at balancing the relativistic self-focusing, defocusing and diffraction. The power trapped in this quasi-stabilized waveguide makes approximately 25% of the total initial power of the beam.

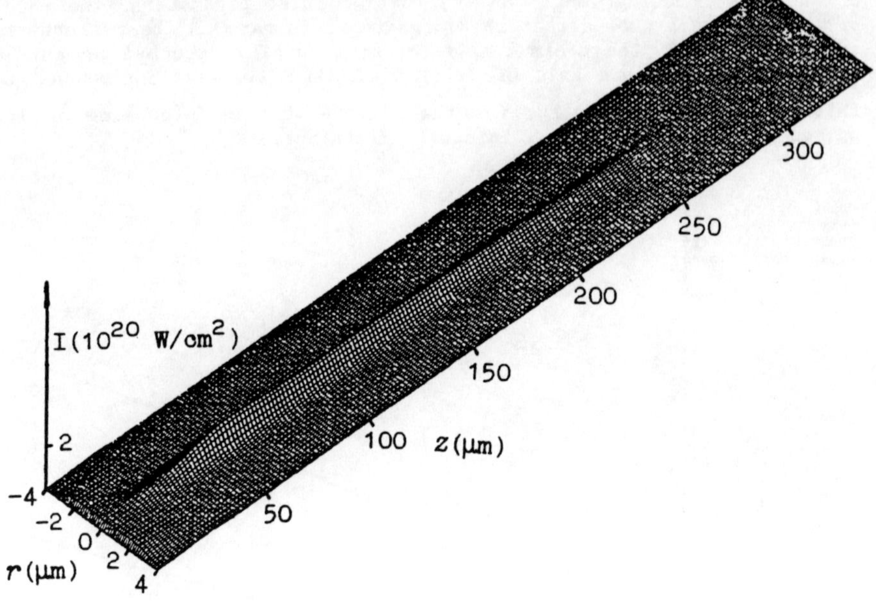

Fig. 6. The intensity distribution $I(r,z)=I_0\cdot|u(r/r_0,z)|^2$ in the case of $u(r_1,0)=exp(-0.5r_1^{\ 8})$, $f(r)=exp\{-(r/r_0)^4\}$.

The spatial dynamics of a beam varies as the defocusing looses or gains strength. The formation of the transverse intensity profile, quasi-stabilized in respect of z gets significantly slower as defocusing decreases.

Fig. 6 presents a graph analogous to those presented in Fig 7,8 for

N_3=4 and $r_*=r_0$ in (15). The amplification of the defocusing causes a quick formation of a transverse intensity profile, quasistabilized in respect of z. No focal spots arise during this process.

It should be noted that the results, presented in Fig. 1-6 in dimensional units for the beam and plasma parameters given by (17), are also valid (by rescaling r, z, I) for any other parameter values, for which a_1, a_2 are the same as in (18).

4. THE SPATIAL AND TEMPORAL DYNAMICS OF THE FIRST FOCUS FOR RELATIVISTIC PULSES WITH GAUSSIAN INITIAL TRANSVERSE INTENSITY DISTRIBUTIONS IN HOMOGENEOUS PLASMA

Following earlier studies [9,10] we have analyzed the evolution of that waves having small perturbations in order to determine the dependence of the self-focusing length on the initial pulse parameters. The results of this analysis showed that the largest growth rate of the perturbations and, consequently, the minimum self-focusing length, occur at an entrance intensity on the beam axis of $I_o(t) = (2/3)I_r$. Computations showed that this inference remains true for the relativistic self-focusing of laser pulses with Gaussian initial intensity distribution.

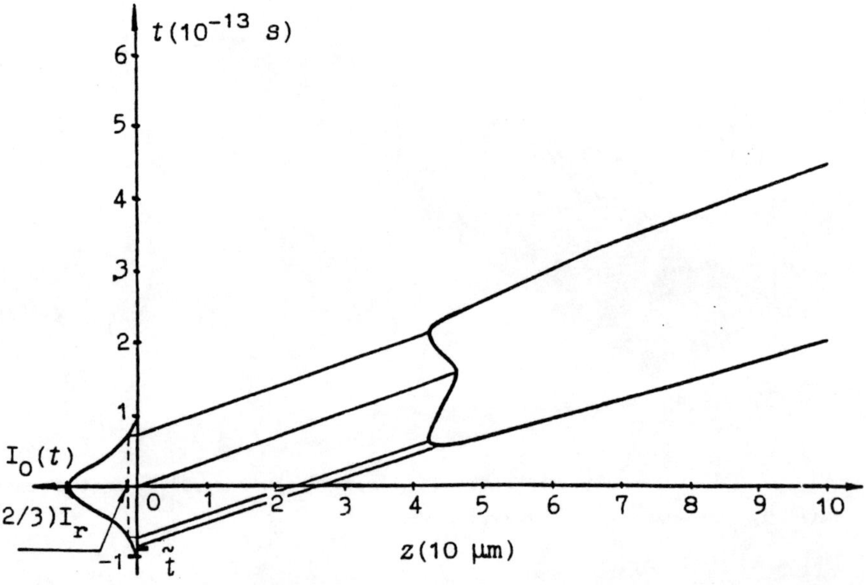

Fig. 7. The (t,z) tracing of the first focus for $I|_{z=0}=I_0(r,z)=I_m exp\{-(t/\tau)^2-(r/r_0)^2\}$, $f(r)\equiv1$.

Fig. 7 shows the locus of the first focus in the (t,z) plane for the entrance signal determined by (12) with $N_1=2$, $N_2=2$, $I_m=6 \cdot I_r \cong 8.04 \cdot 10^{20}$ W/cm^2, $\tau=0.5 \cdot 10^{-13}s$, $r_0=3\mu m$. It reaches the minimal z for $I_o(t) = (2/3)I_r$. If the extreme intensity on the beam axis $I_m \gtrsim I_r$, this locus has three reversal points. Two of them, corresponding to the same z, are due to $I_o(t) = (2/3)I_r$. The central one, corresponding to a greater z, occurs at $I_o(t) = I_m$. In principle this feature establishes a clear diagnostic signature for relativistic self-focusing that enables it to be distinguished from the Kerr nonlinearity which produces a locus with a single point of reversal.

5. CONCLUSIONS

The spatial character of propagation of sufficiently intense subpicosecond ultraviolet pulses in self-generated plasmas is influenced by both relativistic and charge-displacement mechanism. The present treatment has revealed at least three different modes of the relativistic propagation. 1. The relativistic process in a homogeneous plasma can produce self-focusing in the form of a pulsing waveguide involving a structure of alternating rings and focal zones. The location, formation, and nature of the focal regions, including the maximal intensities occurring in them, depend essentially on the properties of the initial transverse intensity distribution. The locus of the first focus in the (t,z)-plane for pulses with Gaussian initial intensity distributions and maximum intensities $I_m \gtrsim I_r$ has three reversal points. For this class of pulses the relativistic self-focusing length is minimal, if the intensity of the initial pulse on the axis equals $(2/3)I_r$. 2. The relativistic propagation in homogeneous plasmas of beams, initially focused by a sufficiently strong lens, results in modes with a single extremely powerful focus. 3. Finally, defocusing is particularly significant for the relativistic mechanism, if the plasma channel in gas has a radius close to that of the beam. In this case it can cause the stabilization of the beam.

LITERATURE

1. A.B. Borisov, A.V. Borovskiy, V.V. Korobkin, and O.B. Shiryaev, Preprint N4, Inst. Gen. Phys., Acad. Sci. USSR, Moscow, 1990 (in Russian).
2. A.B. Borisov, A.V. Borovskiy, V.V. Korobkin, A.M. Prokhorov, C.K. Rhodes, and O.B. Shiryaev, Preprint N61, Inst. Gen. Phys., Acad. Sci. USSR, Moscow, 1990.
3. U. Johann, T. Luk, H. Egger, and C. Rhodes, Phys. Rev. Ser. A. 34, (1986) 1084; T.S. Luk, U. Johann, H. Jara, I.A. McIntyre, A. McPherson, A.P. Shwarzenbach, K. Boyer, and C.K. Rhodes, SPIE, 664, 223.
4. V.I. Lugovoy and A.M. Prokhorov, Usp. Fiz. Nauk 111, (1973) 203 [Sov. Phys. - Usp. 16, (1974) 658].
5. H. Chora, Physics of laser-driven plasmas, Wiley, N. Y., (1981).
6. Gou-Zheng Sun, E. Ott, Y.C. Lee, and P. Guzdar, Phys. Fluids 30, (1987) 526.
7. J.C. Solem, T.S. Luk, K. Boyer, and C.K. Rhodes, IEEE J.

Quant. Electron. 25, (1989) 2423.

8. V. I. Zakharov, V. V. Sobolev, and V. S. Synakh, Zh. Eksp. Teor. Fiz. 60, (1971) 136 [Sov. Phys. - JETP 33, (1971) 77].

9. V. I. Bespalov and V. I. Talanov, Pis'ma Zh. Eksp. Teor. Fiz. 8, (1966) 471 [Sov. Phys. - JETP Lett. 3, (1966) 26].

10. T. B. Benjamin and T. B. Feir, Fluid Mech. 27, (1967) 417.

PRE-PLASMA ELECTRODYNAMICS OF AN IONIZED MEDIUM IN A HYPERATOMIC INTENSITY LASER FIELD

M.Yu. Romanovsky and *L. Sabirov
General Physics Institute, Moscow, USSR
**Samarkand University, Samarkand, USSR*

Introduction

Investigations of laser interaction with matter were mostly devoted to the effects when laser field does not exceed the so-called "inner-atomic" field, i.e. the atomic electrical field in the point of the valence electron. Because of this question about the classical direct ionization of atoms or ions in electrical field did not arise, although the processes of quantum ionization have been well studied.

Now the superbright excimer lasers are changing the situation [1,2]. There comes a new type of laser interaction with matter due to the opened channel of classical direct (above threshold) ionization by means of the appearing free electrons. Here we are interested in the collective dynamics of charged particles in such laser fields, partially in the simplest model of it - when electrons are stimulated to oscillate and impact ions. In this model, stimulated amplitudes and velocities of ions are so small that we will not take them into account and assume them to be in rest. Nevertheless ions will be excited and ionized in several ways.

We will be interesting in the problem of the electrons motion and its influence on the laser field propagation.

We believe that the amplitude of oscillation velocity does not reach relativistic values. This means that the laser intensity does not exceed $10^{19} W/cm^2$.

Let us briefly consider the evolution of system. There is a cold gas, and a strong laser pulse with a very short duration introduced into it. When the field strength exceeds that of the atomic electric field for the valence electron or, perhaps, for one of the inner electrons, all of the above electrons become free and begin to oscillate in the laser field. The necessary time for this is of the order of the pulse front duration. These free electrons can collide with ions and cause their secondary ionization (Ne^{8+} is observed in [1]). The oscillation amplitudes must be more then electrons free tracks. Because of these collisions electrons loose energy acquiring accelerations which are antiparallel their velocity vectors. Electrons reradiate and extinguish the laser field. The laser radiation scatters and its spectrum broadens. This causes a perturbation in the oscillation amplitude during the collisions and, may be, an interruption in the phase. The scattering results from the transversal acceleration. New electrons appear as a result of collisions therefore the electron density increases in a definite point of the gas when the laser pulse passes through it. Electron-electron collisions lead to the stochastization of the electron dynamics and determine a conversion of the pre-plasma state into plasma.

We should not use the plasmas formulae of [2] for calculating the cross-sections because they provide abnormally high values for low temperatures. This is a separate problem. We will calculate accelerations of electrons. There are three types of them. Between the collisions there is only a field acceleration. During one collision the above antiparallel and transverse accelerations.

The major of aim of this article is to and introduce a certain physical value which correctly describes this physical system. For plasma it is temperature. For pre-plasma, it is the velocity of "electron" fluid. We will show that this reasoning is relevant for the pre-plasma system.

NONRELATIVISTIC PRE-PLASMA ELECTRODYNAMICS EQUATION FOR ELECTRONS AND IONS IN THE HYPERATOMIC LASER FIELD

1. Basic equations

To investigate dynamics equations for all electrons in laser field is impossible. It is necessary to average these equations and to introduce physical values which correctly describe the model . Temperature can not be this value because the considered system is strongly unequilibrium.

Only two forces act on the electron between collisions. If m - the electron mass, e - the charge, $\bar{a}$ - the acceleration of the electron in the point $\bar{r}$ in a time moment t, **E** is the laser electric field in this point, then

$$m\bar{a} = eE + MF_{\rho} \qquad (1)$$

The first term in the right part is Lorenz force, F_{ρ} - the force, which is created by all charges in the point of this electron. If $E \sim 10^{10}$ V/cm (laser beam intensity $I \sim 10^{18}$ W/cm^2)

the maximum of oscillation velocity **v** is still much less than the light velocity **c**. The amplitude of oscillations without collisions and zero initial velocity of this free electron is $x \sim 2eE/m\omega^2$ ($\omega=2\pi c/\lambda$, λ - laser wavelength) and if λ=1060 nm $x \sim 5*10^{-6}$cm. The average distance of the electron free pass is $l \sim n^{-1}*\sigma$, n is ions density, σ is ~collisions cross-section and for $n \sim 10^{19}$cm^{-3} and $\sigma>10^{-15}$ $l<x$, i.e. collisions exist, but $x<\lambda$ and the laser beam radius r_o is such that, at least, $r_o>\lambda$. Therefore

$$(E/\nabla_{\perp}E)>>x;\ \nabla_{\perp}=\{\partial/\partial x,\partial/\partial y\}$$

and it can be expected in the first approximation that oscillations occur under constant laser field amplitude. Let us derive the expression for the number of electrons-ions stimulated collisions in a such laser field.

2. The equation for the stimulated collisions frequency in a laser field

We assume that the laser radiation polarization is linear and laser beam propagates along z - axis. Oscillations occur under the constant field amplitude. Let us introduce the laser pulse envelope $f(t-z/c)$, the field amplitude E_m is a function of z in general case. The equation for the electron motion between collisions is:

$$\frac{d^2x}{dt^2} = \frac{eE_m(z)}{m}\cos(\omega t-kz)*f(t-z/c) \qquad (2)$$

Here $k=2\pi/\lambda$. Initial conditions with time $t=t_k$ - the time of the previous collision with number k are $x=x_k$ and $v=v_k$. What is occurring under a collision? There occurs an instantaneous loss of the electron velocity with a parallel laser field vector. Transverse changes of the electrons velocity due to the attraction of electrons by ions also take place. The lost parts of velocities can have various values which, of course, are not more than velocities of electrons at the time of collisions. This means that the electrons motion becomes partially stochastic.

Let us choose the time moment t_o and coordinate z such that the electron velocity is equal to zero (Fig.1). If $f(t-z/c)$ is comparatively smooth ($\partial f/\partial t<<\omega$), we have (not on the pulse front):

$$v \simeq \frac{eE_m(z)}{m\omega}\sin[\omega(t-t_o)-kz] \qquad (3)$$

and this expression is true up to the first collision. We remember that there are several collisions during a laser optical period under such (and larger) laser field amplitudes.

Because the time interval between collisions $<<\omega^{-1}$, it is easy to obtain from (2) that the distance between k and $k+1$ collisions coordinates is

$$x_{k+1} - x_k \simeq (t_{k+1} - t_k)\bar{v}_{k+1}. \quad (4)$$

Here $\bar{v}_{k+1}$ is the velocity right before the $k+1$ collision. Therefore

$$(x_{k+1} - x_k)/(t_{k+1} - t_k) \simeq \bar{v}_{k+1}.$$

If we average this formula in space the left part will be $1\Delta k/\Delta t$ so the average distance between hard particles along electron trajectory is the electron free track length and $\Delta k=1$. The averaging of the right part must give "actual" value of the collided electrons velocity. It means that "actual" value of $\bar{v}_{k+1}$ is the average one of ordinate $v_m \sin\omega t$ within the optical half-period:

$$\overline{v_{k+1}} = 2eE_m(z)/\pi m\omega$$

As a result, using two previous formulas we obtain the collision frequency equation:

$$\frac{dk}{dt} = \frac{2eE_m(z)}{\pi l m\omega} \quad (5)$$

Thus the collisions frequency depends on the coordinates through the field amplitude. Here k is the number of one electron collisions. The value of collisions number of a unit volume is more convenient we multiply the electrons numbers density n on this value and obtain:

$$n_e\frac{dk}{dt} = n_e\frac{2eE_m(z)}{\pi m\omega l} . \quad (5a)$$

The equations (5) or (5a) can be derived also by a very simple way. The "noncollisions" amplitude of electrons oscillations is $\bar{x} = 2eE_m/m\omega^2$, the number of collisions during the period of laser radiation is $\bar{2}x/T_{las}l$, or

$$dk/dt = 2eE_m(z)/\pi m\omega l$$

that coincides with (5).

Average values of electrons motion under consideration are found coinciding with "noncollisions" ones. But it can be seen from Fig.1 and Fig.2 that both values **v** and the oscillations amplitude **x** possess dispersion. The collisions process has Poisson's character in time because the probability of each following collision depends only on the time interval between collisions and does not depend on time. The calculation of the dispersion **v** gives value $\bar{k}\ \overline{\Delta v}$, here $\bar{k}$ is the average number of collisions during optical half-period, Δv is the average velocity loss as a result of collision. The problem is identical to so-called problem of the telegraph signal [3]. Accordingly, the dispersion **x** is $\pi\bar{k}\ \overline{\Delta v}/\omega$. Below we will note only physical followings of nonzero dispersions of **v** and **x**.

Thus we obtained the collisions frequency equation. As the formula (5a) is the averaging formula (5) in space, the convective term in full time derivative d/dt must be: $d/dt = \partial/\partial t + (\bar{v}\,\bar{\nabla})$. Here **v** is the macroscopic average velocity of electrons in this point of space in the considerable time moment. If such macroscopic motion exist in this problem it will be the needed value which describes our physical situation instead of plasma temperature.

3. The macroscopic electrons motion in laser pulse field - ponderomotive forces

If there was a plane monochromatic wave instead of real laser pulse it would be easy to see that any macroscopic electrons motion.Real pulse is a "sausage" which has "washed out" front, sides and "tail": usually from side to side $E(r) \sim \exp(r^2/2r_o^2)$, the envelope $f(t-z/c)$ has been introduced earlier. For example, one picosecond duration laser pulse has the length (along z-axis) of 300 μm. If laser wavelength is 308 nm, there are 974 optical periods in the pulse. It must be well focused for our aims therefore $r_o \sim (10-100)\mu m$. The plane wave approximation is rough both along and across such pulse. Note, that the pulse front length can not be less than several wavelengths. This pulse looks as shown on Fig.1.

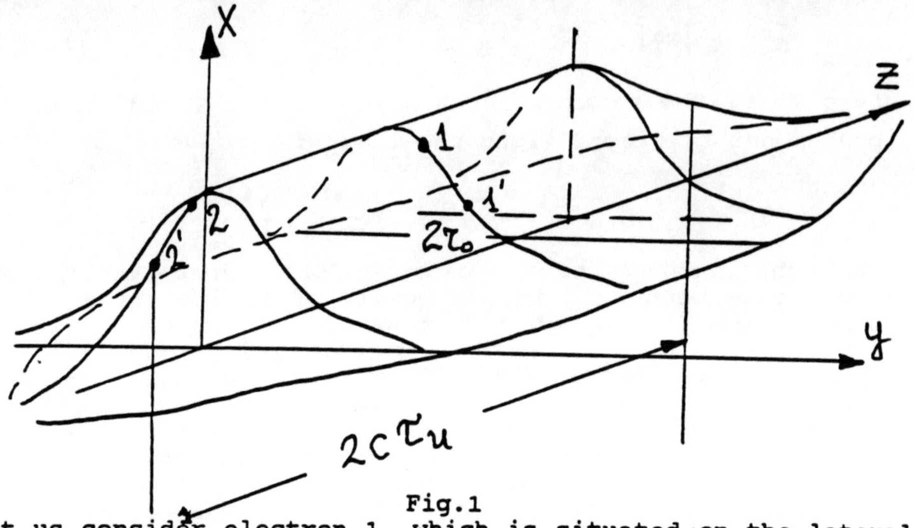

Fig.1

Let us consider electron 1, which is situated on the lateral slope of the pulse. The Lorenz forces which affect this electron in 1 and 1′ positions differ, as the electrical field values are differed in these points. The forces difference $m\,d^2x/dt^2$ is

$$m\frac{d^2x}{dt^2} = eE(x_1) - eE(x_{1'}) = e\bar{x}\frac{\partial E}{\partial x}.$$

We substitute here the average "actual" oscillation amplitude $x = 4eE_m/\pi m\omega^2$ and obtain:

$$m \frac{d^2x}{dt^2} = \frac{4E_m}{\pi m\omega^2} \frac{\partial E_m}{\partial x} ,$$

or for arbitrary polarization of a laser light:

$$m \frac{d^2\bar{r}}{dt^2} = \frac{32e}{m\omega^2} \nabla_\perp I(r) . \tag{6}$$

in case of cylindrically-symmetric problem. Here $I = cE_m^2/8\pi$, $\nabla_\perp=\{\partial/\partial x, \ \partial/\partial y\}$. The value $U_p=32\nabla_\perp I/mc\omega^2$ is the so-called ponderomotive potential. The force (6) is directed so that the electron is constantly pushed out of the area of more strong field.

Let us consider electron 2 which is situated on the back front of the laser pulse. What has happened to it during the oscillation period? The field decreases during the electron oscillation, because the pulse goes away. Field changes are $(\partial E/\partial t)\bar{t}$, $\bar{t}$ is the time of pulse motion relatively this electron, $\bar{t} = \bar{x}/c$. Therefore, the absolute value of this force is

$$|ma| = 32\frac{eI_m}{m\omega^2c^2} \frac{\partial f}{\partial t} . \tag{7}$$

Here I_m is the maximal intensity of laser field. This force pushes out electrons along the z axis:

$$m \frac{d^2z}{dt^2} = -32 \frac{eI_m}{m\omega^2c^2} \frac{\partial f^2}{\partial z} .$$

Thus the presence of macroscopic force which actions along and across beam axis is the result of finite space sizes of laser pulse (it is not a flat wave) and finite amplitude of oscillation (it is not a zero). This is known as a ponderomotive force [1]. Therefore the unit volume of electrons "fluid" which is influenced by electrical field during laser pulse can have macroscopic velocity, and there exists a convection part of the full time derivative from all macroscopic values. The equations of electron "fluid" electrodynamics do not have a pressure term as there is not any big exchange of energies among electrons, but it is necessary to take it into account for more long pulses (see below). Our equations are [cylindrically-symmetric coordinates, $I(r,\phi) = I(r)$]:

$$\frac{\partial v_r}{\partial t} + v_r\frac{\partial v_r}{\partial r} + v_z\frac{\partial v_r}{\partial z} = - \frac{32e}{m\omega^2 c}\,\frac{1}{r}\,\frac{\partial}{\partial r}\, rI(r),$$

$$\frac{\partial v_z}{\partial t} + v_r\frac{\partial v_z}{\partial r} + v_z\frac{\partial v_z}{\partial z} = - \frac{32e}{m\omega^2 c}\,\frac{\partial I}{\partial r}\,. \qquad (8)$$

In the equation (8) we omit the force which arises due to the volume separation of charges: electrons push out the pulse, and ions move much more slowly. The force F_ρ can be found in [1].

The convective derivative also exists in equation (5a), i.e

$$n_e\frac{\partial k}{\partial t} + n_e(\bar{v}\ \bar{\nabla})k = \frac{2en_eE_m}{\pi m\omega l} \qquad (5b)$$

Remember, that in the cylindrical coordinates:

$$(\bar{v}\ \bar{\nabla}) = v_r\frac{\partial}{\partial r} + \frac{v_\phi}{r}\,\frac{\partial}{\partial \phi} + v_z\frac{\partial}{\partial z}\,.$$

Thus the hydrodynamics equations block is exhausted. Now the field dynamics which is connected with electrons motion should be considered.

4. Laser field changes due to electron dynamics.

The energy of stimulated electrons oscillations is taken from the energy of electromagnetic laser field. If electrons density increases, it takes more field energy therefore there exists the critical value of electrons density when all laser energy transits to electrons and the complicated process of reradiation is the result. It is easy to estimate this critical density: the laser pulse energy should equal the oscillation energy of all electrons into the pulse volume:

$$n_e \le n_e^{crit} = \frac{m\omega^2}{2\pi e^2} \qquad (9)$$

It means that, for radiation wavelengths 308 nm, 1060 nm and 10600 nm, critical values of electrons densities will be $2\times10^{22}cm^{-3}$, $2\times10^{21}cm^{-3}$, and $2\times10^{19}cm^{-3}$ respectively. The last value for CO_2-laser is less than the usual (atmospheric) particles number per unit volume and means that CO_2-laser radiation can not propagate in a comparably dense gas. The second number is compared with high-ionized plasma values [4] and only excimer laser pulses can "free" propagate in a dense (as compared to atmospheric one) gas, so light atoms contain few electrons, and densities $n_e \sim 10^{22}cm^{-3}$ can be achieved only under high pressures.

In time of collision the electron acceleration can be to decomposed in two components. One of them is parallel to the electromagnetic field vector (in this moment, of course), the other acceleration is transverse to them. The last component arises due to the ion's attracting during the electron's moving near the ions, i.e. in time of collision (Fig.2).

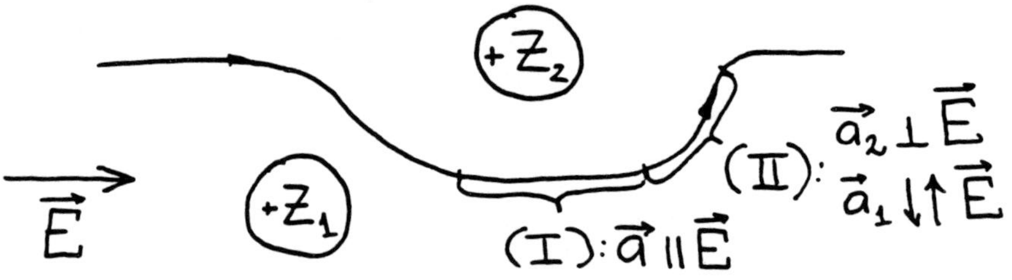

Fig.2

The acceleration is parallel to the field vector on the "free" motion way (I), there are antiparallel and the transverse components under collision (II) apart the first one.

The moving electron creates the electromagnetic field which is described by the LV-potential [5]. The electron which is situated in the point $\bar{r}'$ in the moment t' and possess the velocity $\bar{v}=d\bar{r}'/dt'$ and an acceleration $\dot{\bar{v}}=d^2\bar{r}'/dt'^2$ creates the electric field [17] in the point $\bar{r}$ in the moment t $[|\bar{r} - \bar{r}'|=c(t-t')]$:

$$\bar{E}(\bar{r},t) = e\frac{1-v^2/c^2}{(R-\bar{R}\bar{v}/c)^3}(\bar{R}-R\bar{v}/c) + \frac{e}{c^2(R-\bar{R}\bar{v})^3}[\bar{R}[(\bar{R}-\bar{R}\bar{v}/c),\dot{\bar{v}}]] \qquad (10)$$

Here $R=|r-r'|$.

The first term is the electrostatic field of electron in a nonrelativistic approximation. It is either compensated by the electrostatic field of ions or gives the force $\bar{F}_\rho$ under space charge separation. The second term determines reradiation. It is necessary to sum all these terms to receive the whole electrons reradiation field, when these electrons are situated in the points $\bar{r}_i$ in the time moments t_i'. The value $R/c - (t-t_i)$ can be changed into the time interval from zero to $2\tau_i$ - pulse duration. This sum can be changed on the integral over space of the pulse volume. It is possible to write the full field after a taking away nonrelativistic terms and the stochastic part:

$$\bar{E}(\bar{r},t) = \frac{n_e e^2}{mc^2}\int_0^{2\tau_i}\int_{V_p}\frac{d\bar{r}'dt'}{R^3}[\bar{R}[\bar{R},\bar{E}]]\exp(-r'^2/r^2+ikr'-i\omega t')\times$$

$$\delta(R/c-t+t')\times f(t'-z'/c) + E\,\bar{e}\exp(i\omega t-ikr) \qquad (11)$$

Here $k=2\pi/\lambda$, the second term is unperturbed laser field [5].

The integrand δ-function permits to integrate over t'. The amplitude of reradiation field is the negative addition to the laser radiation amplitude. Its value has been estimated above.

The calculation of this field by the formula (11) is very difficult. Let us do some physical considerations. The first speculation is about the change of laser pulse shape. The pulse front propagates practically without distortions so the electrons reradiation propagates with the same velocity as the initial laser pulse and can not give the same contribution to the front amplitude as compared to the contribution to the another part of the pulse. Accordingly, the reradiation influences the pulse back side stronger and extinguishes it [6]. It is an example of "linear" selfaction.

Let us consider the followings of another two accelerations which exist in time of collisions. If there are several collisions per one laser optics period the oscillated electron has corresponding numbers of accelerations and radiates corresponding times. It means that there appear a noncoherent part of reradiation which is physically determined by electrons scattering on ions. It is the stimulated scattering so we can estimate the average frequency and broadening of this radiation. The most simple estimation of the frequency of this noncoherent reradiation is $\omega\bar{x}/l$ - there are such collisions during one laser period. So the average distance between collisions is l and the dispersion of linear contribution of ions is about $n^{-1/3}$, we can estimate the linewidth of this reradiation by value $\omega\bar{x}n^{-1/3}/l^2 \ll \omega\bar{x}/l$. It is easy to estimate the average side scattering intensity of this field:

$$I_f = \frac{Z^3 e^6 n^3 \tau_1}{m^2 c^2 \omega^2} \frac{r_0}{r} \overline{(\Delta v)}^2 . \qquad (12)$$

Here Z is the average ions charge. If $r=r_0$, $2\tau_1 = 1$ ps, $\lambda=308$ nm, $n=2.67\times10^{19}cm^{-3}$ $I_f = 250$ KW/cm^2, the full fluctuation power $P_f = 150$ W, the averaging frequency of the scattering radiation is equal $\bar{x}\omega/l$.

It was the reradiation which is connected with the electrons scattering on ions. There exists the broadening of coherent part of the reradiation, because there are the dispersions of $\bar{x}$ and $\bar{v}$ (see above). There are not any changes of the reradiation frequency , but the dispersion of $\bar{v}$ provides the broadening as well as in a Doppler case. The lineshape is determined by the loss of energies of collided electrons. It is the sum of two terms; the first of them is a corresponding ionization potential, the second is an initial energy of a pushed electron. The lineshape will be also determined by the possibility of ionization processes. So there are several ionization, and the energy of the second electron + ionization potential is less than the following ionization potential, there will be several lines in a reradiation spectra, which are separated in principle.

Thus the hyperatomic intensity field laser pulse shape will distort under propagation in a medium due to the extinguishing of the pulse "tail". There appears the nonhomogenious broadening in the optical spectra, its linewidths are determined by ionization potentials and the distributions of initial energies of the second electrons, their relative intensities - by the number of electron with necessary energies. There also appears the noncoherent reradiation due to a stimulated scattering of electrons of ions.

5.Processes of the plasma formation

It is sufficient that ions and electrons temperatures are low therefore it is impossible to use convenient plasmas formulas for electron impact ionization cross-sections [2]: $\sigma \sim T^{-2}$ and $\sigma(T=300\ K) \sim 10^{-12} cm^{-2}$! It is also impossible to use the simple gas-kinetic cross-section ($\sigma \sim 10^{-15} cm^{-2}$) because ions attract electrons strongly. The cross-section for top electrons should be more than gas-kinetic one. The impacting electron has to hit the ion in order to make the ionization. This electron has to hit the inner shell in order to ionize it.These rough speculations permit to estimate the inner ionization cross-section: the force which affects electron from the ion should be the same value as the force affecting the electron from the laser field:

$$\frac{Ze^2}{r_0^2} = eE_m \ ; \ \sigma = \pi r_0^2 \sim \frac{\pi Ze}{E_m}$$

Now it is easy to estimate the collisions frequency when an inner electron is pushed out:

$$\frac{\partial k^1}{\partial t} = \frac{1}{\tau_c} = n_e \bar{v} \sigma = \frac{2Z^2 e^2 n}{m\omega}$$

One can see that the collisions frequency depends on the laser field only through the average ions charge Z, and changes in time accordingly.As it was mentioned above, the minimal Z corresponds to the classical above-threshold ionization and the multiphoton ionization on the pumping laser pulse front. Really the first cause gives more contribution than the second, minimal Z is determined by:

$$Z = \frac{\hbar}{me^2} \sqrt[3]{E_m \hbar me} \qquad (13)$$

Here $\hbar$ is the Plank constant.

The minimal number of collisions k^1 during optical laser period are:

$$k^{\imath} = \frac{2\pi}{\omega\tau_{\imath}} = \frac{4\pi n\hbar^2 (E_m \hbar m e)^{2/3}}{m^3 e^2 \omega^2} \qquad (14)$$

For experimental conditions [1] $E_m = 2{,}5\times10^{10} W/cm^2$, $n \sim 10^{19} cm^{-3}$, $\omega = 8\times10^{15} s^{-1}$ and the low boundary of the collisions number with inner ionization can be obtained from the formula (21): $k^{\imath} \simeq 0{,}035$. Thus, about 30 optical periods are necessary for the 100 percent-proof inner ionization ($k^{\imath} \simeq 1$) as a maximal number, here the value of gas-kinetic cross-section ($\sigma \sim 10^{-15} cm^2$) can be used as a minimum. The ionization probability Ω is determined by a multiplication of two values: k^{i} and T^{opt}: $\Omega = k^{i} T^{opt}$, here corresponding k^{i} is necessary for definite Ω.

Let us consider the transition from the regular system behavior - electrons oscillations, coherent reradiation - to plasma. The transition is determined by the fact that the field transversal electrons stochastic velocity component increases and becomes comparable with actual oscillations velocity $\bar{v}$. We also suppose that ions which have a charge Z acquire balanced for this charge temperature [2]:

$$Z = 2(mk_b/2\pi\hbar^2)\exp[-I_{ion}(z)/k_b T]/n \qquad (15)$$

Her k_b is the Boltzmann constant, $I_{ion}(Z)$ - the corresponding ionization potential. This formula permits to nonevidently determine T as a function on Z, and, therefore, due to (13) - E.

The cross-section of a transmit energy from one electron to another above the definite value $\bar{E}$ under collisions is:

$$\sigma_{ee} = 2\pi e^4 [1/\bar{E} - 2/mv^2]/mv^2 \qquad (16)$$

Here v is the collided electron velocity. The time when a transversal electrons velocity components acquire the same value the actual one is t_{plas}:

$$t_{plas} = (n_e \sigma_{ee} \bar{v})^{-1}. \qquad (17)$$

Formulae (15)-(17) solve the problem about the time transition to the plasmas state:

$$t_{plas} = \frac{E_m}{\hbar^2 Z n e^3 \omega} \left[\frac{1}{k_b T(Z)} - \frac{\pi^2 m\omega^2}{2e^2 E_m^2}\right]. \qquad (18)$$

Let us consider experiment [1] conditions: $k_b T \simeq 100$ eV for Z=8, $t_{plas} \le 17$ ps. The ionized medium is not plasma during the pulse time [1] 0,25 ps and can be described by the above supposed method.

Conclusion

While propagating in a medium, the laser radiation of hyperatomic intensity ionizes it. Released electrons start to oscillate with a pump radiation frequency. Oscillating electrons reradiate the electromagnetic field with the pump laser frequency, and this reradiation extinguishes the "tail" of the laser pulse.

Collisions of electrons with ions lead to the nonhomogenious broadening of the full radiation due to the distribution of electrons over velocities. Moreover, there appears the side-scattering radiation which has a multiplex laser frequency.

Electron-electron collisions lead to an increase in the chaotic electrons velocity component which will not be compensated during oscillations. The increase time of this chaotic component to the magnitude of the actual velocity of electrons oscillations will be the time of plasma formation from the initial ionized medium. This time is about (1-10) ps in a wide spectrum of superbright excimer laser experiments.

REFERENCE

[1] A.MacPherson, T.S.Luk, G.Gibson, J.C.Solem and C.K.Rhodes, in the Proceedings of the Seminar on Fundamentals of laser interaction II, February 26 - March 4, Bundessportheim Obergurgl, Austria, Springer-Verlag (1989).

[2] Ya.B.Zel'dovitch and Yu.P.Rayzer. The physics of shock wave and high-temperature hydrodynamics phenomena. Moscow, (1966), p.761.

[3] S.M.Rytov. Introduction in the statistical radiophysics. Moscow,(1966), p.404.

[4] B.J.MacGovan, S.Maxon, C.J.Keane, R.A.London, D.L.Matthews, and D.A.Whelan. J.Opt.Soc.Am.B, 5, (1988) 1858.

[5] L.D.Landau, E.M.Lifshitz. The field theory. Moscow, (1988), p. 510

[6] V.V.Korobkin and M.Yu.Romanovsky.In book of Abstracts XIX Int.Conference on Phenomena in Ionized Gases. July 10-14, Belgrade, Yugoslavia, vol.3,(1989) p.638-639.

STUDY OF THE DYNAMIC EFFECTS OF SEMICONDUCTORS HAVING A NONQUADRATIC DISPERSION LAW UNDER THE ACTION OF ALTERNATIVE ELECTRIC FIELDS

P. Kh. Musayev
Navoiyskei Branch of Bukhara Technology Institute, Navoi, USSR

The study of substance non-linear characteristics such as non-linear susceptibility (non-linear current) allows to obtain valuable information about the structure and properties of the studied object that is difficult to obtain by other methods. The given characterictics may be studied in the optical range with the help of optical quantum generators. In particular, there was a discussion of laser frequency CO_2 doubling in the constant field in JnSb in the literature.

The purpose of the present work is to reveal in quantum-mechanical systems with continuous spectra (electrons in the field of periodic potential) the relation of non-linear current with the substance energy structure as well as to reveal on a number of objects (semi-conductors: indium antimonide, gallium arsenide, as well as semi-conductors with the toroidal dispersion law of the zink winding structure) the information about the object structure which can not be obtained during the study of non-linear responds. The dynamic effects are most explicitly exposed in alternative fields in which the collision frequency ν is by fai less than the field in frequency ω. The matter is that the collision effects are connected with the active part of current while the dynamic effects are revealed also in the reactive part of the current. Under $\omega \gg \nu$, the reactive part of the current may considerably exceed an active part thus allowing to expose the dynamic part (against the background of the dissipative one).

Now let's calculate the non-linear current for the semi-conductors having non-quadratic dispersion law [1]

$$\varepsilon(\vec{k}) = h^2(k_\perp - k_{\perp 0}) /2m_\perp + h^2 k_z/2m_z \qquad (1)$$

Here k_z, $k_\perp$ are the wave vector components along symmetry axis and in the plane perpendicular to it; m_z, $m_\perp$ are the effective mass components in corresponding directions; $k_{\perp 0}$ is a loop radius.

The isoenergy serfaces corresponding to the dispersion law (1) are the tores with the elliptic cross-section in any plane coming through the symmetry axis.

Picus showed that the experiment revealed the realisation of dispersion law (1) in semi-conductors p-InSb.

Let the crystal be under the effect of spacily uniform but alternative in time harmonic electric field $\vec{E}(t)$ polarised along X axis. Under this condition the current density may be computed if the distribution function $f(\vec{p})$ is known

$$j_x = \frac{2e}{(2\pi h)^3} \int \frac{\partial \varepsilon(\vec{k})}{\partial p_x} f(\vec{p}) d^3p \qquad (2)$$

Note that the impulse in the alternative field depends on time. It was shown in [2] that the current calculation according to formula (2) is correct in any optional order in the electric field as far as a single-zone approximation is applicable. This means that the highest degree in expanding the current of the electric field in series should be less than the ratio of interzone transitions $\omega_{nn'}$ frequency to the field frequency ω. Let us consider that in the Maxwell distribution

$$k_0 T_e \ll \Delta \qquad (3)$$

where k_0 is a Boltzmann's constant, T_e is an electronic temperature, Δ is a loop depth.

Thus the current has 2 Fourier's components

$$j_x = \sigma E_0 e^{i\omega t} + \sigma \left(\frac{E_0}{E^*}\right)^2 E_0 e^{i3\omega t} \qquad (4)$$

one having the field frequency (first item) and the other having tripled frequency (second item). Here

$$(E^*)^2 = \frac{\omega^2 m_\perp}{\pi^{1/2} e^2} (\Delta k_0 T_e)^{1/2} e^{-\Delta/k_0 T_e} \qquad (5)$$

In order to estimate E^* let us assume that $\Delta \cong 10^{-4}$ eV; $m_\perp \cong 10^{-27}$ g; $kT = \Delta/2$, then for E^* we obtain the value $\cong$ 10 V/cm. Thus when applying an outer alternative voltage with the amplitude 0.1 V to a sample 0.1 cm, there will be a generation of voltage on the tripled frequency with the amplitude of about 3 mV.

In order to observe the discussed effect the radiation

in the frequency range $4 \cdot 10 - 5 \cdot 10^2$ rad/sec may be used ($\lambda \sim$ 4-0.5 mm).

REFERENCES

[1] G.E.Picus J.of experimental and theoretical physics, (1961),v.41, 1507.
[2] V.N.Genkin, P.M.Mednis J.of experimental and theoretical physics, (1969) v.54, 1138.

NONLINEAR COHERENT OPTICS OF RESONANT LATTICES (BRAGG SOLITONS)

B. I. Mantsyzov and R. N. Kuz'min
Physics Department of Moscow State University
Moscow, 119899, USSR

There is no doubt, that in the nearest future the progress in creation of short-wave lasers will result in the appearance of lasers with the length of waves of radiation in the range of shortest periods of superstructures or even of atom distances in crystal lattices. Therefore the investigation of wave propagation in nonlinear periodic structures is essential now. It have been shown recently, that solitons of self-induced transparency [1,2] and optical solitons [3-7] can exist in nonlinear Bragg structures though their light frequency lies within the forbidden frequency band. However nonlinear coherent optics of resonant lattices remains still insufficiently researched field of physics.

In this work we present the results of theoretical research of dynamics of nonlinear coherent interaction of radiation with a resonant Bragg lattice.

The state of two-level medium and field is described by semiclassical two-wave Maxwell-Bloch equations. We discuss the effects of nonlinear self-induced suppression of Bragg reflection, as well as the process of formation of two-wave solitons in an one-dimensional Bragg lattice under condition of incidence of short field puls on the lattice. Soliton solution are also derived for the Maxwell-Bloch equations, which describe the two-wave diffraction in a two-dimensional Bragg structure. In this case a two-wave stable pulse propagate towards a normal to a reciprocal lattice vector (nonlinear likeness of Borman effect). Dynamics of a stable puls in a deformed lattice in case of small deflection from Bragg conditions is examined. It is shown that in such a case a stable time-dependent two-wave field pulse propagate with the periodically changing velocity and amplitude without energy loss.

THE TWO-WAVE BRAGG SOLITONS

We use the semiclassical picture to describe the interaction of a coherent field with a resonant medium. A discrete one-dimensional periodic resonant medium consists of a set of periodically positioned thin layers containing two-level atoms. To ensure discreteness, the period d of the structure should satisfy the condition $d \geq \lambda$ (λ is the wavelength of the radiation). In two-wave approach this model is described by two-wave Maxwell-Bloch equations for slowly varying amplitudes [2]

$$\pm c\Omega^{\pm}_{x}(x,t) + \Omega^{\pm}_{t}(x,t) = \tau_c^{-2} \sum_i \exp(-ikx_i)\, P(x_i,t)\, \tilde{\delta}(x-x_i),$$

$$P_t(x_i,t) = n(x_i,t)\, [\, \Omega^{+}(x_i,t) \exp(ikx_i) + \Omega^{-}(x_i,t) \exp(-ikx_i),$$

$$n_t(x_i,t) = -\mathrm{Re}\{\, P^{*}(x_i,t)\, [\, \Omega^{+}(x_i,t) \exp(ikx_i) + \Omega^{-}(x_i,t) \exp(-ikx_i)\,]\}, \qquad (1)$$

where $\Omega^{\pm} = 2(\mu/\hbar)E^{\pm}$, $E^{\pm}$ are the complex amplitudes of the electromagnetic field; P is a dimensionless characteristic of the atomic polarization; n is the population inversion of the atoms; τ_c is the cooperative time; the function $\tilde{\delta}(x-x_i)=1$ for $x \in (x_i \pm \lambda/2)$, and is zero otherwise.

The equations (1) are considerably simplified when the Bragg condition is fulfilled exactly, and allow for an analytic solution - Bragg soliton. For the field pulse of width $\tau >> d/c$ Eq.(1) takes the form

$$(\pm\, c\, \partial/\partial x + \partial/\partial t\,)\, \Omega^{\pm}(x,t) = \tau_c^{-2}\, P(x,t),$$

$$P_t(x,t) = n(x,t)\, (\, \Omega^{+} + \Omega^{-}\,),$$

$$n_t(x,t) = -\, P(x,t)\, (\, \Omega^{+} + \Omega^{-}\,). \qquad (2)$$

This set of equations can be reduced to a single equation in the quantity

$$\theta(x,t) = \int_{-\infty}^{t} (\, \Omega^{+} + \Omega^{-}\,)\, dt \ .$$

We obtain the sine-Gordon equation for $\theta(x,t)$:

$$c^2\, \theta_{xx} - \theta_{tt} = 2\tau_c^{-2} \sin\theta\ . \qquad (3)$$

As is well known, Eq.(3) has a complete set of localized soliton solutions, which lead to a set of two-wave Bragg solitons of Eq.(2):

$$\Omega^{\pm}(x,t) = (\, \theta_t \mp c\theta_x\,)/2.$$

The most simple single-soliton solution is

$$\theta(x,t) = 4 \text{ arctg } \exp[(t-x/v)/\tau],$$

$$\Omega^{\pm}(x,t) = \pm [(1 \pm u)/u\tau] \text{ sech}[(t - x/v)/\tau], \qquad (4)$$

where $u = v/c$. The Bragg soliton (4) consists of two field modes and the excitation of the medium and moves along the medium at the constant speed

$$v = c/(1 + 2\tau^2/\tau_c^2)^{1/2} .$$

Thus, solution (4) describes the self-induced transparency of a resonant Bragg structure.

Nonlinear Bragg reflection of the field pulse at resonant lattice we studed by means of a numerical integration of the system of Eq.(1) with boundary conditions

$$\Omega^{+}(x=0,t) = \Omega_o \text{ sech}[(t-t_o)/\tau_o], \quad \Omega^{-}(x=l,t) = 0,$$

$$\Omega^{\pm}(x,t=0) = 0, \quad \theta(x,t=0) = 0,$$

corresponding to an external pulse $\Omega^{+}(0,t)$ incident on the boundary $x=0$ of an unexcited medium of length l. We have shown [2,8] that nonlinear suppression of the Bragg reflection (because of production of a Bragg soliton in the medium) is observed, when the pulse parameters satisfy

$$\theta^{+} = \int_{-\infty}^{\infty} \Omega^{+}(0,t)dt \geq \theta_{cr},$$

$$\theta cr(\tau;\tau_c,\tau_d) = (\pi/4)(\tau_d/\tau_c)^{-1/2}\, \tau + 4\pi^2(\tau_d/\tau_c)^{1/2},$$

where $\tau_d=d/c$. There is an interesting phenomenon which occurs when the external field $\Omega^{+}(0,t)$ produces a two-wave pulse of area $\theta^{+} = 2\pi-\Delta$ ($\Delta \ll 1$) in the medium. The latter propagates through the sample, undergoing an inversion in shape and direction of motion. As a result, most of the external field of the pulse is radiated after some delay which depends on the magnitude of the area decrease. The delayed reflection can be described by a part of breather solution [9].

OSCILLATING FIELD PULSE

Let us examine the dynamics of a field pulse in a deformed lattice in case of small deflection from Bragg condition

$$d = (1+\varepsilon)\lambda , \quad \varepsilon \ll 1 \qquad (5)$$

Under this condition we can write Eq.(1) in nondimensional variables as

$$\Omega'_{tt} - \Omega'_{xx} = 2P'_t \ ,$$

$$P'_t + i\delta P' = n\Omega' ,$$

$$n_t = -(1/2)(P'^*\Omega' + P'\Omega'^*) \ , \qquad (6)$$

where

$$\Omega' = \Omega'^+ + \Omega'^- \ ; \quad \Omega'^{\pm} = \tau_c\Omega^{\pm} \exp[-i\delta(t \mp x)] \ ;$$

$$P' = P\exp(-i\delta t) \ ; \quad \delta = kc\tau_c\mathcal{E} << 1$$

We have obtained the following soliton solution of Eq.(6):

$$\Omega'(x,t) = (2/\tau) \exp[i(\delta/2v)(x-vt)] \ \mathrm{sech}[(x-vt)/\tau].$$

Soliton velocity

$$v = [1+2\tau^2/(1+\tau^2\delta^2/4)]^{-1/2}$$

is limited by the value

$$v_{min} = \delta/2\sqrt{2} \ .$$

Numerical integration of Eq.(1) under conditions (5) shows that, indeed, soliton (7) can propagate in the medium only with enough fast velocity. If the pulse is slow ($v<<1$), then it dynamics becomes similar to zoomeron dynamics [10]: a stable timedependent two-wave field pulse propagate with the periodically changing velocity and amplitude without energy loss (Fig.1). Fig.2 shows the behaviour of a Bloch vector $\vec{R}$ = {ReP, JmP, n} corresponding to oscillating two-wave pulse.

NONLINEAR BORMAN EFFECT

We consider the interaction between a coherent resonant field and a two-dimensional periodic structure in a case of nonlinear two-wave diffraction. Corresponding Maxwell-Bloch equations take the form

$$(\partial/\partial\vec{k}_{o,h} - \partial/\partial t) \ \Omega_{o,h}(\vec{r},t) = P(\vec{r},t) \ ,$$

$$P_t(\vec{r},t) = n \ (\Omega_o + \Omega_h) \ ,$$

$$n_t(\vec{r},t) = - \mathrm{Re}[P^*(\Omega_o+\Omega_h)] \ , \qquad (7)$$

where Ω_o, Ω_h - nondimensional functions of complex amplitudes of incident and diffracted fields; $\vec{k}_o$, $\vec{k}_h$ - wave vectors of incident and diffracted waves.

In a simple case of symmetrical geometry of diffraction the Eq.(7) are written as

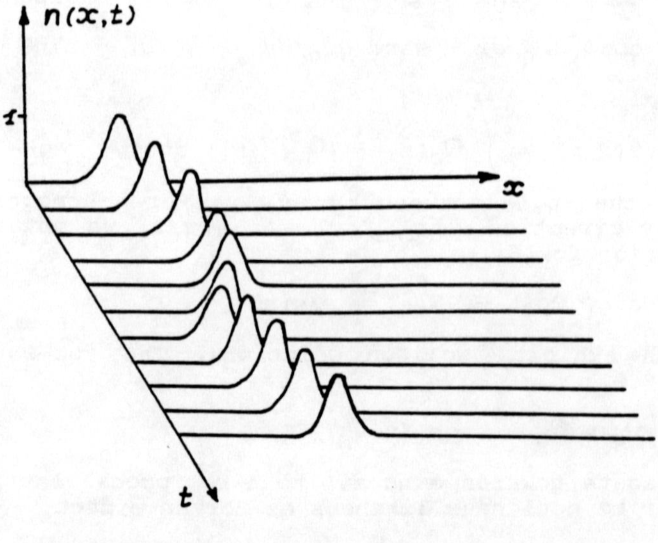

Fig.1. The dynamics of the medium excitation in oscillating two-wave pulse.

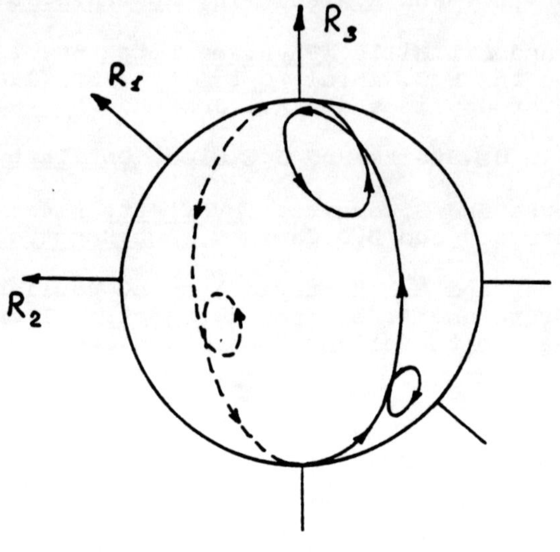

Fig.2. The track of a Bloch vector on a Bloch sphere corresponding to oscillating pulse.

$$\cos\varphi\ \partial\Omega_o/\partial x + \sin\varphi\ \partial\Omega_o/\partial y - \partial\Omega_o/\partial t = \sin\theta \ ,$$

$$\cos\varphi\ \partial\Omega_h/\partial x - \sin\varphi\ \partial\Omega_h/\partial y - \partial\Omega_h/\partial t = \sin\theta \ , \qquad (8)$$

where

$$\theta(\vec{r},t) = \int_{-\infty}^{t} \Omega_o(\vec{r},t')+\Omega_h(\vec{r},t')\ dt' \ ,$$

φ is the angle between $\vec{k}_o$ and $\vec{k}_h$. For a homogeneous field on the y-direction ($\partial\Omega_{o,h}/\partial y = 0$, $\varphi\neq\pi/2$) we obtain sine-Gordon equation for θ:

$$\sin\varphi\ \theta_{xt} - \theta_{tt} = 2\sin\theta \ , \qquad (9)$$

which exhibits soliton solutions. The two-wave soliton of Eq.(8,9)

$$\Omega_o = \Omega_h \sim \mathrm{sech}[(x-vt)/\tau]$$

propagate towards a normal to a reciprocal lattice vector and describe nonlinear likeness of Borman effect.

REFERENCES

[1] B.I.Mantsyzov and R.N.Kuz'min, Sov.Tech.Phys.Lett. 10, (1984) 359.

[2] B.I.Mantsyzov and R.N.Kuz'min, Sov.Phys.JETP 64, (1986) 37.

[3] W.Chen and D.L.Mills, Phys.Rev.Lett. 58, (1987) 160.

[4] J.E.Sipe and H.G.Winful, Opt.Lett. 13, (1988) 132.

[5] D.N.Christodoulides and R.I.Joseph, Phys.Rev.Lett. 62, (1989) 1746.

[6] C.Martijn de Sterke and J.E.Sipe, Opt.Lett. 14, (1989) 871.

[7] A.B.Aceves and S.Wabnitz, Phys.Lett. A141, (1989) 37.

[8] B.I.Mantsyzov and D.O.Gamzaev, Opt.Spectrosc. 63, (1987) 116.

[9] T.I.Lakoba and B.I.Mantsyzov, to be published

[10] F.Calogero and A.Degasperis, Spectral Transform and Solitons, North-Holland Publishing Company, Amsterdam, (1982).

PART IV

X-RAY OPTICS AND SPECTROSCOPY

ON THE EFFICIENCY OF GRAZING INCIDENCE OPTICS THE SPIRAL COLLIMATOR

I.A. Artyukov, A.V. Vinogradov, and I.V. Kozhevnikov
P.N. Lebedev Institute of Physics, Leninsky Prospect 53, Moscow, USSR

Abstract

Geometrical properties and construction of the spiral collimator are discussed. The expression for the efficiency of power conversion of divergent soft X-ray point source into parallel beam is obtained. The efficiency of the spiral collimator is shown to be up to 30-90% in the whole soft X-ray spectral range.

Various X-ray optics to concentrate and collimate parallel and divergent beams ($\lambda \sim$ 10-300 Å) were discussed in [1-5]. Such devices are of interest in many problems: optical pumping of shortwavelength lasers, X-ray lithography, plasma diagnostics, microanalysis, astrophysics,X-ray microscopy [6]. To get maximum efficiency various shapes of mirrors have been discussed: paraboloid [5], ellipsoid (including an off-axis one) [4,7], toroidal surface [8], as well as waveguide [8-10] and a bunch of waveguides [12]. In shortwavelength range the efficiency of all these optical systems drops abruptly due to the decrease of critical angle $\theta_c(\lambda)$ ~ $(0.1 - 0.2)\lambda$ where λ is the wavelength in Å, θ, in degrees.

In this paper an X-ray collimator with the efficiency approaching not 0 but 1 at short wavelengths, will be discussed

If the reflectivity of mirror coating is 1 for all angles of incidence, then the optical system which transports flux density with maximum efficiency is well known. It is a so-called "ideal optical concentrator" considered in [7,12] in connection with solar energetic problems. It has the shape of an off-axis ellipse

Another evident example of optical collimator is paraboloid with point source in its focus, which is also ideal if the coating material is totally reflective.

In X-ray range, as a result of diminishing value of critical angle, the paraboloid efficiency $\nu(\lambda)$ (the part of point source radiation power converted into parallel beam) is extremely small $\nu(\lambda)$ ~ $\theta_c(\lambda)$ (Fig. 1).

Let us put the question: what maximum efficiency $\nu(\lambda)$ is possible, and what is the corresponding X-ray collimator?

The efficiency of X-ray collimator is high enough if the grazing angle of most rays does not exceed critical angle θ_c. As θ_c is small, it is evident that the collimator can utilize the source radiation from large solid angle only if it provides manifold reflections of the rays. On the other hand, it is known [14,15] that it is manifold reflection at grazing angles ("whispering gallery" mo- de) that allows effective turning of X-rays at large angle (up to π).

After these notes let us start constructing an optimal X-ray collimator.

There is a curve for which any ray emitted by point source O falls at the same angle φ_0. This curve is known to be a logarithmic spiral:

$$\rho(\varphi) = \rho_0 \exp \frac{\pi - \varphi}{\mathrm{tg}\ \varphi_0}, \quad \varphi_0 \le \frac{\pi}{2}, \tag{1}$$

where π, φ are polar coordinates; ρ_0, φ_0, spiral parameters. Note that at $\varphi = \frac{\pi}{2}$ the logarithmic spiral degenerates in a circle.

We cut the logarithmic spiral at angle $\varphi = \varphi_0$ (point A), where a tangent is parallel to z-axis. Then the range of polar coordinates is determined by:

$$\varphi_0 \le \varphi \le \pi, \quad \rho_0 \le \rho \le \rho_0 \exp \frac{\pi - \varphi_0}{\mathrm{tg}\ \varphi_0}. \tag{2}$$

3d-collimator is obtained by rotation of the logarithmic spiral around z-axis.

Now let us prove the following statement: any ray going from the source 0 will escape through crosssection AA', so that the angle φ_1 between the escaping ray and z-axis does not exceed φ_0 : $\varphi_1 \le \varphi_0$. In other words, the involved optical system converts point source radiation into a beam with divergency $2\varphi_0$, and the efficiency of this conversion ν is equal to 1. (Mirror surface is regarded as an ideally reflective one).

Firstly, it is clear that the rays emitted by the source O at angle $\varphi : |\varphi| < \varphi_0$ come out through the crosssection AA' without being reflected by surface ADA'. Now take the ray OBhaving initially an angle $\varphi > \varphi_0$ to z-axis. After the reflection from the surface at point B_1 the ray turns by angle $2\varphi_0$, and the angle between the reflected ray B_1B_2 and z-axis will be $\varphi_1 = \varphi - 2\varphi_0 > -\varphi_0$. It is possible to prove that the ray B_1B_2 does not intersect ray OA', and,consequently, the lower part DA' of the collimator.

Hence, after the first reflection the ray OB_1 : (1) either escapes through the crosssection AA' , or (2) falls again on the upper part DA of the collimator at some point B_2. If the ray escapes through AA', then the angle φ_1 to z-axis would not be more than φ_0. If the reflected ray does not get out through AA' then it would fall on the surface at some point B_2 with the angle being less than φ_0. Taking into consideration the auxiliary ray OB_2 enables one to see that the same dilemma keeps true for ray B_1B_2 as well as for the ray OB_1. It either falls on the surface DA again or escapes through crosssection AA' with the angle φ_2 to z-axis, so that $\varphi_0 > \varphi_2 > -\varphi_0$.

Thus, all the rays emitted by the source O have grazing angles not larger than φ_0, and are steered by the colimator into the solid angle $\Omega = 2\pi(1-\cos\varphi_0)$ so that the angle between the outcoming ray and z-axis is not larger than φ_0.

Let us consider the sequence of spiral collimators for which parameter φ_0 tends to 0. Then the ray emitted by the source will fall on the surface at smaller and smaller angles and propagate over the surface due to more and more reflections. In the limit $\varphi_0 \to 0$ each ray will fall on the collimator surface at an angle tending to 0, and, thus, will slide over it ("whispering gallery" effect) with the output divergence $\varphi_0 \to 0$. Hence, this sequence of surfaces represents the sequence of collimators which steer point source radiation into solid angle along z-axis that can be chosen smaller than any given value.

Let us suppose now that the coating material is nonabsorptive ($\mathrm{Im}\ \varepsilon = 0$) and has critical angle θ_c, and take the spiral with the parameter $\varphi_0 \leq \theta_c$. Then any ray on its path inside the spiral has the grazing angle less or equal to θ_c. Hence, this collimator transforms the point source radiation into quasiparallel beam having the divergency $\varphi_0 \leq \theta_c$ with the efficiency equal to 1 though the coating material is not ideally reflective.

Let us take into account the absorption. In shortwavelength range (when θ_c is small) all the rays in the collimator, as was mentioned above, propagate in "whispering gallery" mode. In this case, as shown in [8,14,15], the coating materials exist that provide sufficiently high reflection for the rays being turned by a large angle (up to π). Maximum reflectivity corresponds to the sliding ray with grazing angle equal to 0. Therefore,the collimator with parameter $\varphi_0 \to 0$ will have the maximum efficiency.

The efficiency of the collimator with parameter φ_0 can be easily written in the form:

$$\nu = \sin^2 \frac{\varphi_0}{2} + \frac{1}{2} \int_{\varphi_0}^{\pi} \sin \varphi \prod_{j=1}^{n(\varphi)} R\,(\theta_j(\varphi))\, d\varphi \,, \qquad (3)$$

where φ is the polar angle of a ray emitted by the source; $R(\theta_j(\varphi))$, the Fresnel reflectivity at grazing angle $\theta_j(\varphi)$; $n(\varphi)$, the number of reflections. The first term in expression (3) corresponds to the rays escaping through crosssection AA' of the collimator without any reflection,the second one - to the rays being reflected by collimator walls.

In order to obtain an expression for the maximum efficiency of the spiral collimator consider formula (3) at a very small φ_0. Then, for $\varphi_0 \to 0$ all the rays are in "whispering gallery" mode. In this case, as was shown in [8,14], the reflectivity for the ray being turned by concave mirror does not depend on the shape of the mirror. Consequently, each ray escaping the spiral collimator has the reflectivity given by the known "whispering gallery" mode formula [8,14] :

$$\prod_{j=1}^{n(\varphi)} R\,(\theta_j(\varphi)) \xrightarrow[\Sigma\theta_j = \frac{\pi}{2}]{\varphi_0 \to 0} \exp\,(-\xi\varphi); \quad \xi = 2\mathrm{Im}\,\frac{1}{\sqrt{1-\varepsilon}} \tag{4}$$

By integrating (3) over φ and taking into account (4) we finally obtain the efficiency of transformation of X-ray point source radiation into parallel beam:

$$\nu_{max}(\lambda) = \frac{1}{2}\,\frac{1 + \exp\,(-\xi\pi)}{1 + \xi^2}\,, \quad \xi = 2\mathrm{Im}\,\frac{1}{\sqrt{1-\varepsilon}} \tag{5}$$

The efficiency $\nu_{max}(\lambda)$ is completely defined by optical constants of the mirror material (i.e. the real and imaginary parts of the dielectric constant ε)*). The value $\nu_{max}(\lambda)$ is less than 1 due to the only reason: the X-ray absorption in the collimator walls.

As shown in [8,9] the value of parameter ξ in (5) crucially depends on the material taken. Thus, the following step is to choose optimal materials for mirror coating of the collimator in different spectral ranges. The results of such analysis based on work [16] are shown in figure 1. The best materials (Ru, C, LiH,

---

*) Note that the same parameter ξ from formula (5) defines the efficiency of X-ray waveguides [9].

LiF) provide the efficiency 30-90 % that is 10-10^3 times higher than the efficiency of a paraboloid. In contrast to paraboloid the efficiency of the spiral collimator does not decrease, but rises in short wavelength range, and tends to 1. This results from the fact that spiral efficiency depends not on the critical angle θ_c, but on the ratio of the imaginary and real parts of dielectrical constant ε. This ratio tends to 0 at short wavelength.

Summary

Geometrical properties and construction of the spiral collimator are discussed. The expression for the efficiency of power conversion of soft X-ray point source radiation into parallel beam is obtained. The efficiency of the spiral collimator is shown to be up to 30-90% in all soft X-ray spectral ranges. This efficiency is limited only by a fundamental reason: X-ray radiation absorption in mirror coating of the spiral collimator. In this sense the spiral collimator is more effective than any known X-ray optical collimator system with one or manifold reflections.

At very short wavelengths, when θ_c is small, the collimator becomes more and more stretched. This fact makes difficult the practical use of the spiral collimator. Nevertheless, one can utilize not the whole collimator (as shown in Fig. 2), but part of it. The size of the collimator as well as parameters ρ_0 and φ_0 are determined by the type of a source, manufacturing possibilities and specific problem to be solved. *) To diminish the size of the collimator a set of nestled spiral collimators can be used as in the case of X-ray telescopes.

In the spiral collimator the conversion of divergent beam into a parallel (or quasiparallel) one is carried out through "whispering gallery" modes with very small grazing angles. Naturally, the question arises: how do the microroughness and microinhomogeneities of the mirror coating influence the results obtained. The basic properties of the X-ray mirror in "whispering gallery" mode follows from the limiting angle dependence of interface reflectivity:

$$r(\theta) = 1 + A\theta, \quad \theta \to 0 \tag{6}$$

where $r(\theta)$, the amplitude Fresnel reflectivity coefficient; θ, the grazing angle. It can be shown that similar dependence on the grazing angle keeps true for nonideal surfaces with surface transition layer [17] as well as microroughness [18]. In this case the constant A in formula (6) depends not only on optical properties of the coating material but also on the properties of the transition layer. Thus,the main results of this paper are true for nonideal surfaces too.

We wish to thank A.V.Popov, O.I.Tolstikhin and N.N.Zorev for useful discussions.

---

*) X-ray sources being of interest for applications have the following charactristic size: plasma micropinch, a few mm; laser plasma, 10^2-10^3 μm; X-ray tubes, from 10 μm to several mm; synchrotron sources, from 1 cm to 0.1 mm (after being focused) [19].

References

[1] A.V.Vinogradov "XUV cavity and pumping optics", J.Phys., 47, 10,p.c6-286-c6-297 (1986).

2. J.E.Young, J.J.Macklin, S.E.Harris "Grazing-incidence ellipsoidal reflector for longitudinally pumping short-wavelength lasers", Opt.Lett., 12, 2, p.90-92 (1987).

3. R.Tatchyn, P.Csonka, H.Kilic et al. "Focusing of undulator light at SPEAR with a lacquer-coated mirror to power densities of 10 Watt.cm^{-2}", Proc. SPIE, 733, p.368-378 (1986).

4. A.V.Vinogradov, O.I.Tolstikhin "Soft X-ray concentrators", Trudy FIAN, 196, p.168-181 (1989).

5. A.V.Vinogradov, N.N.Zorev, I.V.Kozhevnikov "On the limitations and possibilities of soft X-ray optics", Trudy FIAN, 176, p.195-210 (1986).

6. "X-ray instrumentation in medicine and biology, Plasma Physics, Astrophysics and Synchrotron Radiation", Ed. R.Benattar. Proc.SPIE, 1140 (1989).

7. V.K.Baranov "Properties of parabolic focons", Optico-Mekhanicheskaya Promyshlennost' , 6, 1-5 (1965) (in Russian).

8. I.V.Kozhevnikov "X-ray optical systems with manifold reflections", Trudy FIAN, 196, p.143-167 (1989).

9. A.V.Vinogradov, I.V.Kozhevnikov "On properties of X-ray waveguide", Sov.Phys.Tech.Phys, 29, 1023 (1984).

10. M.Watanable, I.H.Suzuki, T.Hidakaet et al. "Soft X-ray transmission characteristics of hollow-core fused quartz fibers", Appl.Opt., 24, 23, p.4206-4209 (1985).

11. V.A.Arkad'ev, A.I.Kolomiytsev, M.A.Kumakhov et al. "Broad band X-ray optics with large angle aperture", Sov.Phys.Uspekhi, 157,

3, p.529-537 (1989).

12. W.D.Welford, L.Winston "The optics of nonimaging concentrators", N.Y. : Acad.Press, 1978.

13. A.V.Vinogradov, O.I.Tolstikhin "Properties of an ideal optical X-ray concentrator", Sov.Phys. - Lebedev Institute Reports, 9, p.28-33 (1985).

14. A.V.Vinogradov, V.F.Kovalev, I.V.Kozhevnikov, V.V.Pustovalov "Concave mirrors for turning of X-ray beams", Sov.Phys. Tech.Phys., 30, p.144-150 (1986).

15. A.V.Vinogradov, V.F.Kovalev, I.V.Kozhevnikov,V.V.Pustovalov "The diffraction theory of grazing modes for X-ray mirrors and X-ray cavity", Sov.Phys.Tech.Phys., 30, p.335-344 (1986).

16. B.L.Henke, P.Lee, T.J.Tanaka et al. "Low-energy X-ray interaction coefficients: photoabsorption, scattering and reflection", Atom.Data and Nucl. Data Tables, 27, p.1-144 (1982).

17. A.V.Vinogradov , A.V.Popov "On X-ray reflection at grazing incidence", Opt. Spectrosc., 65, 5, p.1109-1111 (1988).

18. A.V.Vinogradov, N.N.Zorev, I.V.Kozhevnikov, I.I.Yakushkin "On the effect of total X-ray reflection", Sov. Phys. JETP, 62, p.1225-1230 (1985).

19. R.E.Burge, J.H.Beswetterick, M.T.Brown et al. "Scanning X-ray Microscopy" Proc. SPIE, 1140, p.528-529 (1989).

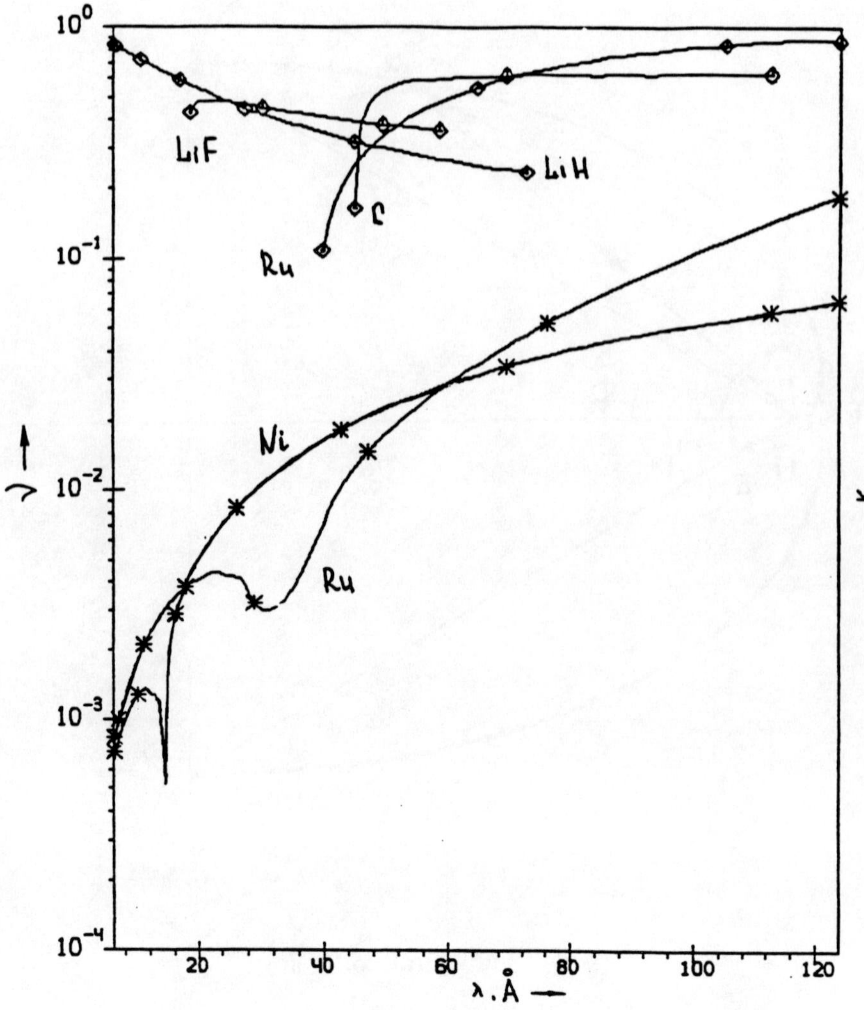

Fig. 1. The efficiency of paraboloid and spiral X-ray collimators with different coating materials:

✳ - paraboloid, formulas for efficiency $\nu(\lambda)$ in terms of Re ε and Im ε are taken from [13].

◊ - spiral collimator, calculations are made acording to formula (5).

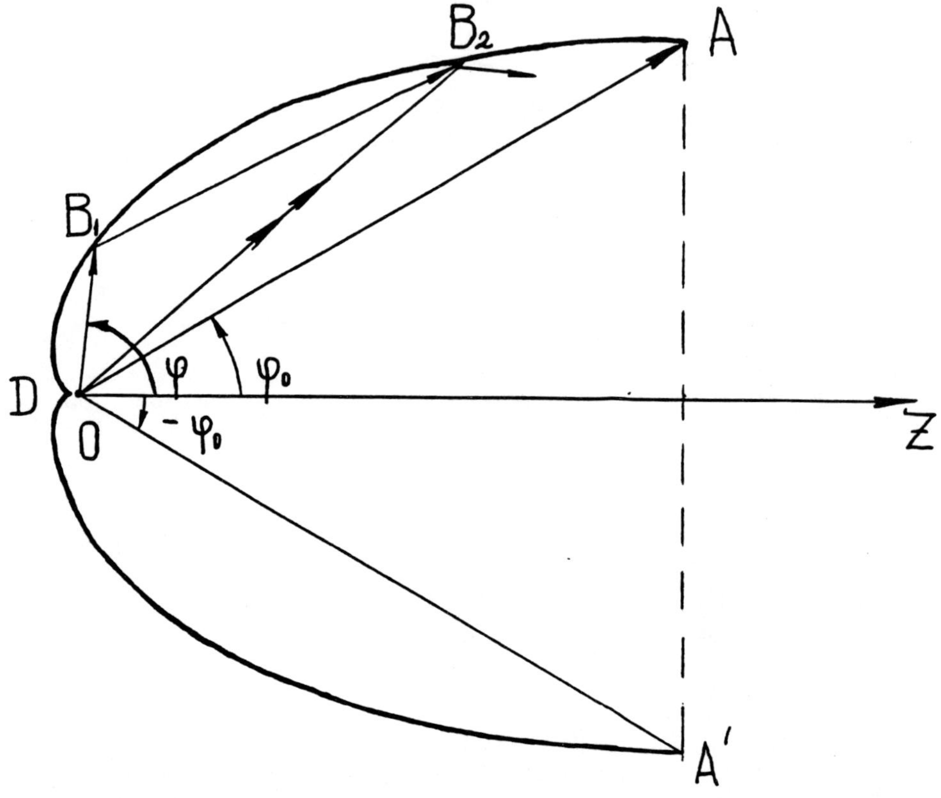

Fig. 2. The spiral collimator.

MEASUREMENT OF THE SOFT X-RAY OPTICAL CONSTANTS OF CARBON IN THE WAVELENGTH RANGE 6-11 NM

A.G. Michette and M. Kühne*
Physics Dept, King's College, London, UK
**PTB, Abbestraße 2-12, Berlin, Germany*

1. INTRODUCTION

Previous work by Tatchyn *et al* [1] has shown that it is possible to measure the soft x-ray optical constants δ and β of a material using the diffraction efficiencies of transmission gratings with rectangular line profiles. The optical constants are defined by

$$\tilde{n} = 1 - \delta - i\beta$$

where $\tilde{n}$ is the complex refractive index; i.e. δ is the refractive index decrement and β is the absorption index. It will be demonstrated here that a similar analysis can be applied to gratings with other line profiles, so long as the profile can be parameterized by a relatively simple function.

2. MOTIVATION

The most comprehensive compilation of soft x-ray optical constant data is that of Henke *et al* [2,3] which covers all elements ($Z = 1 - 92$) and the wavelength range $\lambda = 0.124 - 124$ nm for β and $0.124 - 24$ nm for δ. This compilation is of the atomic scattering factors f_1 and f_2, related to δ and β by

$$\delta = Kf_1 \qquad \beta = Kf_2$$

with

$$K = \frac{r_0\lambda^2}{2\pi}\frac{N_A}{A}\rho$$

where r_0 ($= 2.818 \times 10^{-15}$ m) is the classical electron radius, A is the atomic weight, N_A is Avagadro's number and ρ is the density.

The f_2 values are obtained primarily from measurements of bulk absorptance and the f_1 values are derived primarily from the Kramers-Kronig relationship which requires knowledge of f_2 over the complete wavelength range. This knowledge is not always available and so there is reason to suspect the tabulated values of f_1. Since f_2 is measured directly values of β derived from it for a material of known density can be assumed to be accurate. In order to obtain δ and β from f_1 and f_2 it is necessary to know the density of the material, but the techniques used to make soft x-ray optical components mean that it may not be appropriate to use the bulk densities of the materials in the calculations. In particular, some of the properties of carbon-contamination material suggest that its composition may be close to that of diamond [4]. The values of δ and β obtained from the tabulated f_1 and f_2 may therefore be subject to systematic errors; however, it is important to know these values as accurately as possible in order to predict the performances of the components and to optimize their designs. Thus measurements of the optical constants using the component itself (or one made in the same way) are essential.

The procedure to be described arose from a proposal to use ultrahigh line density transmission gratings (7000–17000 lines mm^{-1}) in the study of laser-produced plasmas. The gratings, typically with apertures of a few tens of micrometres and areas of around 1000 μm^2, are made in carbon by electron-beam contamination lithography [5]. They can be made with nominal bar widths w less than $d/2$, where d is the period, to allow for subsequent coating by a heavier material to enhance the diffraction efficiency [6]. In order to use these gratings for spectroscopy and to allow coating at the correct angles and to the correct thicknesses to optimize the diffraction efficiencies, it is necessary first of all to characterize the original carbon gratings. Preliminary work on this has been carried out in the laboratory of the Physikalisch-Technische Bundesanstalt (PTB) at the Bessy storage ring, Berlin, in the wavelength range 6–11 nm. The experimental arrangement has been described in detail previously [7]; briefly, the beam from a toroidal grating monochromator (TGM) is incident on the component under test and the reflected (for mirrors) or diffracted (for gratings) radiation is detected by a channel electron multiplier (CEM), operated in photon-counting mode. The CEM can be rotated at a fixed distance (200 mm) from the component with a smallest angular step of 0.0072°. The efficiency of the component is obtained by comparing the count rate with the straight-through rate with no component in place. The TGM gives well-characterized radiation at wavelengths between about 5 nm and 40 nm, and the minimum spectral bandpass is about 0.003 nm, i.e. a resolution ($\Delta E/E$) of 5×10^{-4} at 6 nm.

3. MEASUREMENTS OF THE DIFFRACTION EFFICIENCIES

The gratings were made on thin carbon substrates supported by 50 μm diameter electron microscope apertures. The nominal parameters (table 1) were derived from observations during manufacture. The periods were subsequently determined, from the positions of the diffraction peaks in the PTB measurements, to be 110.4±0.1 nm and 89.8±0.2 nm for gratings A and B respectively. During the manufacturing process sideways scattered electrons cause a thin contamination coating over the whole grating area, i.e., the carbon substrate is thickened. The amount of thickening depends upon the conditions during manufacture but will be, to a first approximation, constant over the grating area and can be estimated in the procedure used to fit the measured diffraction efficiencies to the calculated ones (see section 4). Sideways scattered electrons will also cause the grating lines to be broadened and will alter their shapes; these effects can also be taken into account in the fitting procedure.

The gratings were mounted in a simple holder and optically aligned using the support structures (with 5 μm spacing) perpendicular to the grating lines. The angular accuracy of this alignment was about 1–2°.

Table 1. The nominal parameters of the gratings used in the measurements. The grating aperture is the dimension perpendicular to the lines.

Grating	Aperture diameter (μm)	Substrate thickness (nm)	Period (nm)	Line height (nm)	Line width (nm)	Grating aperture (μm)	Grating area (μm ^{2})
A	50	25	110	≈100	≈30	28	945
B	50	35	87.5	≈100	≈30	34	1200

The measured diffraction efficiencies are summarized in table 2. The grating mounting caused a cut-off at large diffraction angles (20–30°). This cut-off was not necessarily symmetric in positive and negative diffraction orders because the mounting was designed for ease of use rather than extreme accuracy, and so the grating was unlikely to be exactly at the centre of the conical exit hole of the mounting. Thus it was not possible to measure to large diffraction orders, and it was not always possible to measure to the same diffraction order in both positive and negative directions. The choice of which orders are called positive and which are called negative is arbitrary, and the convention adopted corresponds to the definition of positive and negative angles in the coordinate system of the beamline.

The incident flux was measured by removing the grating and inserting a 140 μm diameter aperture 300 mm upstream of it, taking the beam divergence into account. The beam divergence was measured as a function of wavelength from the width of the beam at the detector; it increased from 0.21 mr at 6 nm to 0.30 mr at 8 nm and then decreased again to 0.24 mr at 10–11 nm. The zero-order efficiency had then to be estimated from the total signal at the zero-order position, with the signal due to transmission through the substrate surrounding the grating subtracted from it. This subtraction depends upon the value of the absorption index β; the substrates were made of amorphous carbon and therefore it was assumed that the β values calculated from the Henke compilation could be used.

4. FITS TO THE LINE PROFILES AND OPTICAL CONSTANTS

The measured diffraction efficiencies have been fitted, as described below, to the efficiencies $\epsilon_m(k)$ calculated using [8]

$$\epsilon_m(k) = T(k)\frac{1}{d^2}\left|\frac{id}{2\pi m}\left[1-\exp\left(\frac{\mathrm{i}2\pi ma}{d}\right)\right] + \int_a^d \exp\left(\mathrm{i}kz(\delta+\mathrm{i}\beta) + \frac{\mathrm{i}2\pi mx}{d}\right)\mathrm{d}x\right|^2$$

where m is the diffraction order $(0, \pm 1, \ldots)$, $k = 2\pi/\lambda$, $T(k)$ is the transmission of the substrate, the grating is open for $x = 0$ to $x = a$ (i.e. $a = d - w$ where d is the grating period and w is the line width) and $z = z(x)$ describes the line

Table 2. Measured diffraction efficiencies of the gratings.

Wavelength (nm)	Diffraction order	Efficiency	
		Grating A	Grating B
6	0		0.789±0.036
	+1		0.0136±0.0051
	−1		0.0132±0.0059
	+2		0.0008±0.0007
8	0	0.833±0.056	
	+1	0.0183±0.0035	
	−1	0.0181±0.0039	
	+2	0.0030±0.0012	
	−2	0.0023±0.0013	
8.5	0		0.736±0.022
	+1		0.0220±0.0075
	−1		0.0245±0.0077
	+2		0.0013±0.0012
9	0	0.768±0.036	0.645±0.014
	+1	0.0251±0.0041	0.0252±0.0057
	−1	0.0241±0.0038	0.0281±0.0086
	+2	0.0041±0.0006	0.0015±0.0007
	−2	0.0036±0.0005	
	+3	0.00017±0.00016	
9.5	0		0.629±0.021
	+1		0.0240±0.0062
	−1		0.0250±0.0089
	+2		0.0018±0.0016
10	0	0.678±0.006	0.581±0.025
	+1	0.0284±0.0019	0.0341±0.0059
	−1	0.0261±0.0020	0.0285±0.0080
	+2	0.0044±0.0002	0.0013±0.0007
	+3	0.00024±0.00018	
10.5	0		0.505±0.018
	+1		0.0269±0.0080
	−1		0.0245±0.0062
	+2		0.0010±0.0004
11	0	0.655±0.010	
	+1	0.0370±0.0022	
	−1	0.0333±0.0024	
	+2	0.0051±0.0002	

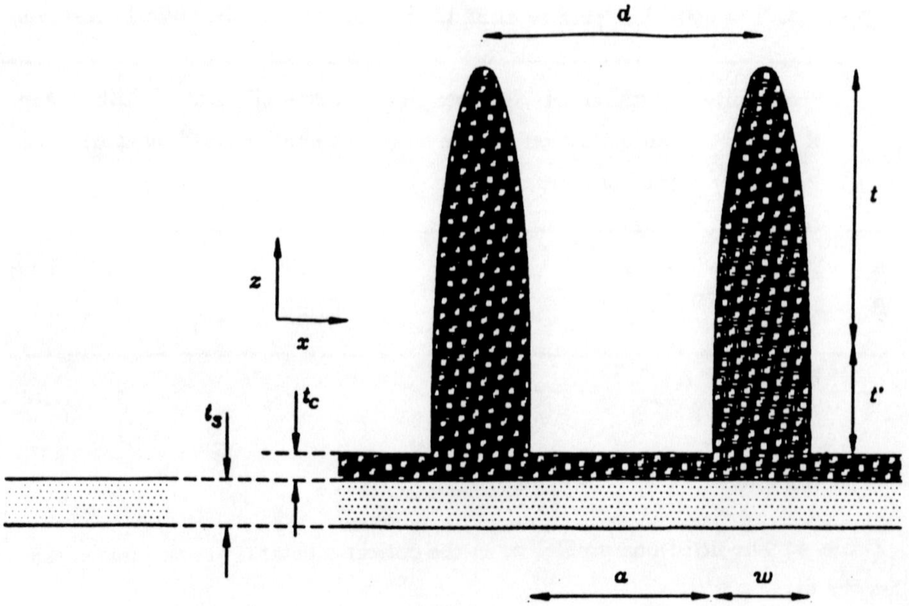

Figure 1. The model used to describe the characteristics of the grating.

profile. The integrations were done numerically to an accuracy of 0.1%, using the computer routine D01AHF from the Numerical Algorithms Group Library which uses Patterson's method [9]. The free parameters were obtained using a non-linear least-squares fit [10,11]. The errors to the fitted parameters were defined by allowing the normalized χ^2 of the fit to vary by unity. The grating model used in the fitting procedure is shown in figure 1. The thin contamination coating over the whole grating area has been discussed in section 3; the thickness t_c of this, which affects $T(k)$, is a free parameter in the fit. It is known from previous observations [5] that the profile of an isolated line drawn by contamination lithography has almost parallel sides with a curved top and an aspect ratio (height/width) of $\approx$5. Lines drawn close to one another are likely to be broadened by proximity effects, but their aspect ratios should still be somewhat greater than unity. A simple description of the line profile is that it has straight edges of height t' with a semi-elliptical top of height t and width w, as shown in figure 1. The line-profile parameters t', t and w are all free parameters in the fit.

4.1 Line profile fits using the Henke data

Using this line profile and the Henke values for δ and β (calculated assuming the density was equal to that of amorphous carbon) fitted values for the free parameters t_c, t', t and w of the two gratings are as shown in table 3. The fit probabilities were calculated from the χ^2 values of the fits. There are several reasons for rejecting these fits; the low probabilities (especially for grating B), the low aspect ratios $(t+t')/w$ and the large thicknesses t_c of the contamination coatings which imply that in each

Table 3. The fitted line profiles using the Henke data for the optical constants.

	Probability of fit (%)	Thickness of contamination coating t_c (nm)	Straight edge height t' (nm)	Semi-elliptical height t (nm)	Width w (nm)	Aspect ratio
A	15	37±6	6±2	49±2	47±1	1.17±0.07
B	≈0	74^{+17}_{-16}	7±6	51±7	61^{+5}_{-7}	0.95±0.19

Table 4. The fitted line profiles when the optical constants are also free parameters in the fit.

	Probability of fit (%)	Thickness of contamination coating t_c (nm)	Straight edge height t' (nm)	Semi-elliptical height t (nm)	Width w (nm)	Aspect ratio
A	65	10±5	7±4	75±4	47±1	1.74±0.12
B	59	18±12	14^{+8}_{-9}	70±11	59^{+5}_{-7}	$1.43^{+0.28}_{-0.26}$

Table 5. **Fitted values for δ and f_1, assuming that the density of contamination material is 1.6±0.3 times that of amorphous carbon.**

λ (nm)	δ $\times 10^2$	f_1
6	$0.52^{+0.23}_{-0.39}$	$1.81^{+0.87}_{-1.41}$
8	$0.84^{+0.22}_{-0.28}$	$1.66^{+0.49}_{-0.62}$
8.5	$0.84^{+0.44}_{-0.50}$	$1.47^{+0.82}_{-0.92}$
9	$1.13^{+0.34}_{-0.49}$	$1.77^{+0.63}_{-0.84}$
9.5	$1.11^{+0.45}_{-0.84}$	$1.57^{+0.69}_{-1.21}$
10	$1.40^{+0.46}_{-0.54}$	$1.77^{+0.65}_{-0.77}$
10.5	$1.48^{+0.41}_{-0.54}$	$1.69^{+0.57}_{-0.71}$
11	$1.74^{+0.13}_{-0.14}$	$1.82^{+0.30}_{-0.32}$

case more than 60% of the deposition would have been due to sideways scattered electrons. Similar results are obtained for other line profiles (e.g. a parabolic top rather than a semi-elliptical one) except that the fit probabilities are lower. If a higher density is used for the contamination material the only significant effect is to decrease the thicknesses (and therefore the aspect ratios) by amounts consistent with the density increase.

4.2 Simultaneous fits to the line profiles and the optical constants

By allowing the values of δ and β at each wavelength, as well as the line profile parameters, to vary freely it is of course possible to obtain fits with higher probabilities. However, if this is done the number of degrees of freedom becomes small and so it is difficult to obtain well-constrained fits. As has been discussed previously, the values of β calculated from the Henke data should be accurate provided that the density of the material is known. Thus the fits can be carried out by constraining all the β values to be a constant multiplicative factor different from the Henke values, this factor representing the difference in density between contamination material and ordinary carbon. When this is done (with the β values for the original substrate kept at the Henke values) the best fits are obtained when the multiplicative factor is 1.6±0.3, which indicates that the density of contamination material is close to that of diamond.

For these fits the line profile parameters are as shown in table 4. The fits have significantly higher probabilities, and the line profiles are much more consistent with those expected. The thicknesses t_c of the contamination coatings now indicate that about 20% of the deposition is caused by sideways scattered electrons. The fitted values for δ are shown in table 5, as are values for f_1 calculated assuming that the density of the material is higher as indicated by the multiplicative factor for β. These f_1 values are on average 42% of the Henke values. The calculated diffraction efficiencies are shown in table 6; these agree well with the measured efficiencies.

5. CONCLUSIONS

It would of course be possible to attempt to fit the data using other line profiles; for example replacing the semi-elliptical top with a parabolic one. In this case neither the line profile parameters nor the optical constants are significantly different to those shown in tables 4 and 5 (the δ values are, on average, less than 1% higher), although the fit probabilities are lower (significantly so for grating B). This is an indication that the measurements are not yet sufficiently precise to allow accurate determinations of the line profiles. The line thicknesses and widths must vary over the grating, the fitted values representing the means. Since the magnitudes of these variations are not known they cannot be taken into account when estimating the errors, and thus it is not worthwhile, at this stage, to attempt to parameterize the line profile by a more complicated function.

The values obtained for the refractive index decrement δ in the wavelength range 6-11 nm are, on average, a factor of 0.66 times those derived from the Henke data for amorphous carbon, with no significant fluctuations from this factor. Assuming that the high (compared to Henke) values of the absorption index β are due to the density of contamination material, the atomic scattering factors f_1 are, on average, a factor of 0.42 times the Henke values. Few other direct measurements of δ appear to have been made in this wavelength range, but the compilation of Hagemann *et al* [12], gives δ values (also derived from a Kramers-Kronig analysis) about 30% lower than the Henke values. Recent reflectance measurements by Windt *et al* [13] at wavelengths of 6.76 nm and 11.4 nm give δ values in agreement with the Henke values, but the method used to extract the optical constants is less direct than that reported here.

At this stage the low values obtained for δ and f_1 should be treated with caution; more extensive measurements over a wider wavelength range are currently being carried out, and will be reported at a later date. In addition, the procedure described above did not take into account the possibility that some unpolymerised oil was present; this tends to congregate between the grating lines. It is likely, however, that any such oil will have been burnt off early in the exposure of the gratings to the synchrotron beam and so will not have affected the grating characteristics.

Once the line profiles and optical constants of the carbon gratings are known, the gratings can be coated with metal; any material which can be evaporated or sputtered is, in principle, suitable. Similar measurements and analysis can then lead to δ and β values for the coating material. Inevitably, this will be more complex than for the straightforward carbon gratings, since the lines will consist of two materials and the profiles will not be so simple to parameterize. Alternatively, the carbon gratings could be replicated into other materials [14] but this is restricted to a narrower range of metals. However, the line profile will be simpler, although that of the carbon grating will not be retained. In addition, the same carbon master could be used to give several metal replica gratings.

The main purpose of the present analysis has been to show that the method of measuring soft x-ray optical constants using transmission gratings is potentially more powerful than originally envisaged by Tatchyn *et al* [1], and that it is a more

Table 6. The diffraction efficiencies calculated using the fitted parameters.

Wavelength (nm)	Diffraction order	Efficiency	
		Grating A	Grating B
6	0		$0.838^{+0.019}_{-0.021}$
	1		$0.0150^{+.0056}_{-.0033}$
	2		$0.0005^{+.0003}_{-.0002}$
8	0	$0.798^{+0.014}_{-0.015}$	
	1	$0.0179^{+.0029}_{-.0022}$	
	2	$0.0028^{+.0005}_{-.0004}$	
8.5	0		$0.689^{+0.034}_{-0.035}$
	1		$0.0176^{+.0057}_{-.0036}$
	2		$0.0007^{+.0004}_{-.0002}$
9	0	0.740±0.017	$0.639^{+0.036}_{-0.037}$
	1	$0.0243^{+.0023}_{-.0020}$	$0.0265^{+.0055}_{-.0043}$
	2	$0.0038^{+.0004}_{-.0003}$	$0.0010^{+.0005}_{-.0003}$
	3	$0.00020^{+.00004}_{-.00003}$	
9.5	0		$0.617^{+0.040}_{-0.041}$
	1		$0.0230^{+.0074}_{-.0047}$
	2		$0.0009^{+.0005}_{-.0003}$
10	0	0.688±0.019	$0.571^{+0.041}_{-0.042}$
	1	0.0285±0.0015	$0.0305^{+.0061}_{-.0050}$
	2	0.0044±0.0002	$0.0013^{+.0006}_{-.0004}$
	3	0.00026±0.00004	
10.5	0		0.543±0.043
	1		$0.0297^{+.0055}_{-.0046}$
	2		$0.0013^{+.0005}_{-.0004}$
11	0	0.632±0.021	
	1	0.0338±0.0019	
	0	0.0051±0.0003	

straightforward and therefore more reliable method of obtaining the refractive index decrement, in particular, than in analyses relying on reflectance measurements. A more detailed analysis of a larger data set will be presented in a forthcoming publication

ACKNOWLEDGEMENTS

The authors are grateful to T. Taguchi (King's College, London) for making the gratings and to numerous colleagues at King's College and at the Max-Planck-Institut für Quantenoptik (MPQ) for useful and enlightening discussions. Part of this work was carried out while AGM was an Alexander von Humboldt research fellow at MPQ.

REFERENCES

1. R Tatchyn, I Lindau, E Källne and E Spiller *Phys. Rev. Letts* **53** 1264–7 (1984)
2. B L Henke, P Lee, T J Tanaka, R L Shimabukuru and B J Fujikawa *At. Data Nucl. Data Tables* **27** 1–144 (1982)
3. B L Henke, E M Gullikson, J C Davis, M Fryer and A Oren *Proc. SPIE* **911** 107–21 (1988)
4. P Charalambous, Ph. D Thesis, London University, (1982)
5. C J Buckley, M T Browne and P Charalambous *Proc. SPIE* **537** 213–7 (1985)
6. A G Michette, E Fill, T Taguchi and M Kühne *Proc. SPIE* **1140** 247–52 (1989)
7. M Kühne and P Müller *Proc. SPIE* **1140** 220–5 (1989)
8. H Braüninger, P Predehl and K P Beuermann *Appl. Opt.* **18** 368–73 (1979)
9. T N L Patterson *Math. Comp.* **22** 847–56 (1968)
10. D W Marquardt *J. Soc. Ind. Appl. Math.* **11** 431–41 (1963)
11. B R Bevington *Data Reduction and Error Analysis for the Physical Sciences* (New York: McGraw-Hill) pp232–45 (1969)
12. H-J Hagemann, W Gudat and C Kunz *DESY Report* **SR-74/4** (1974)
13. D L Windt, W C Cash Jr, M Scott, P Arend, B Newman, R F Fisher, A B Swartzlander, P Z Takacs and J M Pinneo *Appl. Opt.* **27** 279–95 (1988)
14. C J Buckley *The fabrication of gold zone plates and their use in scanning X-ray microscopy* Ph.D. Thesis, University of London (1087)

DISTINCTIVE FEATURES OF NEON ISOELETRONIC SEQUENCE 31-31' AND 2s-2p TRANSITIONS

A.V. Gulov and E.P. Ivanova
Institute of Spectroscopy, USSR Academy of Science, 142092, Moscow Region, Troitsk, USSR

INTRODUCTION

Laboratory X-ray laser investigations are supplemented recently with new ideas concerning a resonant photoexcitation as a pumping mechanism [1-3]. Preliminary estimations of suggested schemes [2] showed that the spectral intensity of the pump line should be more than 10^{10} w/cm^2·mÅ. In this case the plasma electron density amounts to 10^{21}–10^{22} sm^{-3}. Thus, the prediction of ions spectroscopic characteristics becomes more complicated due to the presence of strong electric microfields. The final theoretical model to describe laser enhancement in plasma is not elaborated yet even for the simplest case of collisional excitation for active levels population. The first step is to calculate with a sufficient accuracy the energy levels, transition probabilities and excitation cross-sections for an isolated ion.

The neon-like scheme for lasing proved to be one of the promising schemes [4-7] however, neon isoelectronic sequence distinctive features may play an important role for prediction of population levels kinetics. Namely a strong mixing of certain levels for definite nuclear charge Z may take place which leads to the drastic redistribution of oscillator strengths.

In the present paper wavelengths and probabilities of 3-3 electron transitions and 2-2 vacancy transitions are given for neon-like isoelectronic sequence. Radiative lifetimes of the $2p^5 3l$, $2s2p^6 3l$ (l=0-2) for eight Ne-like ions from CuXX to PbLXXIII are critically compared with the results of other theoretical approaches. This work continued the study of Ne-like spectroscopic characteristics [8,9]. In [8] TiXIII was considered as an example, as for TiXIII the abundant data about wave-lengths and radiative lifetimes were accumulated experimentally by Träbert and Jupén [10]. In [8] we came to the conclusion that there might arise a significant difference between results obtained with non-relativistic (Hartree-Fock-Pauli) and relativistic calculations. The reason of this difference is discussed below. The first 72 excited energy levels of Ne-like ions $2p^5 3l$, $2p^5 4l$, $2s2p^5 3l$, $2s2p^5 4l$ (l=0-2) and probabilities for E1, M1, E2, M2, E3, M3, M4 radiative decays of these states to the ground state $1s^2 2s^2 2p^6$ were presented by us in [9]. Excitations to the n=3 shell were studied for all ions in the range Z=27-92; excitations into the n=4 shell were included up to Z=66. Investigation of each term of the sequence allows one to reveal the extrema in Z-functions of radiative transitions probabilities. In [9] we investigated a strong configurations interaction for those Z values, where two energy levels draw together and change their mutual positions. In above Z region a sharp redistribution of state self vector components takes place, that may result in provoking the irregularity in function energy on Z. The calculation accuracy in such Z region depends much on the method used. The shortcomings of the theoretical method can be obviously revealed while considering the processes cross-sections. Without satis - factory taking into account the correlation effects the calculation error of transition probabilities may reach thousands of percents.

RELATIVISTIC PERTURBATION THEORY WITH MODEL ZERO APPROXIMATION.

The RPTMP method proved to be reliable in calculation of energy levels in various isoelectronic sequences. The detail description of our calculation model is given in [9,11,12]; here we point at the principal foundations of our model.

Successive inclusion of relativistic effects is inevitably based on the electrodynamic theory and the particular calculation scheme must be constructed as a well grounded approximation of the electrodynamic approach. We use one of the possible versions of such an approach so-called energy approach [13]. Well known in nonrelativistic theory is the field form procedure, connected with the diagonalization of secular matrix for the calculation of the energy shift ΔE of degenerate atomic state [14]. In our method this procedure is based upon Gell-Mann and Low adiabatic formula [15] and the perturbation expansion of the scattering matrix; we use the analogous scheme with the electrodynamic scattering matrix [13]. The expansion for energy has only even perturbation powers; the perturbation expansions are usually presented in the form of Feinman diagram contributions [13].

In the consequent electrodynamic perturbation theory, a principally new moment is that already the second-order correction (the first order by interelectron interaction) for the secular matrix elements is a complex value. The imaginary part of the second-order correction is asso - ciated with the possibility of a single-photon decay of the excited state. A complete calculation of the energy and decay probability of degenerate states is reduced to the estimation of the real and imaginary parts of the secular matrix and its diagonalization. The RPTMP repre - sents level width as the sum over virtual states of the atomic system. The sum can be transformed in such a way that separate term of the sum represents decay probability over separate physical channel [13]. The energy approach uses a simple calculation technique for angular parts of the energy shift developed earlier for a nonrelativistic atoms. The real and the imaginary parts of the secular matrix elements differ only in the determination of radial integrals.

Calculation of atomic characteristics with the use of model potentials are winning growing recognition. Both the economic efficiency and simplicity of these calculations and also the possibility of conside -

rable increase in the accuracy of final results seem attractive. Field form perturbation theory with model zero approximation of atomic system was developed in [13]. An important difference and advantage of our approach is the effective inclusion of high order perturbation theory corrections [11,12].

In the calculation made here much attention is paid to the search for the most adequate description of one-particle zero approximation energies and wave functions. One-particle electron and vacancy energies are counted from the core $1s^2 2s^2 2p^6$ energy. By definition the total energy of one particle over the core is the sum of all one-particle diagrams of all orders of the perturbation theory. Precisely determined one-particle energies allow the computation already in the zero-order PT to take into account the major part of the correlation, relativistic and quantum electrodynamic (QED) effects. It should be stressed that the convergence of the perturbation series and also the accuracy of final results are determined to a large measure by the quality of the one-particle basis. For the vacancy one-particle zero order energy is determined by ionization potential of Ne-like ion and by the structure of terms in F-like ion. For the electron, one-particle zero order energy is determined by the first ionisation potential of the Na-like ion and by the term structure in Na-like ion. The structures of energy levels of one electron over the core (Na-like ions) and of one vacancy over the same core (F-like ions) were investigated by one of the authors earlier [16,17]; the way of the model potential construction is also given in [16]. Recently [9] we have specified one-particle energies in the region of large $Z>80$.

The whole calculation procedure can be formed by three steps: i) construction of core potentials along the isoelectronic sequence to determine wave functions of single quasi-particle over the core; ii) correction of one quasi-particle energies to reach ultimately complete accounting for correlation, relativistic and QED effects; iii) estimation of spectroscopic characteristic of ions with few quasi-particles over the core along the isoelectronic sequence. In principle, the way of one-particle zero order wave functions and energies determination is the main difference of our method from well-known self-consistent procedures and fitting methods.

RESULTS OF CALCULATION

The discrepancy between different theoretical transition probabilities values for $2p^5 3p[^1S_0]-2p^5 3s[^{1,3}P_1]$ was discussed by us in [8]. We believe that the reason of discrepancy consists in difference of state vector compositions obtained in different calculations. Note that so strong sensitiveness of self vector composition on the approximation used is the distinctive feature of exactly Ne-like ions. The investigation [12] revealed that while energy matrix elements calculating a significant mutual compensation of some correlation, relativistic and QED contributions may take place. Compensation effect is a characteristic feature of atomic systems containing a vacancy and an electron; for the Ne-like system it seems to be the most strongly pronounced. Provided the high accuracy of all effects contributions calculation the spectroscopic characteristics of Ne-like ions can be predicted satisfactorily. The effect of drawing together of levels with the same parity and the same total moment J, resulting in a strong redistribution of oscillator strengths can be observed at Fig.1, where nonmonotonous changing along

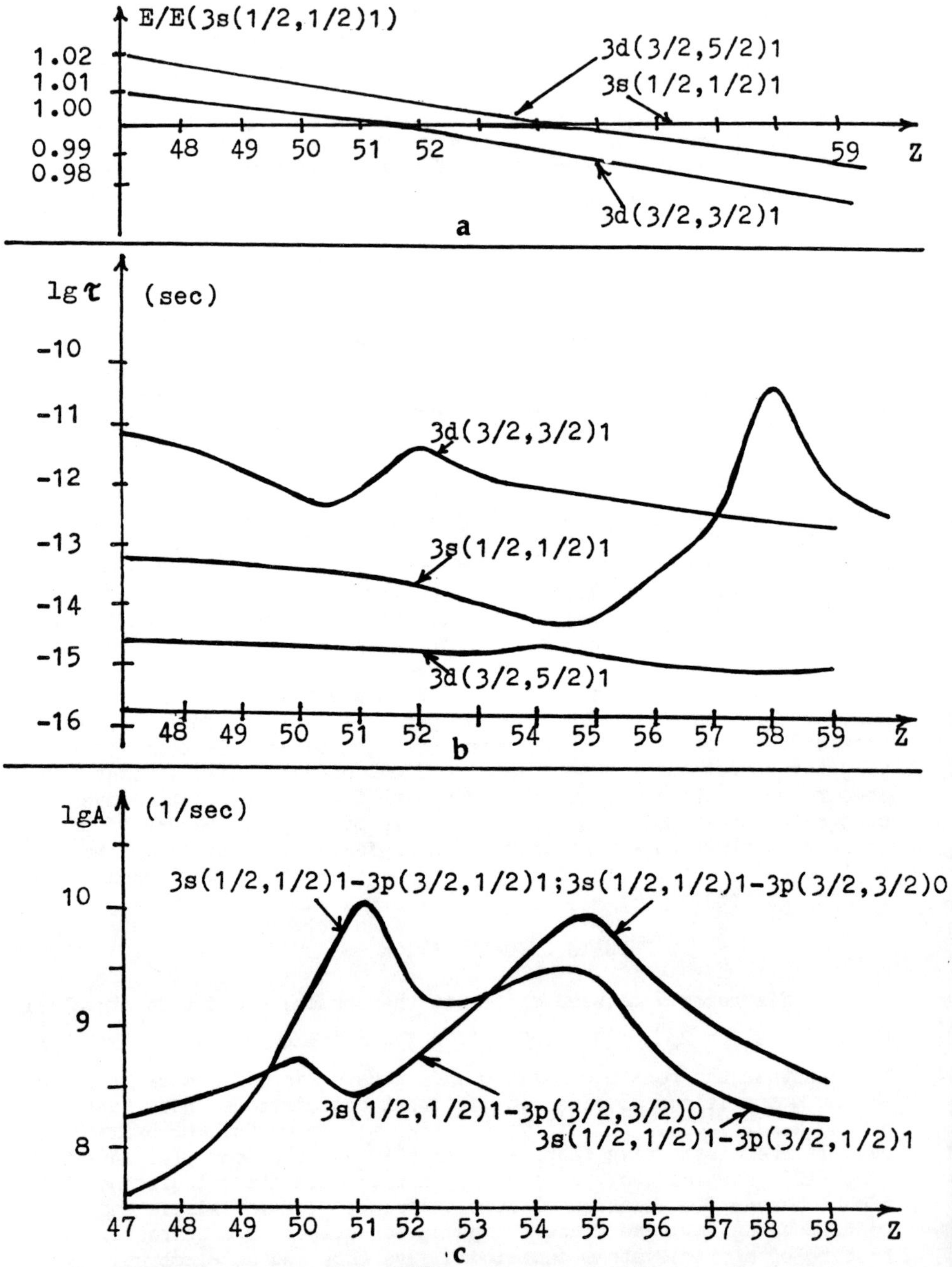

Fig.1 **a**)Changing along Z of $2p^5_{3/2}3d_{3/2}$(J=1), $2p^5_{3/2}3d_{5/2}$(J=1) energy levels positions relative to $2p^5_{1/2}3s_{1/2}$(J=1) energy level; **b**)Lifetimes of these 3 levels as function on Z; **c**) 3-3 transition probabilities on Z.

Z of lifetimes of $2p^5_{3/2}3d_{3/2}$(J=1), $2p^5_{3/2}3d_{5/2}$(J=1) and $2p^5_{1/2}3s_{1/2}$(J=1) levels are shown. The changing along Z of the mutual levels positions are also shown at the top part of Fig.1

The lifetimes of levels showed at Fig.1 are determined mainly by the radiative decay into the ground state $2p^6$; the nonmonotonous behavior of these transition probabilities on Z caused by the effect of level mixing we discussed in [9]. The present calculation shows that the effect of strong level mixing influences appreciably over 3-3 transition probabilities. It is especially pronounced for 3-3 transitions $2p^5_{1/2}3s-2p^53p$, also showed at Fig.1.

Fig.1 manifests the high sensitiveness of transition probabilities to the external perturbation influence. The another distinctive feature of neon sequence is the changing along Z of the mutual positions of le - vels of opposite parties showed at Fig2. At certain Z points a very close drawing together of states $2p^5_{3/2}3p$ and $2p^5_{1/2}3s$ takes place. At certain Z points some of these states may undergo the strong mixing under the influence of an external electric field. Basing upon above properties of neon sequence one may use the varying external electric field as a tool to control active levels population kinetics.

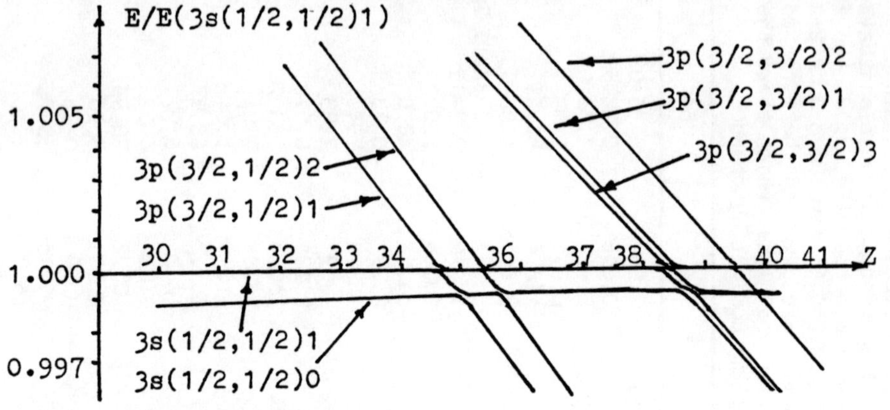

Fig.2. Changing along Z of the mutual positions of levels of opposite parities.

Our final results of neon-like ions spectroscopic characteristics are given in Table 1. Some of the most promising transitions probabilities are also shown at Fig.3. 3p(J=2)-3s(J=1) transition probabilities coincide with corresponding data obtained by Hartree-Fock-Pauli calculation [18,19]. The probabilities for 3p(J=0)-3s(J=1) transition, calculated in [18,19] are appreciably stronger, they are shown at Fig.3 by separate points.

Note the promising features of 2s-2p transitions for lasing under optical pumping conditions. These transitions are of shorter wavelengths in comparison with 3-3 transitions; besides the states $2s2p^63s$(J=0,1) decay very rapidly into $2p^5_{1/2,3/2}3s$(J=1) states, and these channels are dominating decay ones for these states (see Table 1 and Fig.3).

TABLE 1. Radiative lifetimes (τ, in picosecond), dominating 3-3 or 2-2 decay chanels, branch coefficients (k) within dominating channels, and corresponding wavelengths (λ, Å) for $2p^5_{j_v}3l_{j_e}$, $2s_{j_v}2p^6 3l_{j_e}$ states of Ne-like ions. The states $2p^5_{j_v}3l_{j_e}(J)$ are denoted $3l(j_v,j_e)J$, the states $2s_{j_v}2p^6 3l_{j_e}(J)$ are denoted $2s2p^6$ $3l(j_v,j_e)J$. First two states: $2p^5_{3/2}3s(J=0,1)$ have no 3-3 or 2-2 transitions; only lifetimes are given for these states. The dominating channels are given in the first column. The states changing their dominating channel along Z have the fourth additional line, which shows the dominating channel for given Z; the dash means the repetition of the preceding dominating channel.

state Z		29	34	47	54	59	64	74	82
3s(3/2,1/2)2	τ	2.43+06	4.72+05	2.00+04	5.36+03	2.31+03	1.07+03	2.66+02	9.88+02
3s(3/2,1/2)1	τ	0.611	0.264	0.0561	0.0296	0.0196	0.0134	6.64-03	3.94-03
3s(1/2,1/2)0	τ			8.41+02	4.56+02	2.63+02	1.39+02	4.00+02	15.5
-3p(3/2,1/2)1	k			0.97	0.99	0.996	0.998	0.999	1.0
	λ			118.6	51.25	31.70	20.77	10.01	6.000
3s(1/2,1/2)1	τ	0.825	0.397	0.0666	4.50-03	1.14	0.0940	0.0304	0.0174
-3p(3/2,3/2)1				-	3p(3/2,3/2)0	3p(3/2,3/2)1	3p(3/2,1/2)2	-	-
	k			1.8-05	2.2-05	8.6-04	1.2-04	1.3-04	2.0-04
	λ			207.4	133.7	47.21	20.73	9.993	5.994
3p(3/2,1/2)1	τ	145.	110.	64.5	48.8	39.1	30.1	14.8	7.00
-3s(3/2,1/2)2	k	0.61	0.69	0.79	0.78	0.74	0.68	0.46	0.28
	λ	340.5	264.1	163.8	134.1	118.0	104.7	83.98	71.25
3p(3/2,1/2)2	τ	145.	85.8	17.9	7.83	4.51	2.70	1.08	0.571
-3s(3/2,1/2)2	k	0.42	0.33	0.12	0.070	0.047	0.033	0.018	0.012
	λ	311.8	247.8	158.4	130.7	115.5	102.8	83.03	70.77
3p(3/2,3/2)3	τ	118.	69.7	16.3	7.01	3.79	2.03	0.580	0.215
-3s(3/2,1/2)2	k	1.0	1.0	1.0	1.0	1.0	1.0	1.0	1.0
	λ	287.4	211.8	101.3	68.77	52.31	39.85	23.36	15.42
3p(3/2,3/2)1	τ	128.	73.1	16.3	6.97	3.77	2.01	0.574	0.213
-3s(3/2,1/2)1	k	0.69	0.77	0.82	0.82	0.83	0.83	0.83	0.83
	λ	296.9	220.7	105.2	70.94	53.71	40.75	23.73	15.60
3p(3/2,3/2)2	τ	104.	54.6	9.70	3.98	2.16	1.20	0.383	0.157
-3s(3/2,1/2)2	k	0.44	0.39	0.30	0.28	0.29	0.30	0.33	0.37
	λ	266.1	197.1	95.95	65.75	50.32	38.55	22.80	15.13
3p(3/2,3/2)0	τ	133.	77.0	16.8	7.10	3.82	2.04	0.580	0.215
-3s(3/2,1/2)1	k	0.73	0.87	1.0	1.0	1.0	1.0	1.0	1.0
	λ	236.0	168.3	80.93	56.58	44.03	34.31	20.92	14.15

TABLE 1. (continued)

state Z		29	34	47	54	59	64	74	82
3p(1/2,1/2)1	τ	171.	128.	69.6	52.3	42.4	34.0	20.3	12.1
-3s(1/2,1/2)1	k	0.65	0.66	0.64	0.49	0.59	0.56	0.45	0.34
	λ	329.5	262.0	167.1	134.2	123.4	109.5	88.40	75.47
3p(1/2,3/2)1	τ	117.	68.1	15.3	6.31	3.17	1.39	0.249	0.190
-3s(1/2,1/2)0		-	-	-	-	-	-	3s(3/2,1/2)2	3s(1/2,1/2)0
	k	0.64	0.64	0.62	0.59	0.55	0.43	0.51	0.58
	λ	280.8	208.4	100.6	68.61	52.30	40.01	6.428	15.29
3p(1/2,3/2)2	τ	103.	53.9	9.45	3.86	2.12	1.17	0.374	0.154
-3s(1/2,1/2)1	k	0.86	0.77	0.58	0.44	0.55	0.58	0.64	0.71
	λ	285.3	210.1	100.5	67.36	52.27	39.77	23.32	15.39
3p(1/2,1/2)0	τ	146.	107.	61.0	46.0	36.8	28.4	14.4	6.76
-3s(1/2,1/2)1		-	-	-	-	-	-	3s(3/2,1/2)1	-
	k	0.55	0.66	0.81	0.64	0.76	0.70	0.52	0.72
	λ	213.5	177.6	121.4	99.59	91.16	81.21	7.894	5.048
3d(3/2,3/2)0	τ	56.7	38.8	13.1	6.60	3.91	2.26	0.721	0.282
-3p(3/2,1/2)1	k	0.67	0.79	0.93	0.96	0.98	0.98	0.99	0.997
	λ	227.4	174.7	90.64	63.66	49.31	38.16	22.87	15.27
3d(3/2,3/2)1	τ	5.34	3.29	7.44	0.632	0.125	0.0444	0.0109	4.85-03
-3p(3/2,1/2)1	k	0.041	0.045	0.39	0.070	0.024	0.015	0.012	0.014
	λ	222.0	169.9	87.91	61.79	47.98	37.24	22.46	15.06
3d(3/2,5/2)2	τ	64.6	46.0	24.5	16.4	11.7	7.96	3.46	1.73
-3p(3/2,3/2)1	k	0.36	0.33	0.51	0.53	0.51	0.47	0.39	0.34
	λ	244.9	192.2	113.7	89.90	76.23	64.73	46.65	35.86
3d(3/2,5/2)4	τ	68.2	52.4	27.6	19.1	14.3	10.6	5.57	3.20
-3p(3/2,3/2)3	k	1.0	1.0	1.0	1.0	1.0	1.0	1.0	1.0
	λ	240.7	191.0	117.4	92.67	78.51	66.58	47.86	36.69
3d(3/2,3/2)3	τ	57.0	39.0	13.1	6.64	3.93	2.27	0.722	0.282
-3p(3/2,1/2)2	k	0.86	0.88	0.94	0.96	0.97	0.98	0.99	0.994
	λ	223.0	168.7	87.11	61.44	47.84	37.21	22.49	15.09
3d(3/2,3/2)2	τ	59.4	43.2	13.6	6.72	3.97	2.29	0.725	0.283
-3p(3/2,1/2)2	k	0.34	0.30	0.45	0.48	0.49	0.49	0.50	0.50
	λ	214.7	161.9	85.37	60.21	46.94	36.56	22.17	14.91

TABLE 1. (continued)

state Z	29	34	47	54	59	64	74	82
3d(3/2,5/2)3	τ 68.1	52.4	27.5	19.0	14.2	10.5	5.44	3.10
-3p(3/2,3/2)2	k 0.80	0.80	0.80	0.80	0.79	0.79	0.78	0.78
	λ 236.5	187.3	114.9	90.79	76.98	65.35	47.10	36.19
3d(3/2,5/2)1	τ 0.063	0.0193	2.98-03	2.11-03	9.50-04	6.60-04	3.49-04	2.26-04
-3p(3/2,3/2)1	–	–	3p(3/2,3/2)0	–	–	–	–	–
	k 2.4-04	1.2-04	4.6-05	4.0-05	3.1-05	3.0-05	3.1-05	3.5-05
	λ 201.6	155.4	127.9	99.80	87.35	73.90	52.83	40.25
3d(1/2,3/2)2	τ 56.9	39.0	13.1	6.64	3.94	2.28	0.723	0.282
-3p(1/2,1/2)1	k 0.84	0.88	0.94	0.96	0.97	0.98	0.99	0.994
	λ 220.8	167.2	86.31	60.87	47.39	36.86	22.28	14.95
3d(1/2,5/2)2	τ 67.8	52.0	26.6	17.3	12.3	8.48	3.62	1.67
-3p(1/2,3/2)1	k 0.84	0.87	0.86	0.80	0.76	0.68	0.52	0.46
	λ 237.9	188.4	115.3	90.78	76.65	64.56	48.83	36.54
3d(1/2,5/2)3	τ 68.1	52.4	27.5	18.9	14.2	10.4	5.36	3.01
-3p(1/2,3/2)2	k 0.99	0.99	0.996	0.99	0.99	0.98	0.96	0.94
	λ 236.9	187.7	115.1	90.82	76.92	65.23	46.89	35.96
3d(1/2,3/2)1	τ 0.024	0.0128	3.61-03	2.01-03	1.35-03	8.99-04	6.54-04	3.72-04
-3p(1/2,1/2)0	k 1.4-04	1.4-04	1.6-04	1.9-04	2.2-04	2.4-04	5.8-04	8.7-04
	λ 270.7	195.7	94.50	65.60	50.71	39.35	23.23	15.54
$2s2p^6$ 3s(1/2,1/2)1	τ 16.2	9.34	1.85	0.726	0.371	0.194	0.0556	0.0179
-3s(3/2,1/2)2	k 0.67	0.71	0.79	0.81	0.82	0.82	0.82	0.83
	λ 77.09	57.93	28.24	19.22	14.64	11.17	6.654	4.415
$2s2p^6$ 3s(1/2,1/2)0	τ 16.1	9.26	1.84	0.718	0.364	0.186	0.0494	0.0178
-3s(3/2,1/2)1	k 0.80	0.85	0.95	0.98	0.99	0.99	0.997	0.999
	λ 74.48	56.08	27.54	18.83	14.38	11.01	6.537	4.363
$2s2p^6$ 3p(1/2,1/2)0	τ 14.8	8.65	1.79	0.709	0.362	0.185	0.0493	0.0177
-3p(3/2,1/2)1	k 0.54	0.69	0.91	0.96	0.97	0.99	0.995	0.998
	λ 76.02	57.46	28.19	19.21	14.64	11.19	6.617	4.407

TABLE 1. (continued)

state Z		29	34	47	54	59	64	74	82
$2s2p^{6}$ 3p(1/2,1/2)1 -3p(3/2,1/2)2	τ	0.834	0.273	0.0427	0.0231	0.0188	0.0385	1.50-03	1.20-03
	k	0.033	0.020	0.018	0.026	0.042	0.16	0.025	0.055
	λ	77.33	58.05	28.26	19.23	14.64	11.17	6.629	4.410
$2s2p^{6}$ 3p(1/2,3/2)2 -3p(3/2,3/2)3	τ	14.2	8.19	1.65	0.652	0.333	0.170	0.0456	0.0164
	k	0.50	0.53	0.60	0.62	0.63	0.64	0.64	0.65
	λ	76.94	57.81	28.19	19.20	14.62	11.18	6.610	4.402
$2s2p^{6}$ 3p(1/2,3/2)1 -3p(3/2,3/2)2	τ	0.161	0.0747	0.0161	8.50-03	5.74-03	4.01-03	2.14-03	1.39-03
	k	3.2-03	2.8-03	3.6-03	4.8-03	6.4-03	9.0-03	0.018	0.033
	λ	77.59	58.32	28.40	19.32	14.71	11.23	6.635	4.416
$2s2p^{6}$ 3d(1/2,3/2)1 -3d(3/2,3/2)1	τ	12.5	7.47	1.61	0.647	0.333	0.171	0.0462	0.0167
		–	–	3d(3/2,3/2)2	–	–	–	–	–
	k	0.20	0.23	0.32	0.36	0.37	0.38	0.39	0.39
	λ	76.71	57.71	28.67	19.47	14.80	11.29	6.656	4.426
$2s2p^{6}$ 3d(1/2,3/2)2 -3d(3/2,3/2)3	τ	12.7	7.52	1.47	0.549	0.274	0.139	0.0379	0.0141
	k	0.29	0.35	0.47	0.49	0.50	0.51	0.53	0.55
	λ	78.09	58.53	28.33	19.24	14.64	11.18	6.609	4.401
$2s2p^{6}$ 3d(1/2,5/2)3 -3d(3/2,5/2)4	τ	13.0	7.86	1.72	0.690	0.355	0.182	0.0489	0.0176
	k	0.42	0.47	0.57	0.61	0.62	0.63	0.64	0.64
	λ	77.49	58.16	28.29	19.24	14.65	11.19	6.615	4.404
$2s2p^{6}$ 3d(1/2,5/2)2 -3d(3/2,3/2)3	τ	10.2	5.20	0.85	0.347	0.188	0.103	0.0318	0.0127
		–	3d(3/2,5/2)3	3d(3/2,5/2)2	–	–	–	–	–
	k	0.12	0.11	0.14	0.17	0.19	0.21	0.25	0.28
	λ	75.01	57.65	27.84	19.02	14.52	11.11	6.585	4.391

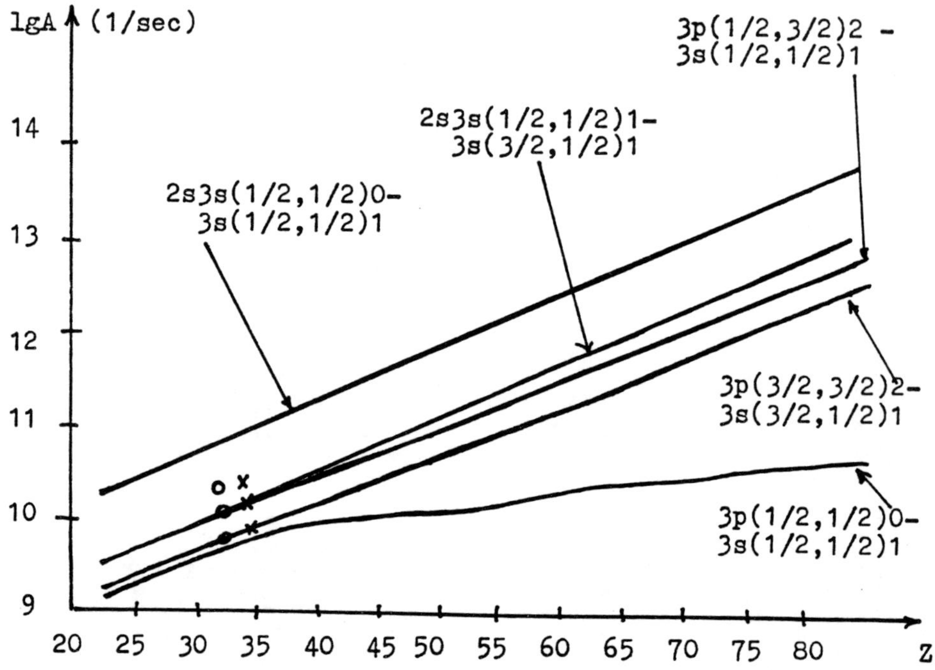

Fig.3. Probabilities along Z of the most promising 3-3 and 2-2 transitions for neon-like ions scheme lasing. o - [18], x - [19].

REFERENCES

1. J.Nilsen, Phys.Rev.A 40 (1989) 5440.
2. J.Nilsen, Optics Commun. 72 (1989) 371.
3. B.N.Chichkov, Reports' Theses, International symposium "Short wave - length lasers and their applications", Samarcand (1990) p.47.
4. A.V.Vinogradov and V.I.Shlaptsev, Kvantovaya Electron. 7 (1980) 1319; 10 (1983) 516.
5. D.L.Matthews et. al. Phys. Rev. Lett. 54 (1985) 110.
6. T.N.Lee, E.A.McLean and R.G.Elton, Phys. Rev. Lett. 59 (1987) 1185.
7. B.J.MacGowan et. al. J. Appl. Phys. 61 (1987) 5243.
8. A.V.Gulov and E.P.Ivanova, Phys. Lett. A 140 (1989) 39.
9. E.P.Ivanova and A.V.Gulov, AT. DATA NUCL. DATA TABLES (in print)

10 E.Träbert and C.Jupén, Phys. Scr. 36 (1987) 586

11 E.P.Ivanova, L.N.Ivanov, A.V.Glushkov, A.E.Kramida, Phys. Scr. 32 (1985) 513.

12 E.P.Ivanova, A.V.Glushkov, JQSRT 36 (1986) 127.

13 M.N.Driker, L.N.Ivanov, Sov. J. Opt. Spectrosc. 49 (1980) 209.

14 V.V.Tolmachev. Adv. Chem. Phys. 14 (1969) 421.

15 M.Gell-Mann and F.Low, Phys. Rev. 84 (1951) 350.

16 L.N.Ivanov, E.P.Ivanova, At. Data Nucl. Data Tables, 24 (1979) 95

17 M.N.Driker, E.P.Ivanova, L.N.Ivanov and A.F.Shestakov, JQSRT 28 (1982) 531.

18 A.K.Bhatia, U.Feldman, and J.Seely, At. Data Nucl. Data Tables 32 (1985) 435.

19 E.Biemont and J.E.Hansen, At. Data Nucl. Data Tables 37 (1987) 1.

THE FACTOGRAPHIC DATABASE "SPECTR-2" FOR SHORT-WAVE LENGTH LASER STUDIES

V.Yu. Bugaev, V.G. Pal'chikov, I.Yu. Skobelev, and A.Ya. Faenov
NPO VNIIFTRI, Mendeleevo, Moscow Region, 141570 USSR

Introduction. The atomic constants of multycharged ions (level energetic structure, wavelength and radiation transition probability) are the main object of studies in the fields of high-temperature plasma physics, quantum radio physics & non- atmospheric astronomy. It's mainly connected with three circumstances: First, the emission spectra of such ions carry fundamental information about the matter properties in the extreme conditions of super high temperature and pressure, Second, the knowledge of atomic constants is necessary for numerical plasma modeling of both laboratory and astrophysical sources, and Third, for theoretical and experimental studies with in the programmes of X-ray lasers development

The data about the spectra characterictics of multicharged ions may be obtained by three methods: subsequent, quantum-mechanical calculation, experimentally, by means of semi-empirical methods, which use both experimental and theoretical information. Until now large material has been accumulated in this field which, however, is available in many original works and accessible to a very narrow circle of experts. A number of compilations of the US National Institute of Standards and Technology which are available for separate chemical elements ion spectra doesn't improve the situation. Due to this fact the task of systematic accumulation and distribution of information on multicharged atomic constants is especially vital at the modern stage and it is obvious the information concentration in the data base DB must mean its critical estimation. It is natural that such DB should be built as automated base with the use of modern computing facilities, since the volume of information is very large and it is not possible to handle it "manually".

"Spectr-2" automated data base ADB is an extension of "Spectr-1" database developed in VNIIFTRI for storage and rapid search of multicharged ions atomic characteristics.

The development of a new ADB version is linked with the necessity of including information about cross-section and excitation speeds of ions by different particles in ADB. On the other hand, large volume of information (more than 200000 transitions) accumulated on spectral characteristics (level energetic structures, wavelengths, radiation transition probabilities etc) as well as the experience of utilizing "Spectr" ADB led to the necessity of changing formats of data presentation in both entry documents and files loaded for automatic search. The process of ADB servecing has also changed. A number of service packges providing a more flexible technology to support data base has been created. There is a possibility to obtain integral charactaristics of the data base to estimate the ADB informational completeness and it's current state.

The present article is devoted to the functional capabilities of the "Spectr-2" database second version.

"Spectr-2" DB information structure. The main unit of information presentation in "Spectr-2" ADB is a document. Every document contains segments which incorporate data or texts. In order to distinguish one data from another they are given a unique tree-symbol label. The data having one label in "Spectr-2" DB may differ by its position within the segment.

Combinations of practically any symbols may serve as data for "Spectr-2" DB. One data from another is separated by at least one blank. Terms indicating element chemical symbols, energy level configuration, nucleous charge levels energy, transition wave length, radiation probabilities, etc serve as data in "Spectr-2" ADB.

The documents contained in "Spectr-2" ADB may be found with the help of terms which are referred to as search terms. The search terms fully identify the information contained in the document. For this purpose every document should feature informational completeness property, i.e. the document should contain all terms by means of which it can be found. For instance, the document "Electronic transitions" should contain information about the element chemical symbols and isoelectronic sequence, the configuration of the initial & final levels, as well as about the transition wavelength. The lack of at least one of the above listed terms makes the document uncertain.

The document structure in "Spectr-2" ADB is oriented into the use of "DIALOG-2" (West analogues "Stairs") as the information search system.Every document consists of records having the length of 80 bites. The first three bites are left for a label, 69 bites serve to locate data or texts, while the last 8 string bites are not used. The labels in the documents are placed in the increasing order. The change in the label's increasing order points to the beginning of the next document.

The documents of the following types are supported in "Spectr-2" DB:

1- "Ionization potential"
2- "Electronic Transitions"

3- "Excitation cross-section"
4- "Excitation rate"
5- "Information supplier"
6- "Bibliography"
7- "Commentaries".

The document of different type are interfaced at a level of similar terms. For exemple, if reference number in the "Bibliography" document coincides with the reference number of the "Electronic transition" documents, they are interfaced at the logical level. The documents of types 4 and 5 which were absent in the "Spectr" ADB previous version, feature specific information presentation. Since excitation cross-sections and speeds may be presented either by tables or by approximation formulas, additional agreements on the method of locating information have been worked out.

All information available in "Spectr-2" DB is coded in and has a symbolic character (including a numerical information) "Spectr-2" DB. Input documents are prepared with the help of programmes-convertors proceeding from the data coming from various sources and presented in practically optional formats.

Interaction of the Terminal User with "Spectr-2" DB and the obtaining of output forms.

The terminal user is provided an opportunity to obtain information from "Spectr-2" ADB in a dialogue mode. The dialogue mode is realised in the electronic computer in the local TV access network by EC-7920 equipment and its analogues under OC MVT 6.1 operational system control and higher in the TV acces system.

The searching prescript is compiled from terms or part of terms truncated from the right and logical operators between them. The result of fulfilling searching prescipts is the number list of documents containing information given in the conditions. This list may be processed by three methods to obtain the so called output forms: by means of report generators and the output of the resulting tables to the printer (type EC-7934) in interactive mode; by means of obtaining tables built on the results of the search in the packeting mode; by means of information output into the intermediate file for further processing and use in application programmes.

In the first case standard facilities of "DIALOG-2" system are realised. They provide the generation and on-line inspection of reports due to the search results. The report formats are created as required and may be stored and adjusted in a dialogue mode. The generated report is output onto the display or printer during the system operation. Unfortunatelly, the printing string length in this case is limited by a screen string length (80 bites) and the use of printer does not provide a sufficient rate of outputting large volumes of information. To eliminate these setbacks a packaging mode for processing found documents is used and it's realised with the help of programmes developed and included in "Spectr-2" ADB.

These programmes automatically form tables from the documents after every request. The forming of tables is carried out after completing a dialogue session. The programme of obtaining output form looks through the lists of output documents twice, sorting them out by request numbers and user's identificattors in advance.

During the first inspection it determines a number of output colums according to the number of labels in the output documents & the data length in columns. If the information in the column doesn't change, it is taken out of the table. Table size & column width are determined after the first inspection. During the 2-nd inspection of the output documents list the resulting list is output to the printer. If the output table string length is bigger than the automatic printer string length, some columns are not printed & the user is informed about the necessity to alter the number of the input data types.

During "Spectr" ADB operation the necessity to obtain output forms not only in the form of various prints, but also in the form suitable for use in data processing applied programmes was revealed (for example in Fortran programmes). For these proposes the third form of data output has been realised. D1A057A, included in DIALOG-2 system is used.

"Spectr-2' ADB Information accumulation.

At present ADB "Spectr-2" provides data about spectral characteristics of ions of isoelectronic sequences H-Ag (theoretical calculations, observation data, experimental results compilation). The total volume of information is about 200000 documents for which, because LS-communication diagrams are mostly spread and traditionally obvious, the classification by extreme LS-terms is used. The possibility to analyse statistical weights of the corresponding wave functions in the framework of multiconfigurational approximation is reserved for complex spectr.

The theoretical calculations of spectra and radiation transition probabilities in hydrogen-like ions were made by methods of Dirac relativity theory with the subsequent account of quantum-electrodynamic effects and adjustments linked with the nuclear structure.

For helium isoelectronic series the theoretical calculations of atomic spectral characteristics were made by two alternative methods: the method of relativity theory of excitations by 1/z and within the framework of semi-empirical approach which uses experimental approximations of transition wavelengths along isoelectronic sequence.

The energy of levels in Li-like ions were calculated in the similar way, through obtained by the method of model potential in the relativistic version was also used. The calculations of oscillator forces and radiation transition probabilities were curried out by the method of Fuse model potential which mode it possible to simplify mathematical computations radically due to the simple analitical form of wave functions. The calculations of level energy for Be-like ions were made both by the method of excitation theory by 1/z & Hartry-Fock approximation. In order to increase accurracy of computations combination of methods of self-consistent field and semi-empirical dependencies along Z were used which helped to perform overall adjustment to study their matching with the experimental data.

The main volume of theoretical data concerning spectral characteristics of other isoelectronic sequences (O-, F-, Ne-,Na- like ions) was obtained by the relativistic method of self-consistent field (in Hartry-Fock-Pauli approximation) which allows to determine the values of atomic constants with the accuracy close to the experimental one.

Spectra calculations of H- and Na- like ions were made in the various institutes of the USSR (Spectroscopy Institute, USSR Academy of Sciences, Institute of Physics, USSR Academy of Sciences, Institute of Physics, Lithuania SSR Academy of Sciences, Voronezh State University,NPO VNIIFTRI). Besides, for the above mentioned types of ions practically all presently known experimental data about the spectral lines in the area of λ < 100 A have been entered into the database.

"Spectr-2" ADB contains the results of spectrum studies for ions of Mg- to Ag- like isoelectronic sequences in the region λ < 100 A published prior to 1990.

References

[1] Bugayev V.Yu. & others "Spectr" automated data base on multicharge ions atomic constants: Scientific works collection NPO VNIIFTRI, Moscow - 1988-C.4-19.

[2] Dialogue Information & Search Systam DIALOG-2. in 4 vol. Moscow; MTsNTI.-1986.-36 issue.

SCALING LAWS AND PERFORMANCES FOR X-RAY LASERS

S. Jacquemot, A. Decoster, M. Cornille*, and J. Dubau*

Commissariat à l'Energie Atomique, Centre d'Etudes de Limeil-Valenton, BP 27, F94195 Villeneuve-St-Georges Cedex, France

**Observatoire de Paris, URA CNRS 0812, F92195 Meudon Cedex, France*

Abstract : At the present time, a systematic investigation of atomic parameters along the neon isoelectronic sequence and many precise interpretations of experiments are available. Then, it is possible to analyze the behavior of the gain with the ionic charge z. For the lasing transitions 3p $J=1 \rightarrow$ 3s $J=1$. This study allows to predict the "performances" of a neon X-ray laser versus the laser-target parameters and to extrapolate from MoXXXIII to "water-window" X-ray laser.

INTRODUCTION

The idea of a neon-like collisional excitation X-ray laser was just investigated as early as 1975 |1| : in a laser-produced plasmas, the Ne-like $(1s^2 2s^2 2p^5)3p$ levels are strongly populated by electron-impact excitation : (A) From the 2p ground state (along a dipole-forbidden line E0) and by dielectronic recombination (plus cascades), (B) From the F-like continuum; population inversions can then occur, via a selective decay mechanism, between these upper levels and the lower 3s $J=1$ ones (which present fast resonant E1 deexcitations to the ground state) |Fig. 1|.

Soft X-ray laser action was first observed at Livermore (LLNL) during 1984 |2| in 3p $J=1 \rightarrow$ 3s $J=1$ lines at 206 and 210 A arising from Se XXV plasmas.

These experiments and most of the following ones (in particular those performed at Limeil-Valenton |3|) are based on the two-sided illumination of exploding-foil targets (lasing material vapor-deposited at a given column density m on a plastic substrate) with green light (λ_L = 0.53nm) by a short-duration (FWHM Δt_L = 500 ps) gaussian pulse (total irradiance I_{tot} reached at t_L = 750 ps) in an elongated line focus geometry (L_y = 250 nm, L_z = 0.8, 1.7, 2.2 cm), from a high-power optical laser (Nova 2-beam or Phebus). This design |Fig. 2| is generally used because of its relative simplicity and because it provides a more uniform temperature and density cylindrical plasma (as compared to, for example, massive slabs).

Our theoretical efforts have been directed towards the analysis of the past experiments |Tab. 1| and the scaling of the scheme (amplification duration Δt_o, effective gain g) to shorter wavelengths.

MODELING METHODOLOGY

The physics of such soft X-ray lasers can be classified in four main areas :

- laser-matter interaction and target hydrodynamics (CHIVAS |8|),
- atomic physics (SUPERSTRUCTURE, DW+JJOM+CBe, AUTOLSJ |9|),
- ionization and inversion kinetics (LASIX |10| : includes the effect of trapped line radiation and provides local gain coefficients α),
- X-ray optics (OPTIQX |11| : accounts for radiation propagation and spectrograph recordings |12|).

The effective theoretical gain g is then deduced from the variation of the line intensity I with the plasma length L_z, according to the formula :

$$I \propto (e^{gL_z}-1)^{3/2} / (gL_z \, e^{gL_z})^{1/2}.$$

DISCUSSION

Three "nominal" experimental cases (Ge, Se, Mo) have been specifically studied and represent privileged steps in the isoelectronic scalings to higher nuclear charges Z_N ($z = Z_N$-a). We have focussed our attention on the temporal behavior of the hydro parameters : electronic temperature T_e and density D_e |Fig. 3| (main factors in the α small-signal gain coefficients resolution) and on the spatial dependence of the last one |Fig.-4) which is repsonsible for plasma refraction. Final results |Table 2, Fig.-5| exhibit quite good agreement between observations and simulations which allows us to suggest qualitative trends of the 3p J=2 → 3s J=1 scheme (From Se → Mo interpolations) :

* design |Table 1| : $I_{tot} \propto z^6$

  $m \propto z^{7/2}$

* atomic physics (along the X-ray laser line |Fig. 6, 7|) :

  $\lambda \propto Z_N^{-2}$

  $A_r \propto \lambda^{-2}$

* hydro (time $-t_o-$ and space $-x_o-$ resolved) |Table 2| :

  $T_e \propto z^{7/2}$

  $T_i \propto z^2$

  $N_e \propto z^4$

* kinetics : $\alpha_{max} \propto z^{1/2}$

TABLE 1

Summary of the considered (3p $^3P_2$ - 3s $^1P_1$) experimental data

XX	m (mg/cm²)	substrate	I_{tot} (w/cm²)	λ_{exp} (A)	G_{exp} (cm⁻¹)
Ge |4|	27	1500 A	$6\ 10^{13}$	232	2.0
Se |5|	36	1500 A	$7\ 10^{13}$	206	4.0
Sr |6|	80	1000 A	$1.4\ 10^{14}$	164	4.4
Mo |7|	88	1400 A	$4.4\ 10^{14}$	131	4.1

TABLE 2

XX	λ_{theo} (Å)	g_{max} (cm⁻¹)	I/LR (cm⁻¹)	Δt_o (ps)	T_e (ev)	T_i (ev)	N_e (10^{20}cm⁻³)
Ge	233	2.2	0.8	300	816	270	5.7
Se	207	4.4	1.2	250	948	308	7.0
Mo	133	4.8	1.9	100	2559	595	22.9

(I/L_R represents the gain lowering by refraction effects : $I/L_R = \alpha_{max} - g_{max}$, and T_i the ionic temperature)

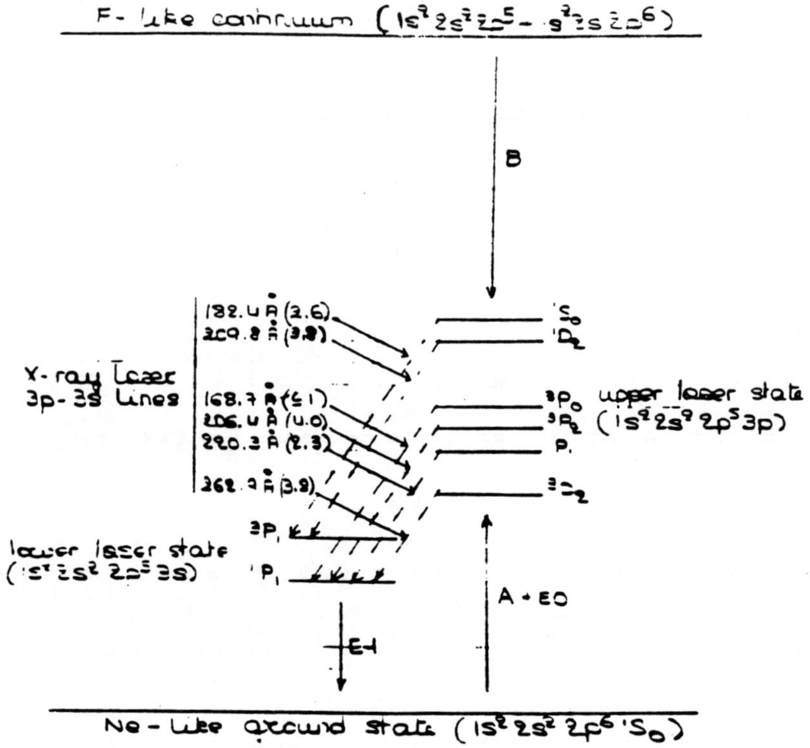

Figure 1

Simplified Grotrian diagram for Ne-like ions (measured wavelengths and effective gain coefficients -in cm^{-1}- are given for each laser transition, in Se XXV).

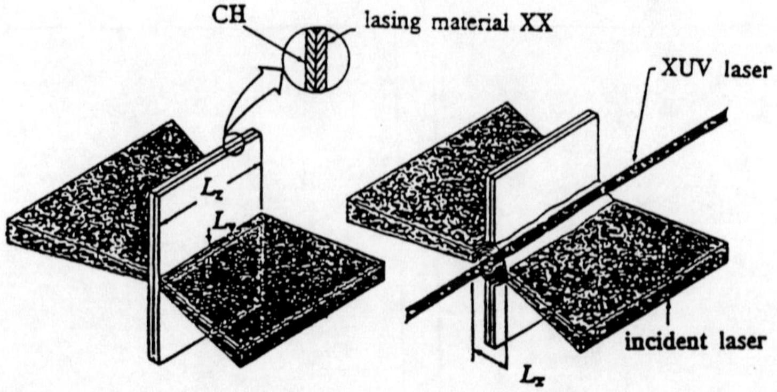

Figure 2

Schematic of the exploding-foil target used in the Ne-like experiments.

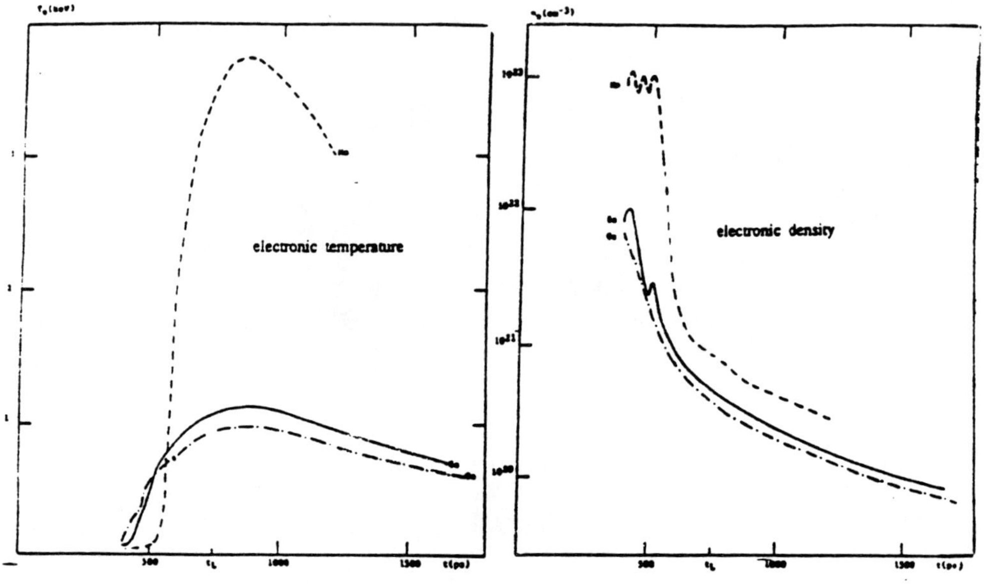

Figure 3

Temporal evolution of the hydro. parameters (T_e, n_e) for a given position in the losing plasma (i.e. at the initial center of exploding foil X_o).

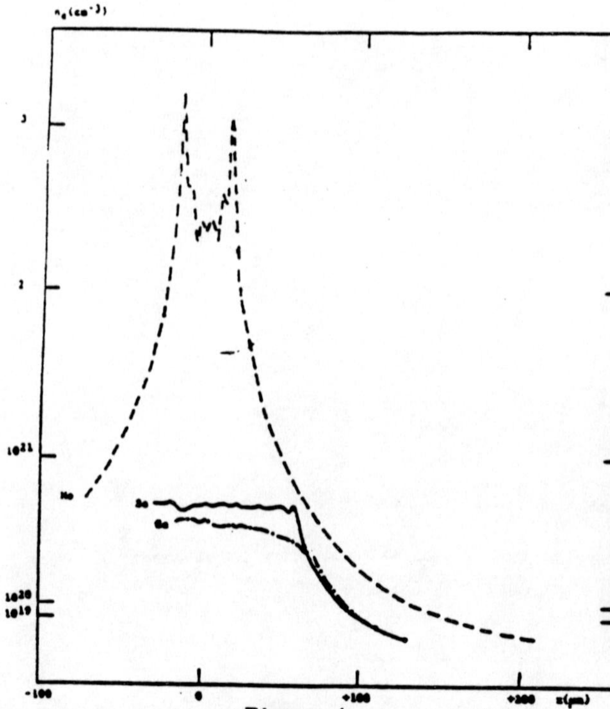

Figure 4

Spatial profile of the dielectronic density
(drawn when the simulated effective gain coefficient reaches its maximum value g_{max} : $t = t_o$).

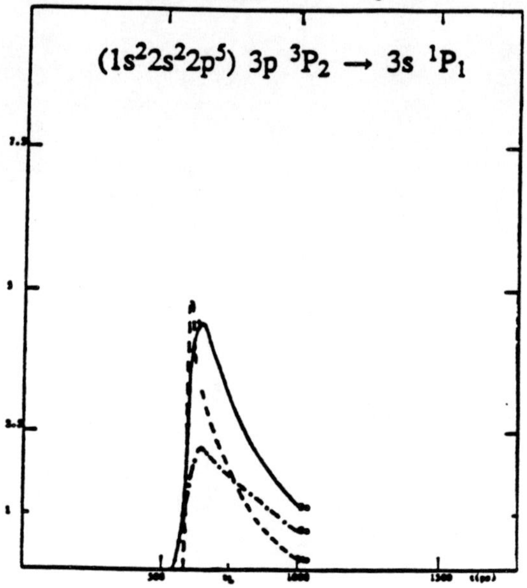

Figure 5

Time dependence of the theoretical effective gain coefficient.

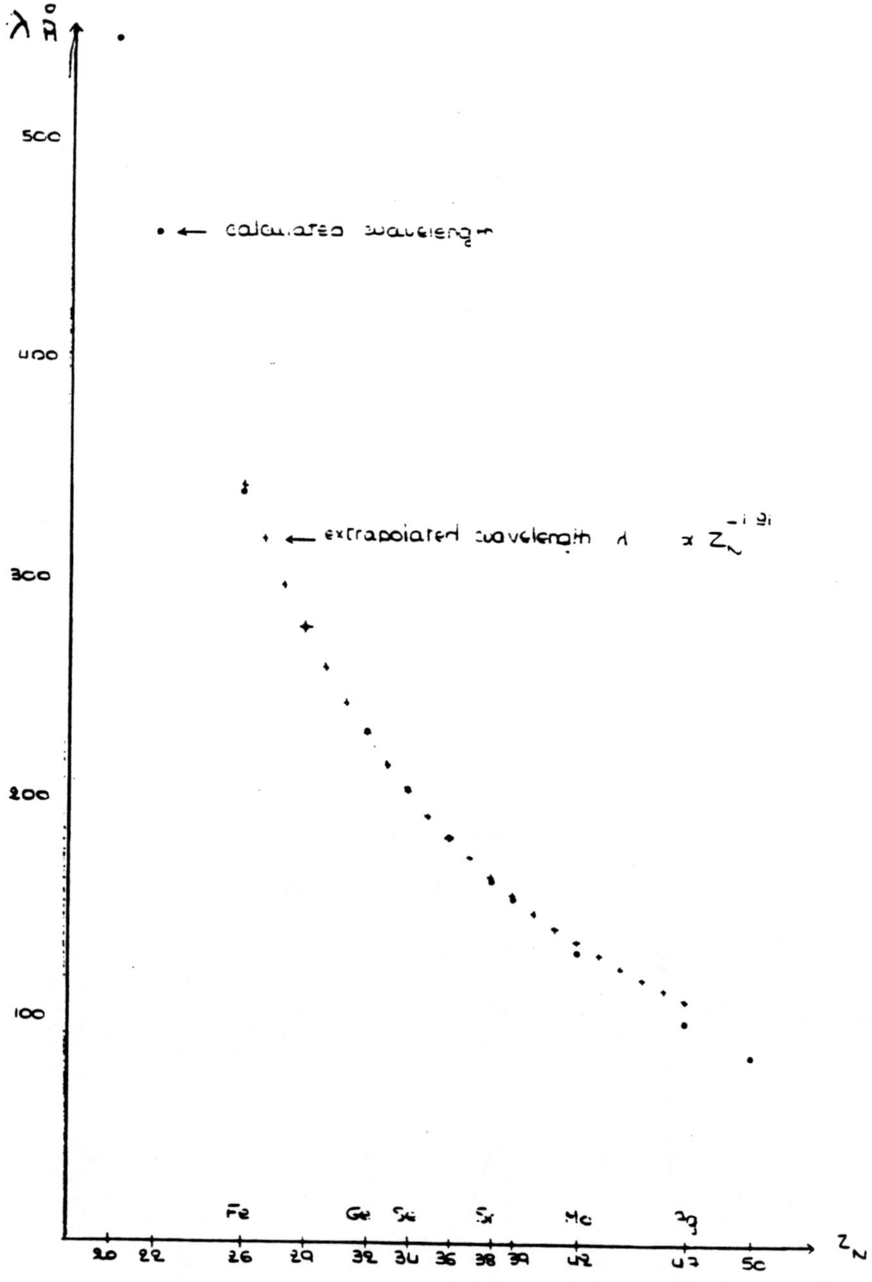

Figure 6

X-ray laser wavelenght λ (in A) versus nuclear charge Z_N

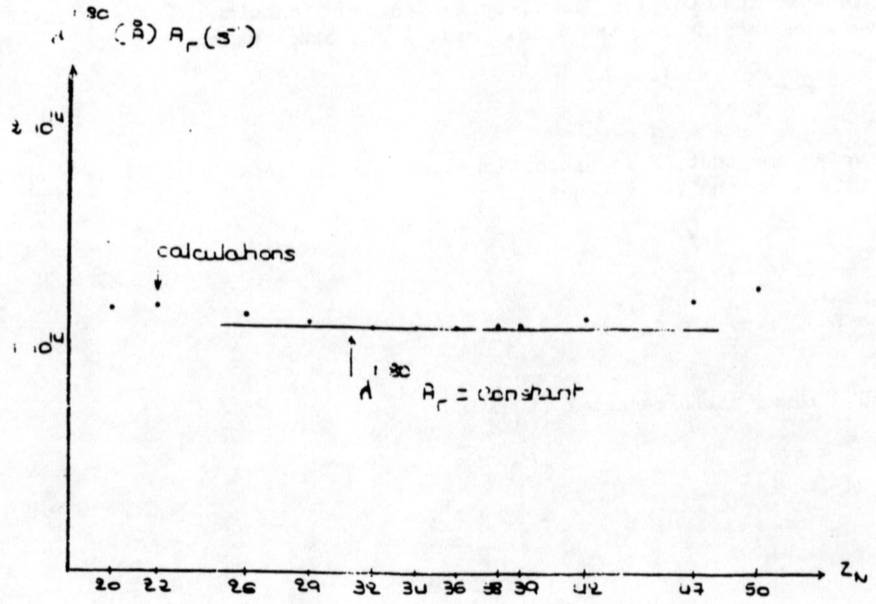

<u>Figure 7</u>

Decay rate of the X-ray laser transition versus nuclear charge.

(this scaling law can be in fact reproduced by a very simple model which expresses the Doppler gain coefficient in fonction of the population inversion density ΔN, the atomic mass A and other parameters :

$$\alpha \; \alpha \; \Delta N \; \frac{A}{T_i} \; \lambda^3 A_r \; ;$$

if we assume that ΔN is proportional to the ion density ($\Delta N \; \alpha \; n_i = \frac{n_e}{z}$) and if we mix up the two charges,

$$\alpha \; \alpha \; n_e \; T_i^{-1/2} \; z^{-5/2}$$

($A \; \alpha \; Z_N$), i.e. $\alpha \; \alpha \; z^{1/2}$)

* XUV- laser performances : $\Delta t_o \; \alpha \; z^{-3}$

$1/L_R \; \alpha \; z^{3/2}$

$g_{max} \; \alpha \; z^o$

CONCLUSION

We may conclude first that while, for moderate Z_N materials, it is easy to desing an effective X-ray laser, refraction requirements become more severe for higher Z_N targets (I/L_R increases as $z^{1/2}$). Secondly , this scheme seems to be unfortunately not appropriate for producing a "water-windows" laser ($\lambda \leq 44$ Å) because while it is still possible to reach 100 Å in Ag-Foil experiments ($m \cong 156$ Mg/cm², $I_{tot} = 8.6\ 10^{14}$ w/cm² |13|) with significant gains (4-5 cm⁻¹), extrapolations ($Z_N = 64$, $\lambda \cong 43$ Å) are just prospective ($I_{tot} > 10^{16}$ w/cm²).

REFERENCES

[1] A.N. Zherikhin, K.N. Koshelev, V.S. Letorkhov, Sov.J.Quant.Electron. 6,82 (1976).

[2] D.L. Matthews et al., Phys.Rev. Lett. 54,110 (1985).

[3] D. Naccache et al. (submitted to J.de Physique).

|4| D. Naccache et al. (submitted to Phys.Rev.).

|5| C.J. Keane et al., J.Phys.B 22, 3343 (1989).

|6| C.J. Keane et al. (submitted to Phys.Rev.)

|7| B.J. Mac Gowan et al., J.Appl.Phys. 61,5243 (1987).

|8| A. Decoster, S. Jacquemot, Rapport des activités laser CEA/CEL (1988), p.298.

|9| M. Cornille et al., J. de Physique 49(3), C1-95 (1988).

|10| E. Berthier et al. in ref. 8, p.304.

|11| J.L. Ferrer CEL-V internal report (1988).

|12| R.A. London Phys.Fluids 31,184 (1988).

|13| R.S. Walling (Communication 20th ECLIM (Schliersee 1990)).

A PERTURBATIONAL METHOD TO OBTAIN IMPROVED ATOMIC ENERGY LEVELS

P. Faucher, P. Lautard, and J. Dubau*

Observatoire de la Côte d'Azur, B. P. 139 06003 Nice Cedex, France

**Observatoire de Meudon, 92195 Meudon Principal Cedex, France*

Abstract. A perturbational method for improving calculated energy levels of atomic elements ($Z \leq 30$) at different ionization stages is presented. The method uses as a starting point the Superstructure code developed at the University College London and uses scaled Thomas-Fermi-Dirac potentials. The perturbation theory is applied up to the second order energy correction and introduces the contribution of the complete basis, i.e. including all eigenstates of the zero-order operator. As an illustration the method is applied to the determination of energy terms of some iron ions and the results are compared to the corresponding ones obtained by the MZ method.

1. INTRODUCTION.

The increasing resolution of the crystal and grating spectrometers which are used to detect the XUV emission in laboratory and astrophysical hot plasmas allows to obtain very complex spectra. For plasma diagnostics from spectra analysis, it is necessary to have atomic data codes which give accurate numerous data. Unfortunately, quality usually does not go with quantity. This large number of data is very important for modelisation and diagnostics of laser plasmas where, due to high density, many levels are excited and often, the important levels for laser production are not the lowest levels. In particular, it is important to have accurate theoretical wavelengths, therefore energy levels, for the analysis of spectra which present lines which have never been identified before, and also in the case of blended lines corresponding either to different atomic elements or different ionisation stages of the same element.

Many atomic codes were developed in order to obtain atomic energy levels with increasing precision (1,2,3,4,5,6). In particular, the Superstructure code(1,2) developed at University College of London calculates, in a very reasonable computing time, a lot of atomic physic parameters with a good precision, owing to a good minimisation procedure.It gives simultaneously many excited levels with an accuracy in relative wavelength ($\Delta\lambda/\lambda$) of the order of 10^{-3}. However, if the wavelength data obtained from this program are good enough to localise and interpret intense lines usually well identified from experiments or observations, the precision is sometimes not sufficient to identify blended lines or less intense isolated lines.

In order to increase the precision of these critical atomic energy levels, we propose a perturbational method which uses the Superstructure code as a starting point. This method takes into account the contribution of the continuum eigenstates of the central operator used in Superstructure in the calculation of the atomic energy levels of the total hamiltonian. The relative importance of this contribution and the quality of the method will be shown from the calculation, in LS coupling, of energy terms of some iron ions.

2. THEORY.

2.1 The Superstructure Code.

Superstructure is a code developed by Eissner, Jones and Nussbaumer (2) which gives with a good accuracy, large quantities of atomic parameters : non relativistic energy terms, relativistic corrections to energy levels, transition probabilities (allowed and forbidden),etc.. . It computes non relativistic multiconfigurationnal wavefunctions for a N electron system from one electron wavefunctions obtained in some central scaled Thomas-Fermi-Dirac potentials $V(\lambda_l, r)$. In this method the hamiltonian **H** of the system is separated in a central operator $\mathbf{H_{cen}}$ which is a sum of independent monoelectronic operators, and an interacting part $\mathbf{H_{int}}$.

$$\mathbf{H} = \mathbf{H_{cen}} + \mathbf{H_{int}} \tag{1}$$

$\mathbf{H_{cen}}$ includes the potentials $V(\lambda_l, r)$ which are different for different electron angular momenta l, and which simulate, for a given electron, the average effect of the N-1 other electrons as well as the nucleus of electric charge Z, the scaling parameters being obtained by a variationnal procedure upon the energy . The expression of this central operator is given by, in atomic units :

$$\mathbf{H_{cen}} = \sum_{i=1}^{N}\{-\frac{\nabla_i^2}{2} - V(\lambda_l, r_i)\} \tag{2}$$

where :

$$rV(\lambda_l, r) \longrightarrow Z \quad \text{for} \quad r \longrightarrow 0 \tag{3}$$

$$rV(\lambda_l, r) \longrightarrow (Z - N + 1) \quad \text{for} \quad r \longrightarrow \infty \tag{4}$$

The interaction operator can be divided in two parts, a non relativistic part and a relativistic part. In this work, we consider only the non relativistic part which is given by :

$$\mathbf{H}_{\mathbf{int}}^{\mathbf{nr}} = \sum_{i=1}^{N}\{V(\lambda_l, r_i) - \frac{Z}{r_i} + \sum_{j>i}^{N}\frac{1}{r_{ij}}\} \tag{5}$$

In the Superstructure program a truncated basis (finite dimension n) of eigenstates of $\mathbf{H_{cen}}$, which includes the terms issued from the the most important bound configurations, is defined. In order to obtain non relativistic wavefunctions the hamiltonian ($\mathbf{H_{cen}}$ + $\mathbf{H_{int}^{nr}}$) is diagonalized in the subspace generated by this truncated basis .

2.2 Present Method.

In order to increase the accuracy of the results of the Superstructure program, a possibility is to increase, in the limit of the computer capacity , the number of configurations of the truncated basis. Unfortunately, this technique does not allow to include all the fine electronic correlation effects: all the bound states and the continuum eigenstates of $\mathbf{H_{cen}}$ must be included in order to have a complete basis. These effects can be introduced by a perturbation theory treatment which takes the Superstructure code as a starting point.

A complete basis of N electrons Slater determinants $| \varphi >$ can be defined from the central operator $\mathbf{H_{cen}}$ of the expression (2).

$$\mathbf{H_{cen}}|\varphi_i >= \epsilon_i|\varphi_i > \qquad \text{for the bound states} \tag{8}$$

$$\mathbf{H_{cen}}|\varphi_\epsilon >= \epsilon|\varphi_\epsilon > \qquad \text{for the continuum states} \tag{9}$$

The states $| \varphi >$ satisfy to the following orthogonality conditions:

$$< \varphi_i|\varphi_{i'} >= \delta_{ii'} \qquad \text{for the bound states} \tag{10}$$

$$< \varphi_\epsilon|\varphi_{\epsilon'} >= \delta(\epsilon - \epsilon') \qquad \text{for the continuum states} \tag{11}$$

The non relativistic multiconfigurational wavefunctions $| \varphi >$ are defined from one-electron wavefunctions for which the radial part satisfies the radial equation :

$$(\frac{d^2}{dr^2} - \frac{l(l+1)}{r^2} + 2V(\lambda_l, r) + \epsilon)F(r) = 0 \tag{12}$$

where for the bound wavefunctions

$$F(r) = P_{nl}(r) \tag{13a}$$

$$P_{nl}(r) \longrightarrow 0 \quad \text{for} \quad r \longrightarrow 0 \tag{13b}$$

$$P_{nl}(r) \longrightarrow 0 \quad \text{for} \quad r \longrightarrow \infty \tag{13c}$$

$$\int_0^\infty P_{nl}(r)P_{n'l}(r)dr = \delta_{nn'} \tag{13d}$$

and for the free wavefunctions $k^2 = \epsilon$

$$F(r) = F_{\epsilon l}(r) \tag{14a}$$

$$F_{\epsilon l}(r) \longrightarrow 0 \quad \text{for} \quad r \longrightarrow 0 \tag{14b}$$

$$F_{\epsilon l}(r) \longrightarrow (\frac{2}{\pi k})^{\frac{1}{2}} sin(kr + \delta) \quad \text{for} \quad r \longrightarrow \infty \tag{14c}$$

where

$$\delta = -\frac{l\pi}{2} + \frac{z}{k} log(2kr) + Arg\Gamma(l+1 - \frac{iz}{k}) + \tau_{kl} \tag{14d}$$

Γ is the Gamma EULER function
$z = Z - N + 1$ is the ionic charge
τ_{kl} is the phase shift introduced by the central potential
The orthogonality relation for the continuum wavefunctions is given by:

$$\int_0^\infty F_{\epsilon l}(r) F_{\epsilon' l}(r) dr = \delta(\epsilon - \epsilon') \tag{14e}$$

In order to have orthogonality relations (13d) and (14e), and the orthogogality of the bound and free radial functions, the scaling parameters λ_l are the same for all radial functions with the same l.

Then this space of N electrons Slater determinants is divided in two subspaces $\mathcal{P}$ and $\mathcal{Q}$. The subspace $\mathcal{P}$ is similar to the space of dimension n of Superstructure which corresponds to the truncated basis defined above. The subspace $\mathcal{Q}$ includes all the other states, bound states (infinite basis), and continuum states. Two projection operators associated with the two subspaces are defined.

$$\mathbf{p} = \sum_{|\varphi>\in\mathcal{P}} |\varphi><\varphi| \tag{15}$$

$$\mathbf{q} = \sum_{|\varphi>\in\mathcal{Q}} |\varphi><\varphi| \tag{16}$$

These operators satisfy to the relations:

$$\mathbf{p}^2 = \mathbf{p} \qquad \mathbf{q}^2 = \mathbf{q} \qquad \mathbf{pq} = \mathbf{qp} = 0 \qquad \mathbf{p} + \mathbf{q} = \mathbf{I} \tag{17}$$

Then a zero order operator $\mathbf{H_0}$ and a perturbation operator $\mathbf{W}$ can be defined :

$$\mathbf{H_o} = \mathbf{pHp} + \mathbf{qH_{cen}q} \tag{18}$$

$$\mathbf{W} = \mathbf{H} - \mathbf{H_0} = \mathbf{pH_{int}q} + \mathbf{qH_{int}p} + \mathbf{qH_{int}q} \tag{19}$$

If we note $| \Phi >$ the complete basis of eigenstates of $\mathbf{H_0}$:

$$\mathbf{H}_0|\Phi_i>=\bar{\epsilon}_i|\Phi_i> \qquad \text{for} \qquad |\Phi_i>\in \mathcal{P} \tag{20}$$

where the eigenvalues $\bar{\epsilon}_i$ correspond to those obtained in Superstructure from the truncated basis, and

$$\mathbf{H}_0|\Phi_i>=\epsilon_i|\Phi_i> \qquad \text{for} \qquad |\Phi_i>\in \mathcal{Q} \tag{21}$$

$$\mathbf{H}_0|\Phi_\epsilon>=\epsilon|\Phi_\epsilon> \qquad \text{for} \qquad |\Phi_\epsilon>\in \mathcal{Q} \tag{22}$$

$|\Phi_i>$ and $|\Phi_\epsilon>$ are identical to $|\varphi_i>$ and $|\varphi_\epsilon>$ the eigenfunctions of $\mathbf{H_{cen}}$ with the same eigenvalues ϵ_i and ϵ .

We are interested by the determination of the energy levels of the states of the subspace $\mathcal{P}$ (in other words we are interessed by the improvement of the results of the Superstructure code). The perturbation theory applied to the state $|\Phi_i>$ of $\mathcal{P}$ at the second order gives:

$$E^i = E_0^i + E_1^i + E_2^i + ... \tag{23}$$

where :

$$E_0^i = \bar{\epsilon}_i \tag{24}$$

$$E_1^i =<\Phi_i|\mathbf{W}|\Phi_i>=0 \tag{25}$$

$$E_2^i = \sum_{|\Phi>\in\mathcal{Q}}{}^{*} \frac{|<\Phi_i|\mathbf{W}|\Phi>|^2}{(\bar{\epsilon}_i-\epsilon)} \tag{26}$$

The summation $\sum_{|\Phi>\in\mathcal{Q}}{}^{*}$ in the expression of E_2 includes all the bound and continuum states of $\mathcal{Q}$.($\sum_{|\Phi>\in\mathcal{Q}}{}^{*} = \sum_{\text{bound states} \in \mathcal{Q}} + \int_{\text{continuum states} \in \mathcal{Q}} d\epsilon$)

2.3 Comparison with the MZ Method.

The results obtained by this method will be compared to those obtained by the MZ method developed at the Lebedev Institute of Moscow and at the Institute of Troitsk by Vainshtein and Safronova.The corresponding code uses the Z-expansion method, i.e. perturbation theory from hydrogenic monoelectronic wavefunctions, and gives very accurate results for systems with few electrons.

With the present method , it is possible to reproduce the MZ results by replacing the potential $V(\lambda_l, r)$ by the nuclear coulombic potential and to check the accuracy of the numerical results.

The MZ method offers some qualities compared to the present one . For an isoelectronic sequence, it obtains results in one calculation while it is necessary, in the present method, to repeat the calculation for each ion , and in particular,with MZ, the second order E_2 is constant. In the calculation of the second order energy this method has to consider not only two-electron operators $(1/r_{ij})$ as in MZ, but also one-electron operators $(Z/r - V(\lambda_l, r))$.

3. RESULTS.

The present results are preliminar results which do not use all the possibilities of the Superstructure program. The central potential is a Thomas-Fermi-Dirac potential (i.e. all the parameters λ_l are equal to 1), and, as already mentionned in (2.2) the results are obtained in LS coupling, i. e. the relativistic effects are not taken into account.

Table 1. Comparison of the two first energy terms (in Ryd) of some He-like ions.

(a) results obtained in this work using an hydrogenic potential
(b) results obtained by the MZ method.

Z		$1s^2\ ^1S$	$1s2s\ ^3S$
6	(a)	-64.81531	-42.83973
	(b)	-64.81533	-42.83974
12	(a)	-273.31531	-175.58466
	(b)	-273.31533	-175.58467
26	(a)	-1319.81532	-835.32280
	(b)	-1319.81533	-835.32282

In the present calculations, the subspace $\mathcal{P}$ is restricted to the terms corresponding to the considered configuration. In order to test the program, table 1 compares for the two first energy terms of some He-like ions, the results obtained by our method using an hydrogenic potential with the corresponding MZ ones. The agreement is very good along the isoelectronic sequence since the difference corresponds to a precision of the order of $10^{-7} - 10^{-8}$. It is a fondamental interest to do this comparison for any system.

Contributions to the second order energy perturbation are separated in three parts :

(i) contribution due to bound states.
(ii) contribution due to continuum states with one continuum electron.
(iii) contribution due to continuum states with two continuum electrons.

As mentioned above the interaction operator includes one- and two-body operators, so there is no contribution from free states with more than two continuum electrons. In the numerical application, the program calculates exactly these contributions for the configurations with a principal quantum number $n \leq 7$ and over a limited energy integration interval. These contributions correspond to E_2^0, E_2^+ and E_2^{++}. E_2^{asymp} represents all the other contributions obtained by extrapolation and is generally small compared to the other ones. Table 2 shows the results obtained for some energy terms of the He-like Fe XXV ion. It is interesting to note the importance of continuum state contributions and the necessity to include them to improve the calculation of the energy terms.It can be mentionned that the relative importance of the three different contributions is not distributed of the same manner for the different energy terms.

Table 2. Relative importance of the different contributions(in Ryd) to the second order energy E_2 for some energy terms of the Fe XXV He-like ions, using a Thomas-Fermi-dirac potential.

E_2^0 : Contribution due to bound states
E_2^+ : Contribution due to free states with one continuum electron
E_2^{++} : Contribution due to free states with two continuum electrons.
E_2^{asymp} : (see text)

$$E_2 = E_2^0 + E_2^+ + E_2^{++} + E_2^{asymp}$$

Term	E_2^0	E_2^+	E_2^{++}	E_2^{asymp}	E_2
$1s^2\ ^1S$	-0.00678	-0.02350	-0.06101	-0.00144	-0.09273
$1s2s\ ^3S$	-0.01809	-0.06755	-0.00066	-0.00195	-0.08826
$1s2s\ ^1S$	-0.08021	-0.04419	-0.01205	-0.00165	-0.13810

Calculations relative to energy terms for some ions of the iron element are shown in table 3.Column (a) represents the results obtained by the present method using a Thomas-Fermi-Dirac potential. They are compared with the corresponding MZ ones (column (b)). E represents the summation of the three first energy contributions ($E = E_0 + E_1 + E_2$). The agreement in total energy E is quite good between the two methods, it is often better than 10^{-5} but it is decreasing with the number N of electrons of the considered ions. The comparison between the second order energy contribution of the two methods is more significative. Thus, for each energy term calculation, the contribution given by the present method is much smaller than the corresponding MZ one and, in more, the relative importance of this contribution decreases versus the number of electrons N : a factor between 1.5 and 3 for the 2-electron system and a factor 13 for the 5-electron system.

Table 3. Relative importance of the second order energy contribution E_2 in the determination of some energy terms for different iron ions.(Energies in Ryd) .
(a) results obtained in this work using a Thomas-Fermi-Dirac potential
(b) results obtained with the MZ method.

terms	E (a)	E (b)	E_2 (a)	E_2 (b)
$1s^2\ ^1S$	-1319.81433	-1319.81533	-0.09273	-0.31533
$1s2p\ ^3P$	-833.40615	-833.40813	-0.08984	-0.14598
$1s^22s\ ^2S$	-1468.62897	-1468.63054	-0.13325	-0.8164
$1s^22p\ ^2P$	-1465.19434	-1465.19384	-0.13798	-1.0572
$1s^22s^2\ ^1S$	-1610.08560	-1610.07374	-0.26220	-1.7658
$1s^22p^2\ ^3P$	-1602.67736	-1602.67342	-0.23071	-2.4314
$1s^22s2p^2\ ^4P$	-1738.64511	-1738.72880	-0.30956	-3.9852
$1s^22s2p^2\ ^2S$	-1734.83293	-1734.89564	-0.35375	-4.4176

4. CONCLUSION.

To have a good perturbation approach, it is necessary to obtain only small contributions. The proposed method satisfies this criterion. For example, for the term $1s^22s2p^2\ ^2S$,the second order energy contribution is only : $E_2 = -0.35375$ compared to the corresponding value given by the MZ method : $E_2 = -4.4176$.

The inclusion of a Thomas-Fermi-Dirac potential has also decreased the variation of E_2 versus the number of electrons N. This is not surprising because this potential includes the average electron contribution. For example,the energy ratio : $E_2(1s^22s2p^2\ ^2S)/E_2(1s^2\ ^1S)$ is equal to 4 from our method and is equal to 14 with MZ.

The present calculations were done in LS coupling. Relativistic effects have now been included in the method.

REFERENCES.

(1) Eissner W, Nussbaumer H, 1969, J. Phys. B, **2**, 1028.
(2) Eissner W, Jones M and Nussbaumer H 1974 Comput. Phys. Commun. **8** 270.
(3) Desclaux J P, 1975, Comp. Phys. Comm. **9**, 31.
(4) Vainshtein L A, Safronova U I,1978,Atomic Data and Nuclear Tables, **21**,49.
(5) Grant I P, McKenzie B J, Norrington P H, Mayers D F, Pyper N C, 1980, Comp. Phys. Comm. **21**, 207.
(6) Bhalla C P, Tunnell T W, 1984, J. Quant. Spectr. Transfer, **32**, 141.

SIMPLE ESTIMATES FOR PLASMA BROADENING AND SHIFT OF NON-HYDROGENIC ION LINES

N. Konjević and M. S. Dimitrijević[†]
[*]Institute of Physics, P.O. Box 57, 11001 Beograd, Yugoslavia
[†]Astronomical Observatory, Volgina 7, 11050 Beograd, Yugoslavia

1. INTRODUCTION

For radiative transfer calculations of laser produced plasmas, electron-impact broadening parameters for a large number of spectral lines of various elements are required. In spite of the possibility of quantum mechanical and semiclassical theories to provide data of higher accuracy, they are often not suitable, since considerable computational efforts and knowledge of numerous atomic data are needed. For Stark broadening parameter calculations for a large number of lines, as e.g. in the case of radiative transfer, simple approximative approaches with good average accuracy (e.g. [1-11]) are more appropriate.

This contribution deals with the modified semiempirical approach [4,5,9,10] which is tested several times [5-8, 12-21] and gives on average a satisfactory agreement with experimental data.

2. MODIFIED SEMIEMPIRICAL APPROACH

According to the modified semiempirical approach [4,9,10], the half-halfwidth w and the shift d of an ion spectral line broadened by Stark effect are given by the following expression:

$$w+id = N\frac{4\pi}{3}\frac{\hbar^2}{m^2}\left(\frac{2m}{\pi kT}\right)^{1/2}\frac{\pi}{\sqrt{3}}\sum_{j=i,f}\{\vec{R}^2_{k,k+1}[\tilde{g}(x_{k,k+1})+$$

$$+i\varepsilon_j\tilde{g}_{sh}(x_{k,k+1})]+\vec{R}^2_{k,k-1}[\tilde{g}(x_{k,k-1})-i\varepsilon_j\tilde{g}_{sh}(x_{k,k-1})]\}+$$

$$+\sum_{j'}(\vec{R}^2_{jj'})_{\Delta n\neq 0}[g(x_j)+i\varepsilon_j g_{sh}(x_j)]-2i\varepsilon_j\Big[\sum_{\Delta E_{jj'}<0}(\vec{R}^2_{jj'})_{\Delta n\neq 0}g_{sh}(x_{jj'})\Big]. \quad (1)$$

In this expression $R^2_{j'j}$ (in units of the Bohr radius a_0^2) is the square of the coordinate operator matrix element summed over all components of the operator and the magnetic substates of the total angular momentum J', and averaged over the magnetic substates of J.

Within the hydrogenic approximation, i.e. assuming that the principal effective quantum number for the same n and different ℓ are nearly the same, we can perform the summation of matrix elements for $\Delta n \neq 0$ obtaining

$$\sum_{j'} (\vec{R}^2_{jj'})_{\Delta n \neq 0} = \left(\frac{3n^*_j}{2Z}\right)^2 \frac{1}{9}(n^{*2}_j + 3\ell_j^2 + 3\ell_j + 11) \qquad (2)$$

One must take into account that in the case of low lying levels of more complex elements this approximation is invalid. This approximation is very crude also for the shifts which are more sensitive to the details of calculations than widths. In such cases it is better to use the more correct expression

$$R^2_{jj'} = \left(\frac{3n^*_\ell}{2Z}\right)^2 \frac{\ell}{2\ell_j+1}(n^{*2}_\ell - \ell^2)\phi^2(n^*_{\ell-1}, n^*_\ell, \ell). \qquad (3)$$

Here, the Bates and Damgaard [23] factor ϕ^2 (tabulated e.g. in Ref. [24]) gives the correction for the deviation of hydrogenic model. The cases where the one electron model (i.e. only one energy level for each $n\ell$ electron) assumed in equations (2) and (3) is not satisfied, are analysed in detail in Ref. [25], e.g. for a multiplet as a whole, $\vec{R}^2_{j'j}$ should be multiplied by $R^2_{mult.}(2\ell+1)/(2L+1)$. The multiplet factor R^2_{mult} can be found in tables of Shore and Menzel [26]. In Eq.(1) the parameter $\varepsilon_j = +1$ if $j = i$, and -1 if $j = f$, $k = \ell_j$ (the orbital quantum number of the valence electron), i and f denote the initial and final energy states,

$$x_{jj'} = 3kT/2|\Delta E_{jj'}|$$

$$x_j = 3kTn^{*3}_j/4Z^2E_H.$$

E_H is the hydrogen ionization energy, E_{ion} is the appropriate spectral series limit. In case that the term in question belongs to a series converging to an excited state of the resulting ion, the excitation energy of this state is added to the usual ionization energy. The residual ionic charge is denoted by Z, N is the electron density, $\Delta E_{j'j} = E_{j'} - E_j$, $\ell = \max(\ell_{j'}, \ell_j)$, and

$$n^*_j = [E_H Z^2/(E_{ion} - E_j)]^{1/2}$$

the effective principal quantum number.

All Gaunt factors g, $\tilde{g}$, g_{sh} and $\tilde{g}_{sh}$ are given in Table 1.

TABLE 1. Gaunt factors

	x	≤1	2	3	5	10	30	100
	g	0.2	0.2	0.24	0.33	0.56	0.98	1.33
	$\tilde{g}$	0.7-1.1/Z+g						
	g_{sh}	0.2	0.25	0.32	0.45	0.66	0.82	0.87
Z = 2	$\tilde{g}_{sh}$	0.35	0.40	0.47	0.58	0.70	0.82	0.87
Z = 3	$\tilde{g}_{sh}$	0.53	0.54	0.57	0.62	0.70	0.82	0.87
Z = 4	$\tilde{g}_{sh}$	0.62	0.63	0.63	0.65	0.70	0.82	0.87
Z > 4	$\tilde{g}_{sh}$	0.88-1.1/Z+0.01+0.01x/Z.						

At high temperatures, say $3kT/2\Delta E>50$, all Gaunt factors in equation (1) may be calculated in accordance with the GBKO high temperature limit [27], viz.

$$\tilde{g}_{j'j} = g_{j'j} = \frac{\sqrt{3}}{\pi}\left[\frac{1}{2} + \ln\left(\frac{2ZkT}{n_j^{*2}\Delta E_{j'j}}\right)\right] \tag{4}$$

$$\tilde{g}_{sh} = g_{sh} = \sqrt{3}/2.$$

The above expression is suitable only if high temperature limit is valid for principal perturbing levels in the whole temperature range of interest, or for rough estimates. If a very close perturbing level exists, it is better to take into account also the Debye screening effects by modifying Gaunt factor for close levels as [28]

$$g_{j'j} = \frac{\sqrt{3}}{\pi}\left\{\frac{1}{2} + \ln\left[\frac{2ZkT}{n_j^{*2}\max(\Delta E_{j'j}, \omega_p/\hbar)}\right]\right\}$$

$$g_{sh} = \frac{\sqrt{3}}{\pi}\left[\frac{\pi}{2} - b\left(\frac{\omega_{jj'}}{\omega_p}\right)\right] \tag{5}$$

$\omega_{jj'}$ is the frequency separation between the level j and the close perturbing level j', $\omega_p = (4\pi Ne^2/m)^{1/2}$ is the plasma frequency, and b is the GBKO [27] Stark broadening function for the shift tabulated in Ref. [2]). One must take into account that shift is very sensitive to the calculation procedure, and the proposed correction is appropriate only if the $b(\omega_{jj'}/\omega_p)$ term is small compared to $\pi/2$. If not, it is better to use the expression

$$g_{sh} = \frac{\sqrt{3}}{\pi} b\left(\frac{\Delta E_{jj'} n_j^{*2}}{eZkT}\right). \tag{6}$$

Equation (1) is obtained assuming the LS coupling approximation, separating the transitions with $\Delta n = 0$ and $\Delta n \neq 0$, and supposing that the nearest perturbing level in the $\Delta n \neq 0$ group may be obtained in the hydrogenic aproximation as:

$$|\Delta E_{n,n+1}| \approx 2Z^2 E_H / n^{*3}.$$

In the cases when perturbing levels exist which strongly violate the assumed approximations, i.e. if there are levels with $|\Delta E_{j',j}| << |\Delta E_{n,n+1}|$, the correction of equation (1) may be done by summing equation (7) and equation (1). Here m' is the number of such perturbing levels and $\varepsilon_j = +1$ if $j = i$ and -1 if $j = f$

$$w_c + id_c = N \frac{4\pi}{3} \frac{\hbar^2}{m^2} \left(\frac{2m}{\pi kT}\right)^{1/2} \frac{\pi}{\sqrt{3}} \sum_{m'} R^2_{j',j} \left[g(x_{j',j}) - g(x_{n_i,n_{i+1}}) \pm \right.$$

$$\left. \pm i\varepsilon_j (g_{sh}(x_{j',j}) \mp g_{sh}(x_{n_i,n_{i+1}}))\right]. \qquad (7)$$

For all cases when $3kT/2|\Delta E_{j',j}| \leq 2$ one can use the values $g = g_{sh} = 0.2$ and $\tilde{g} = \tilde{g}_{sh} = 0.9-1.1/Z$.

3. SIMPLE ESTIMATES

For perturber energies E comparable or lower than $\Delta E_{j',j}$ where j' is the nearest perturbing level, the modified semiempirical approach can be considerably simplified. For all cases when $x_{jj'} \leq 2$, $g = 0.2$ and $g = 0.9-1.1/Z$. Furthermore, one can put $\phi^2 = 1$ into Eq. (1) which is a reasonable assumption for $\Delta n = 0$ (see e.g. Ref. [2], p.31), since exact values of ϕ^2 usually range between 0.8 and 1. If one performs the summation in Eq. (1), it is easy to obtain [9]:

$$w(\mathring{A}) = 0.4430\ 10^{-8} \frac{\lambda^2(\text{cm}) N(\text{cm}^{-3})}{T^{1/2}(\text{K})}$$

$$\sum_{j=i,f} \left[\vec{R}^2_j + \frac{\tilde{g}_{th} - g_{th}}{g_{th}} \left(\frac{3n^*_j}{2Z}\right)^2 (n^{*2}_j - \ell^2_j - \ell_j - 1)\right] \qquad (8)$$

$$\vec{R}^2_j = \sum_{j'} (\vec{R}^2_{jj'})_{\Delta n \neq 0}.$$

Since the contributions to the total line width of transitions with $\Delta n \neq 0$ does not exceed 25% and it is compensated by assuming $\phi^2 = 1$, we can neglect them and finally obtain

$$W(\mathring{A}) = 2.2151\ 10^{-8} \frac{\lambda^2(\text{cm}) N(\text{cm}^{-3})}{T^{1/2}(\text{K})}$$

$$\left(0.9 - \frac{1.1}{Z}\right) \sum_{j=i,f} \left(\frac{3n^*_j}{2Z}\right)^2 (n^{*2}_j - \ell^2_j - \ell_j - 1). \qquad (9)$$

Taking into account that $\tilde{g}_{sh} \equiv \tilde{g}$ and g_{sh} near the treshold, and applying the same procedure as for the width, we obtain similarly in the case of shift

$$d \approx 1.1076\ 10^{-8} \frac{\lambda^2(\mathrm{cm})N(\mathrm{cm}^{-3})}{T^{1/2}(\mathrm{K})}(0.9-\frac{1.1}{Z})\frac{9}{4Z^2}$$

$$\sum_{j=i,f} \frac{n_j^{*2}\varepsilon_j}{2\ell_j+1}\{(\ell_j+1)[n_j^{*2}-(\ell_j+1)^2]-\ell_j(n_j^{*2}-\ell_j^2)\}. \qquad (10)$$

If all levels $\ell_{i,f}\pm1$ exist, an additional summation may be performed in equation (10) to obtain

$$d \approx 1.1076\ 10^{-8} \frac{\lambda^2(\mathrm{cm})N(\mathrm{cm}^{-3})}{T^{1/2}(\mathrm{K})}(0.9-\frac{1.1}{Z})\frac{9}{4Z^2}$$

$$\sum_{j=i,f} \frac{n_j^{*2}\varepsilon_j}{2\ell_j+1}(n_j^{*2}-3\ell_j^2-3\ell_j-1). \qquad (11)$$

4. DISCUSSION

In order to test the modified semiempirical approach, selected experimental data for 36 multiplets (7 different ion species) of doubly and 7 multiplets (4 different ion species) of triply-charged ions were compared with theoretical linewidths. The averaged values of the ratios of measured to calculated widths are as follows [4]: for doubly charged ions 1.06±0.32 and for triply-charged ions 0.91±0.42. Furthermore, in order to test the equations obtained for low temperature limit of modified semiempirical formula, comparison is made between the linewidth results from equations (1) and (9) and previously mentioned selected experimental results, in the case when equation (9) may be applied (32 multiplets). The average ratio of experimental and calculated values from equations (1) and (9) are 1.01 and 1.04, respectively [9]. Comparison is made also with the results of comprehensive width calculations for doubly and triply ionized atoms [5]. The discrepancy did not exceed ±30% on the average [9].

Generally, the possibilities of theory in the case of shifts are considerably smaller than for the linewidths. Reliable experimental data on ion lines Stark shifts exist almost exclusively for singly charged ions. The most extensive set of experimental data is for Ar II lines, where there are about 170 experimental results for line shifts within 47 allowed and forbidden multiplets. In Ref. [18] the obtained mean ratio of measured and calculated Ar II Stark shifts for 165 line shift measurements is 1.38±0.60 (with 6 results with opposite sign of the shifts).

TABLE 2. Average ratios of measured (w_m) and calculated linewidths for various doubly and triply ionized atoms. $N_e=10^{17}$ cm^{-3}, w_{SEM} - modified semiempirical approach; w_{SE} - semiempirical approach

Element	T_e |K|	w_m/w_{SEM}	w_m/w_{SE}
C III	60000	1.29	1.21
N III	24300	0.92	1.71
O III	25400	1.05	1.90
Si III	25600	0.67	1.08
S III	28500	1.16	1.65
Cl III	24200	1.01	1.68
A III	21100	0.99	1.57
average ratio:		1.06	1.53
C IV	60000	1.50	2.57
Si IV	25600	0.66	1.15
S IV	28500	0.88	1.65
A IV	20750	0.76	1.24
average ratio:		0.91	1.56

TABLE 3. Stark widths (FWHM) for heavy elements. $N_e=10^{17}$ cm^{-3}, w_{SEM} - modified semiempirical, w_{SE} - semiempirical and w_{SC} - semiclassical approach.

Ion	Transition (mult.no.) Wavelength(Å)	T(K)	w_{SEM} (Å)	w_{SE} (Å)	w_{SC} (Å)
Ti II	$a^4F-z^4D^o$	2500	0.238	0.230	0.228
	(5)	5000	0.168	0.163	0.164
	3079.4 Å	10000	0.119	0.115	0.125
		20000	0.0840	0.0815	0.0996
Mn II	$a^5D-z^5P^o$	2500	0.264	0.166	0.266
	(3)	5000	0.174	0.117	0.190
	3464.0 Å	10000	0.123	0.0830	0.144
		20000	0.0870	0.0587	0.144

There is another field where approximative theories can be useful. That is the case of heavier elements. Atomic data for heavy element spectra are often insuficient for sophysticated semiclassical calculations. Moreover, the accuracy of usually applied approximations (e.g. Coulomb approximation) is lower for complex spectra. Additional time consuming sophystications (e.g. Thomas-Fermi model) are needed in order to achieve the same accuracy as in the case of lighter atoms. In such cases a reliable approximative method may be a useful tool for quick estimate of needed Stark widths.

As a test for applicability of the modified semiempirical approach to heavier elements, here are presented the results for one TiII and one MnII multiplet from Ref. [3]. The agreement between simple modified semiempirical approach and more sophysticated semiclassical calculations is very encouraging and indicates that this simple method can be used for estimation of electron widths of heavier elements.

When reliable data do not exist, the knowledge on regularities and systematic trends of line broadening parameters can be used for quick acquisition of new data especially when high accuracy of each particular value is not needed (see e.g. Refs. [29,30] and references therein). Similarities and systematic trends can be expected between line broadening parameters of the lines within a given spectrum (multiplet, supermultiplet, transition array), within homologous atoms and ions and within an isoelectronic sequence.

REFERENCES

[1] H.R.Griem, Phys.Rev. 165, (1968) 258.
[2] H.R.Griem, Spectral Line Broadening by Plasmas, Academic Press, N.Y., (1974).
[3] J.D.Hey, and R.J.Bryan, JQSRT 17, (1977) 221.
[4] M.S.Dimitrijević, and N.Konjević, JQSRT 24, (1980) 451.
[5] M.S.Dimitrijević, and N.Konjević, In Spectral Line Shapes, Vol.1 (Ed. B.Wende), W. de Gruyter, Berlin, New York, (1981) p. 211.
[6] M.S.Dimitrijević, and N.Konjević, Astron.Astrophys. 102, (1981) 93.
[7] M.S.Dimitrijević, and N.Konjević, JQSRT 25 (1981) 387
[8] M.S.Dimitrijević, and N.Konjević, Astron.Astrophys. 163, (1986) 297.
[9] M.S.Dimitrijević, and N.Konjević, Astron.Astrophys. 172 (1987) 345.
[10] M.S.Dimitrijević, and V.Kršljanin, Astron.Astrophys.165, (1986) 269.
[11] M.J.Seaton, J.Phys.B 20 (1987) 6431.
[12] M.S.Dimitrijević, In Sun and Planetary System, (Eds. W.Fricke, G.Teleki), D.Reidel, Dordrecht, (1982) p.101.
[13] M.S.Dimitrijević, In The Physics of Ionized Gases (Inv. Lectures of SPIG-82)- (Ed. G.Pichler), Inst.Phys.Univ. Zagreb, (1982) p. 397

[14] M.S.Dimitrijević, Astron.Astrophys. 127, (1983) 68.
[15] M.S.Dimitrijević, In Physics of Formation of FeII Lines Outside LTE. (Eds. R.Viotti, A.Vittone, M.Friedjung), D.Reidel, Dordrecht, (1988) p. 211.
[16] M.S.Dimitrijević, Bull.Obs.Astron.Belgrade, 139, (1988) 31.
[17] N.Konjević, M.S.Dimitrijević, and W.L.Wiese, J.Phys.Chem. Ref.Data 13, (1984) 649.
[18] V.Kršljanin, and M.S.Dimitrijević, Z.Phys.D 14, (1989) 273.
[19] T.Lanz, M.S.Dimitrijević, and M.C.Artru, Astron. Astrophys. 192, (1988) 249.
[20] M.A.El-Farra, and T.P.Hughes, JQSRT 30, (1983) 335.
[21] U.Ackermann, K.H.Finken, and J.Musielok, Phys.Rev.A 31 (1985) 2597.
[22] V.Kršljanin, and M.S.Dimitrijević, Bull.Obs.Astron.Belgrade 140, (1989) 7.
[23] D.R.Bates, and A.Damgaard, Phil.Trans.R.Soc.London Ser. A 242, (1949) 101.
[24] G.K.Oertel, and L.P.Shomo, Astrophys.J.Suppl. 16, (1968) 175.
[25] M.S.Dimitrijević, Astron.Astrophys. 112, (1982) 251.
[26] B.W.Shore, and D.H.Menzel, Astrophys.J.Suppl. 12, (1965) 187.
[27] H.R.Griem, M.Baranger, A.C.Kolb, and G.K.Oertel, Phys. Rev. 125, (1962) 177.
[28] M.S.Dimitrijević, and V.Kršljanin, SPIG-90, Dubrovnik (1990)
[29] N.Konjević, and M.S.Dimitrijević, In Spectral Line Shapes, Vol. 1, (Ed. B.Wende), W.de Gryuter, Berlin, New York, (1981) p. 241.
[30] W.L.Wiese, and N.Konjević, JQSRT 28, (1982) 185.

THERMAL AND RADIATION STABILITY OF MULTILAYER X-RAY MIRRORS Mo-Si AND $MoSi_2$-Si

V.V. Kondratenko, Yu.P. Pershin, O.L. Poltseva, A.I. Fedorenko, E.N. Zubarev, and S.A. Yulin
Kharkov Polytechnic Institute, Kharkov, Ukraine

I.V. Kozhevnikov, S.I. Sagitov, V.A. Chirkov, V.E. Levashov, and A.V. Vinogradov
P.N. Lebedev Institute of Physics, USSR Academy of Sciences, Moscow, USSR

INTRODUCTION

Multilayer systems of alternative thin films of two or more substances with different values of dielectric constant may be used as effective optical elements for spectrometry and application of soft X-ray radiation of numerous sources like plasma, synchrotrons, X-ray lasers, space,etc [1]. Being consisted of ultrathin layers of different materials X-ray multilayer mirrors with abrupt interfaces represent usually nonequilibrium thermodynamic systems, having a tendency to time degradation, gradual change of structure and optical characteristics. Instability of multilayers is especially dramatic at elevated temperatures and under the load of power fluxes of photons and particles. This feature of multilayers limits the scale of their application in many fields of science and technique. Therefore,improving the thermal and radiation stability of multilayers is important for development of X-ray optics.

There are several mechanisms of aging of multilayer optical characteristics:agglomeration of planar layers by diffusion of atoms along interfaces of multilayer; interdiffusion mixing of layers with the formation of solid or liquid solutions and compounds; deformation, crackling or fracture of multilayer as a result of stress relaxation in a system "multilayer-substrate" [2,3]. To exclude or at least to weaken the destroying action of the above noted mechanisms of multilayer ageing one has to choose multilayer materials taking into account many requirements. The pair of the materials chosen has to be in thermodynamic equilibrium with the environment inside of the layers and at interfaces; to have melting temperatures as high as possible; to form refractory eutectic with each other; to have equal thermal and radiation expansivities; to wet each other; not to undergo structure and phase transformations at specific conditions of multilayer applications. The difficulties of simultaneous meeting of these requirements are obvious, but one has to aspire to satisfy, at least the most important of them to improve stability of multilayer functioning.

The aim of this study is to find out how the approaching of thermodynamic equilibrium of two materials at their interfaces improves thermal and radiation stability of multilayer. It is known [4] that thermodynamic equilibrium of two phases at their interfaces may be achieved only if they are situated at the ends of one common conoda, i.e. only if they are neighbours in the phase diagram of binary alloy system. For example, according to phase diagram of binary Mo-Si alloys [5] thermodynamic equilibrium at interfaces of widely used multilayers of pure Mo and Si is impossible because Mo is neighbouring with silicide Mo_3Si whereas Si is the neighbour of silicide $MoSi_2$. Therefore, formation of silicides on interfaces of Mo-Si multilayer is inevitable as at leads to thermodynamic equilibrium. To exclude this mechanism of ageing one has to use $MoSi_2$ and Si instead of pure Mo and Si as a pair of materials for construction of multilayer. From thermodynamics $MoSi_2$-Si multilayers are expected to be much more stable in original state in comparison with Mo-Si ones. The melting temperatures of Mo, $MoSi_2$, Si and eutectic $MoSi_2$-Si are 2898, 2293, 1688 and 1673 K, respectively [5]. So, from diffusion kinetics Mo-Si multilayers have some advantage over $MoSi_2$-Si ones because Mo has higher melting temperature than $MoSi_2$.

Thus, thermodynamic and kinetic considerations lead to different conclusions about thermal and radiation stability of Mo-Si and $MoSi_2$-Si multilayers. The results of experimental study of this subject will be given in the following sections.

EXPERIMENTAL

Mo-Si and $MoSi_2$-Si multilayers were prepared by dc-magnetron sputtering of Mo, $MoSi_2$ and Si targets at pressure of $\sim 3.10^{-3}$ mbar of argon atmosphere. The materials were deposited onto polished flat substrates of float-glass and single crystal silicon or polished spherical substrates of quartz glass. The substrates were mounted on carrier frame rotated over two working magnetron sources. The grazing angle diffraction of CuK_α-radiation with wavelength λ=0.154 nm was

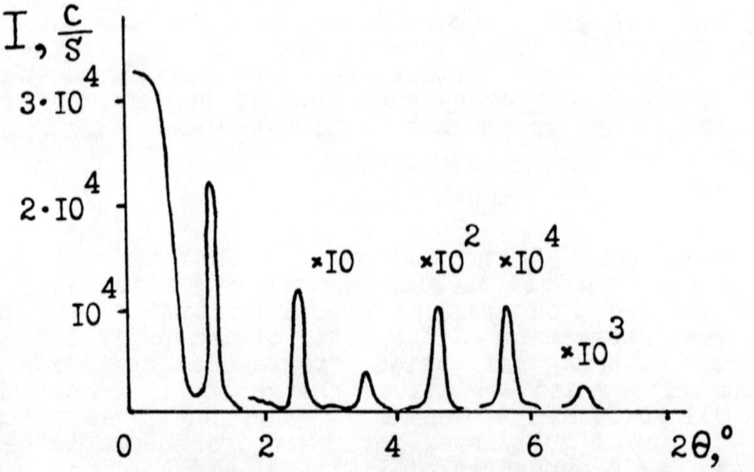

Fig.1. 2θ detector scan of monochromatic (0.154 nm) radiation diffracted by Mo-Si multilayer with period H=8.7 nm and number of periods N=20 (sample 1a).

Table 1. Characteristics of samples.

sample number	materials	$d_{Si}/d_{Mo,MoSi_2}$	period H, nm	number of periods, N
1a	Mc - Si	2 : 1	8.7	20
1b			4.2	40
2a	Mo - Si	1 : 1	7.4	20
2b			4.1	40
3a	Mo -- Si	1 : 2	7.6	20
3b			4.0	40
4a	$MoSi_2$-Si	2 : 1	8.6	20
4b			3.9	40

used for characterization of period and reflectivity of multilayer both in original state and after annealing at 570-1070 K in vacuum $\sim 10^{-7}$ mbar or after irradiation by ions He^+. The crystal structure and phase composition of Mo, $MoSi_2$,and Si layers were studied by X-ray diffractometry of CuK_α-radiation and electron microscopy methods. Monochromatization of CuK_α-radiation wasprovided by its reflection from

(III) planes of silicon single crystal. An example of diffraction spectra of multilayer is shown in Fig. 1. The number of Bragg reflections certificates the multilayer quality. Table 1 contains the characteristics of periods N, ratio of thicknesses of spacer and absorbing materials

THERMAL STABILITY

The intensities I and the angle positions of all Bragg reflections in the diffraction spectra of a multilayer were measured as the functions of annealing time τ at constant temperature. Decreasing of the ratio of intensity $I(\tau)$ of the first order of Bragg reflection after annealing to its initial value $I(0)$ and relative change of period $\Delta H/H = (H(\tau)-H(0))/H(0)$ after annealing were chosen as criterions of degradation of multilayer X-ray mirrors due to temperature treatment. The dependences of $I(\tau)/I(0)$ and $\Delta H/H$ on τ at different temperatures for Mo-Si and $MoSi_2$-Si multilayers are shown in Fig.2 and Fig. 3. One can see that both the reflectivity and period of Mo-Si multilayers are decreasing with τ monotonically at temperatures as low as 570 K. The degradation rate of Mo-Si multilayers grows with increasing of annealing temperature. So, after annealing at 770 K during only

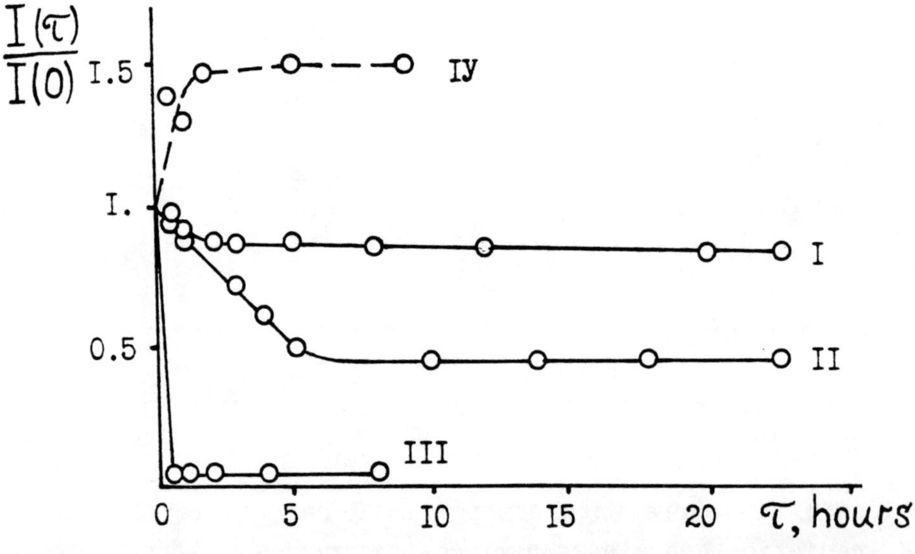

Fig.2. The relative reflectivity in the first Bragg peak of Mo-Si (curves I,II,III) and Mo-Si (curve IV) multilayers with period ~8nm as the functions of annealing time τ at different temperatures (I - 570K; II- 670K; III,IV - 770K).

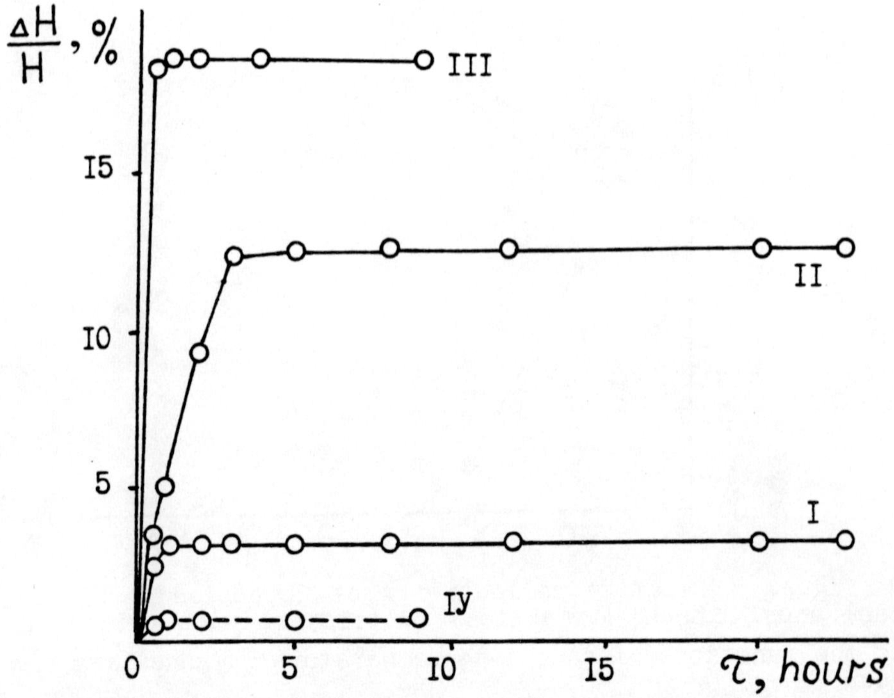

Fig.3. Relative decrease of Mo-Si (curves I,II.III) and $MoSi_2$-Si (curve IV) multilayer periods as the functions of annealing time τ at different temperatures (I-570K; II-670K; III,IV - 770K). All multilayers have 20 periods with H~8nm.

~0.5 hour the value of I(τ)/I(0) of Mo-Si multilayer with H=8.7 nm decreases to as low level as ~0.03-0.05 (curve III in Fig.2) that means practically complete destruction of X-ray mirror.

At the beginning of annealing at all temperatures the degradation rate of Mo-Si multilayers is increased by such processes of reconstruction of the layer structure as migration of grain boundaries, dislocations and other imperfections of polycrystal Mo-layers and crystallization of amorphous Si-layers. After stabilization of Mo and Si-layer structures the degradation rate of Mo-Si layers decreases to a small constant value (Fig.2 and Fig. 3).

Higher Bragg reflections in contrast to the first one show nonmonotonic dependence of relative intensities I(τ)/I(0) on the time of annealing τ. This means that Mo and Si concentration profiles in multilayer cross section does not just tend to sinusoidal shape, but suffer some complicated evolution

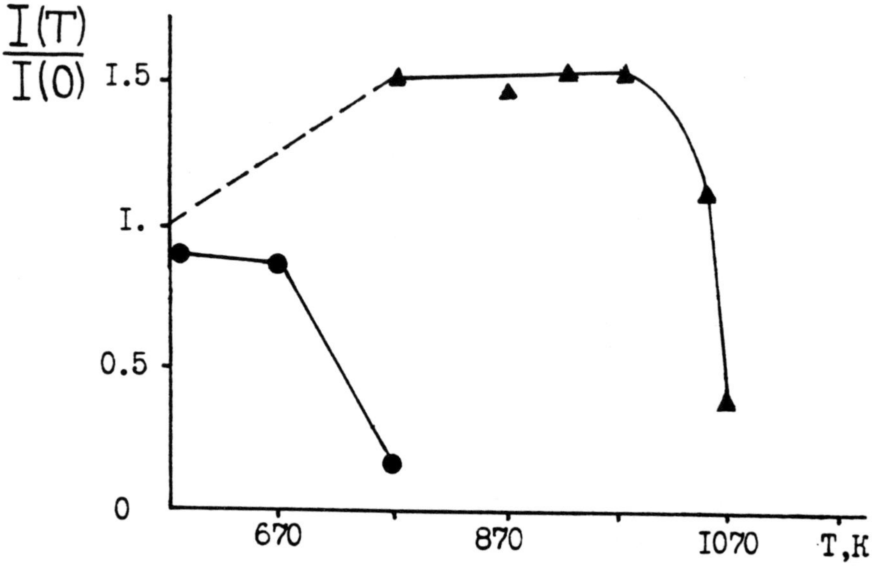

Fig.4. Relative reflectivity of 20-period Mo-Si (circles) and $MoSi_2$-Si (triangles) multilayers with period ~8 nm as the functions of annealing temperature T (annealing time 1 hour).

connected with formation of compound $MoSi_2$ at the interfaces.

Formation of hexagonal Mo-Si_2 on interfaces of Mo and Si layers was revealed by electron microscopy after annealing of Mo-Si multilayer at temepratures higher than 700K. Mo-Si_2 is of less specific volume than original pure Mo and Si [6]. Therefore, the formation of $MoSi_2$ is the main reason for as large as 18% decrease of Mo-Si multilayer period after annealing of the mirrors at 770K (curve III in Fig. 3).

Straightforward application of Hilliard's model [7] to kinetic dependence of $I(\tau)/I(0)$ on τ similar to the ones shown in Fig.2 gives the following values of diffusivities D of Mo and Si layer intermixing at 570K: D= $1.9.10^{-25}$; $2.7.10^{-25}$; $6.2.10^{-25} m^2/s$ for Mo-Si multilayers with period H=2.4; 4.1 and 8.2 nm respectively.

In contrast to Mo-Si mirrors the ratio $I(\tau)/I(0)$ for $MoSi_2$-Si multilayers increases by 20-50% at the beginning of annealing at temperatures below ~1000 K (curve IV in Fig.2), that means a reduction of roughness of $MoSi_2$ and Si interfaces. After the first few hours of annealing the relative reflectivity $I(\tau)/I(0)$ of $MoSi_2$-Si mirrors is stabilized at some higher level than the original one(curve IV in Fig.2) whereas the change of the period $\Delta H/H$ does not exceed some percent (curve IV in Fig.3).

The degradation of $MoSi_2$- Si multilayer (the decrease of $I(\tau)/ I(0)$ relative to its initial value 1) is observed only at temperatures higher than 1000K (see Fig.4). We consider this to be a result of crystallization of amorphous Si-layers, recrystallization of polycrystalline $MoSi_2$-layers and transformation of hexagonal $MoSi_2$ to its tetragonal form. All these high-temperature processes promote agglomeration of $MoSi_2$ and Si layers and development of roughness of their interfaces.

Much more high thermal stability of $MoSi_2$-Si mirrors in comparison with with Mo-Si ones is illustrated by Fig. 4. As a result of improving of thermodynamic equilibrium of the layer materials the working temperature of $MoSi_2$-Si X-ray mirrors is ~1000 K, which is much higher than ~600 K for Mo-Si mirrors (Fig. 4).

The choice of multilayer materials based on the principle of interface thermodynamic equilibrium grants good opportunity for the improvement of thermal stability of X-ray mirrors owing to a decrease in the driving force for diffusion intermixing and formation of chemical compounds of the layer materials.

RADIATION STABILITY

The Mo-Si and $MoSi_2$-Si mirrors were irradiated by He^+ ions with energy 10-30 keV at room temperature using linac as ion source. The dependences of relative intensities $I(P)/I(0)$ of all Bragg reflections and relative change of period $\Delta H/H$ of multilayer on irradiation dose P were measured for the samples listed in Table 1. The most relevant dependences are shown in Fig.5-Fig.8, where one can see that similar to thermal stability the radiation stability of $MoSi_2$-Si mirrors is much better than Mo-Si ones.

Analysis of Figs.5-8 shows the following main effects induced by He^+ irradiation of multilayers: (a) nonmonotonic change of intensities at higher orders of Bragg reflections with increasing of the irradiation dose (Fig. 5); (b) essential decrease of period H of multilayers (Fig. 7); (c) dependence of degradation rate of Mo-Si mirror reflectivity on period H (Fig.5); (d) similarity of dose behaviour of III-order Bragg reflection of sample 1a and II order Bragg reflection of sample 1b (Fig.5), which differ in H by a factor ~2.

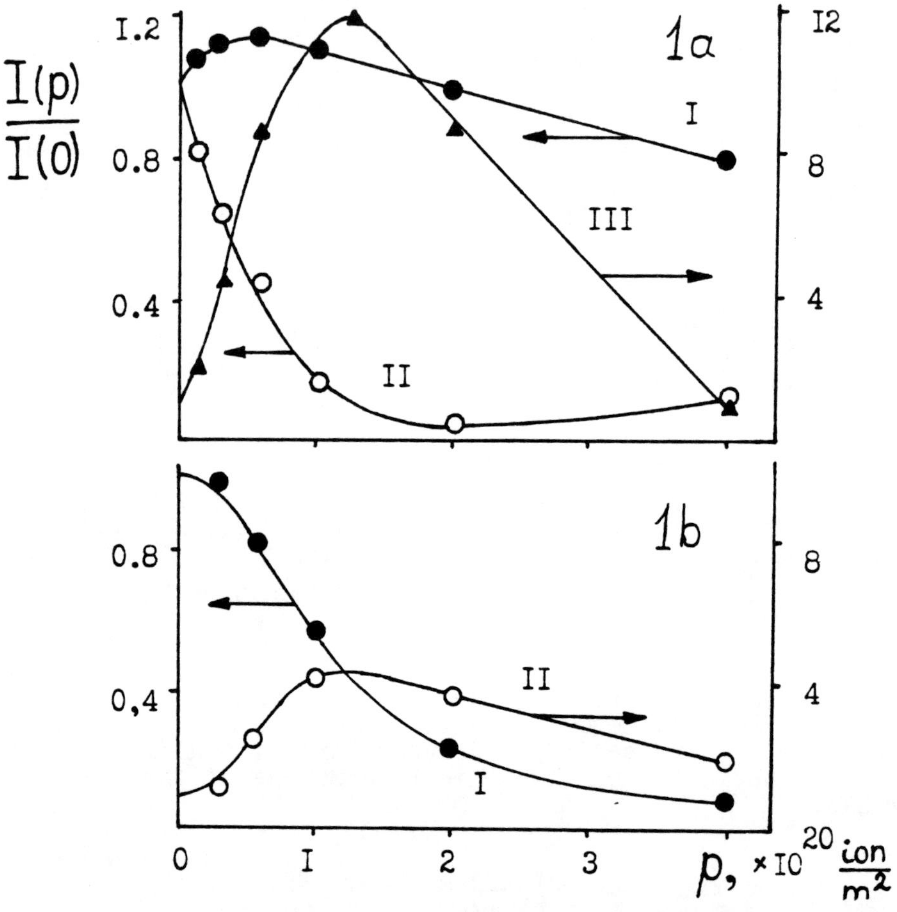

Fig. 5. Relative intensities of Bragg reflections of Mo-Si multilayers with periods H = 8.7 nm (sample 1a) and H = 4.2 nm (sample 1b) as functions of irradiation dose of He^+ with energy E=30 keV (I,II,III-orders of Bragg reflections).

All the effects can be understood if we adopt that the main mechanism of radiation degradation of Mo-Si multilayer is ballistic intermixing of layers in collision cascades induced by He^+ ions nearby interface accompanied by nonhomogeneous formation of $MoSi_2$.

Radiation induced nonhomogeneous growth of silicides like $MoSi_2$ develops the roughness of interfaces of Mo and Si lay-

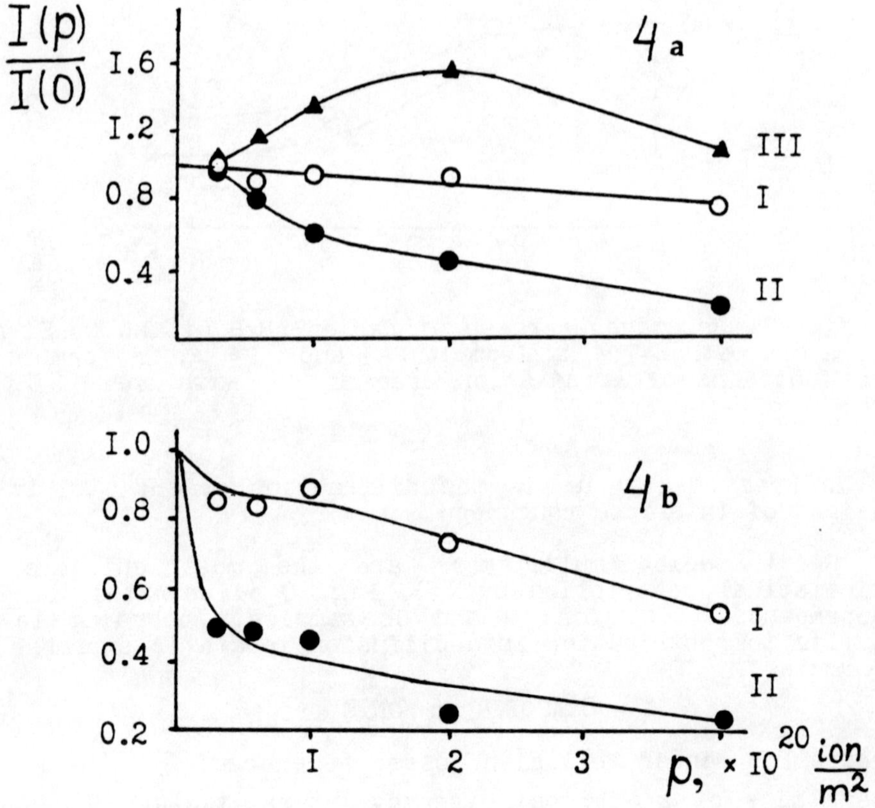

Fig. 6. The relative intensities of Bragg reflections of $MoSi_2$-Si multilayers with periods H=8.6 nm (sample 4a) and H=3.9 nm (sample 4b) as functions of irradiation dose of He^+ with energy E=30 keV (I,II,III - orders of Bragg reflections).

ers, that decreases the peak reflectivity R(g) of multilayer according to the law:

$$R(g) = R_0 e^{-g^2\sigma^2}, \qquad (1)$$

where R_0 is the peak reflectivity of multilayer with ideally smooth interfaces;

$$g = \frac{4\pi}{\lambda}\sin\theta = \frac{2\pi n}{H},$$

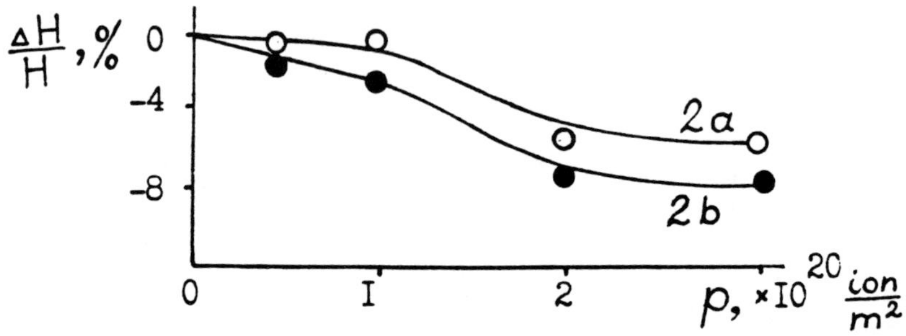

Fig. 7. Relative decrease of period ΔH/H of two Mo-Si multilayers with H=7.4 nm (sample 2a) and H = 4.1nm (sample 2b) as functions of irradiation dose of He^+ with energy E = 30 keV.

θ is Bragg angle; n is the diffraction order; σ is the height of interface roughness.

Short period multilayers are the most suitable for kinematical description by (1). Fig. 8 illustrates the dose dependence of σ^2 for 1b and 3b samples. According to [7] ballistic contribution into diffusion length is expressed by formula:

$$D\tau = 0.42\sigma F_D R_c^2 \phi / \sigma N_0 E_d \qquad (2)$$

where F_D, linear collision losses of energy; R_c, minimum linear size of stable pair vacancy-interstitialcy; N_o, atomic density of system; E_d, mean energy for atom displacement; $\Phi = P/\tau$, fluence.

Table 2. Characteristics of ballistic diffusion intermixing of Mo-Si multilayers.

sample	$\frac{D\tau}{\Phi}$, 10^{-4} nm^4	D, 10^{-23} m^2/s
1b	12	5
2b	20	8
3b	20	8

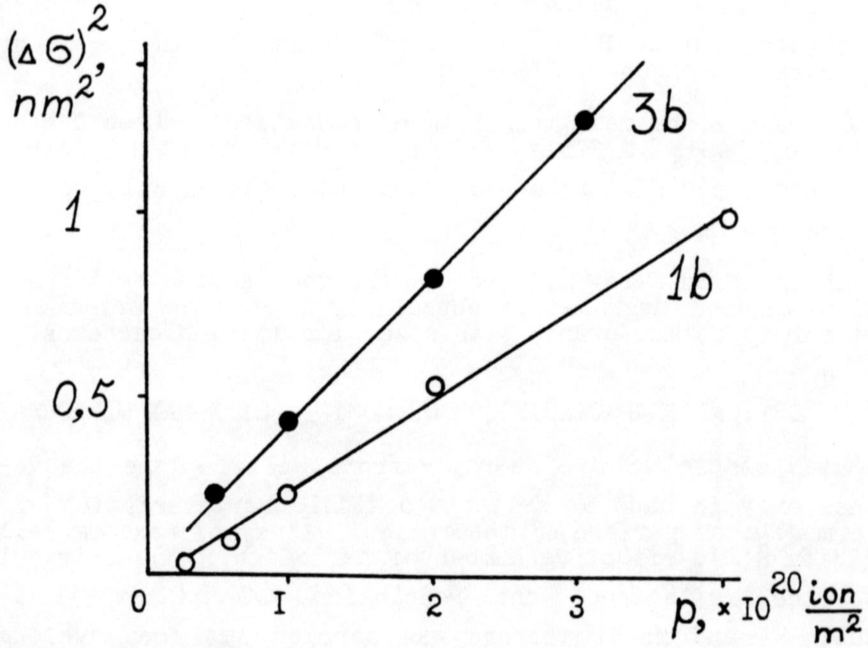

Fig. 8. Increase of interface roughness of two Mo-Si multilayers with periods H = 4.2 nm (sample 1b) and H=4.0 nm (sample 3b) as functions of irradiation dose of He^+ with energy E=30 keV.

Linear dependence of σ^2 on ϕ corresponds to cascade mechanism of layer intermixing. Using that

$$\sigma^2/2 = D\tau \qquad (3)$$

parameter $D\tau/\phi$ and ballistic diffusivity D of ballistic intermixing of Mo and Si layers were determined for samples 1b,2b,3b (Table 2).

Larger concentration of silicon may explain the observed difference of characteristics of ballistic intermixing of sample 1b in comparison with the samples 2b and 3b.

Fig. 6 illustrates that the role of degradation of $MoSi_2$-Si mirrors was much lower than Mo-Si ones (see Fig.5). Really at dose 4.10^{20} ion/m^2 $MoSi_2$-Si mirror with H=3.9 nm has relative reflectivity I(P)/I(0) ~ 0.5 (Fig. 6) whereas similar Mo-Si mirror with H=4.2 nm has I(P)/I(0) ~ 0.08

(Fig. 5) Another advantage of $MoSi_2$-Si mirrors is practically unchanged period H at dose 4.10^{20} ion/m^2 whereas H of Mo-Si mirrors decreases by 6-8 %.

Therefore the main mechanism of radiation induced degradation of $MoSi_2$-Si X-ray mirrors is development of interface roughness and disturbance of $MoSi_2$ stoichiometry by collision cascades.

Shortperiod X-ray mirrors are the most subjected to radiation induced degradation especially when the period of the mirror is commensurable with size of collision cascades.

OPTICAL CHARACTERISTICS OF $MoSi_2$ - Si X-RAY MIRRORS

High stable $MoSi_2$-Si X-ray mirrors will replace the Mo-Si ones only in case of suitable optical characteristics of the former. A comparison of theoretical values of maximum reflectivity $R(\lambda)$, effective number of periods $N_{eff}(\lambda)$ contributing to the reflectance and resolution $\lambda/\Delta\lambda \simeq (\pi/2)N_{eff}$ of $MoSi_2$-Si and Mo-Si mirrors was carried out for wavelengths

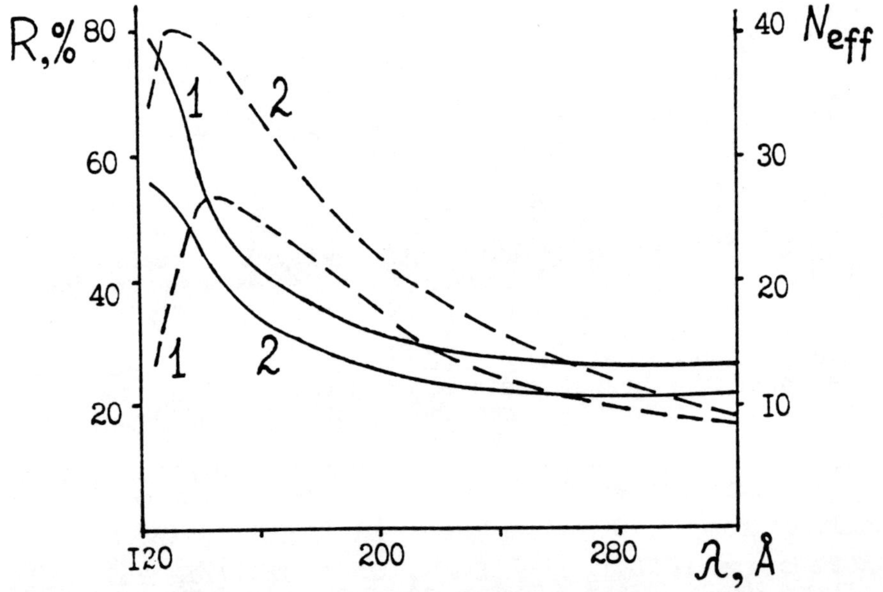

Fig. 9. Maximum reflectivity $R(\lambda)$ (solid lines) and effective number of periods N_{eff} (dotted lines) of Mo-Si (curves 1) and $MoSi_2$-Si (curves 2) X-ray mirrors as the functions of soft X-ray wavelength λ.

$\lambda>12.4$ nm using formulas [8]. In Fig.9 one can see that at $\lambda>15$-20 nm $MoSi_2$-Si and Mo-Si mirrors have practically coinciding values of $R(\lambda)$, $N_{eff}(\lambda)$ and $\lambda/\Delta\lambda$. Only nearby the end of photoabsorption band of silicon ($\lambda=12.4$nm) the Mo-Si mirrors have 10-20 % superiority over $MoSi_2$-Si ones in optical properties (Fig.9).

Note that the resolution of the mirrors shown in Fig.9 is rather low from 40-60 to 10-15. This is due to the fact that calculated structure is optimized to have maximum peak reflectivity. If necessary resolution can be improved by reducing of $MoSi_2$ layer thickness [9].

Experimental measurement of reflectivity of $MoSi_2$-Si mirrors at normal incidence of soft X-rays was carried out by a method described in [9,10].$MoSi_2$-Si mirrors were deposited on spherical quartz substrates with curvature radius 10 cm. Spectrograph of grazing incidence DFS-6 with slit of spatial resolution was used for comparison of two X-ray spectra radiated by laser plasma, and its image created by spherical multilayer mirror [10,11]. Table 3 contains the results of measurements. Much more high reflectivities of $MoSi_2$-Si mirrors in comparison with similar Mo-Si ones indicates that $MoSi_2$-Si interfaces are much smoother than Mo-Si interfaces. Annealing of $MoSi_2$-Si spherical mirror at 500K during 1 hour did not change its reflectivity at $\lambda\approx14$ nm (Table 3). Therefore, the optical characteristics of $MoSi_2$-Si X-ray mirrors are high enough to provide good competability for them in X-ray optics.

Table 3. Experimental reflectivities of spherical Mo-Si and $MoSi_2$-Si focusing mirrors at normal incidence of soft X-rays.

mirror materials	period H, nm	number of periods	λ,nm	R, %	$\lambda/\Delta\lambda$
Mo-Si	7.5	40	14.5	35±4	36
$MoSi_2$-Si	7.5	40	14.3	52±5	41
$MoSi_2$-Si annealed for 1 hour at 500 K	7.7	40	14.9	52±5	43

CONCLUSIONS

1. Construction of multilayers from two materials of one binary alloy system with thermodynamic equilibrium on their interfaces essentially improves thermal and radiation stabilities of multilayer elements of X-ray optics.

2. Thermal and radiation degradation of Mo-Si multilayer is connected with formation of compound $MoSi_2$ at the interfaces. Multilayers with higher thermal and radiation stability have been produced of $MoSi_2$ and Si which is a pair of phases being in thermodynamic equilibrium.

3. Calculated soft X-ray reflectivity of $MoSi_2$-Si multilayer mirror is only slightly lower than that of Mo-Si multilayer. But in practice the prepared $MoSi_2$-Si mirrors appeared to give even higher reflectivities than Mo-Si mirrors.This means that $MoSi_2$-Si interfaces are smoother than Mo-Si ones as a result of lesser mutual solubility of $MoSi_2$ and Si and better their mutual alignment at their interface as compared to Mo and Si.

4. High thermal and radiation stabilities and high reflectivities of $MoSi_2$-Si mirrors at normal incidence of soft X-rays provide good perspective for their application as elements of X-ray optics under heavy radiation load of plasma, synchrotrons, etc.

REFERENCES

[1] X-ray instrumentation in medicine and biology, plasma physics, astrophysics, and synchrotron radiation. (Ed. R.Benattar). Proc.SPIE, Vol.1140 (1989).

[2] A.L.Greer and F.Spaepen, In Synthetic modulated structures (Eds.L.C.Chang and B.C.Giessen), Academic Press, N.Y., (1988) p.419.

[3] K.Hooloway, K.B.Do and R.Sinclair, J.Appl.Phys., 65, (1989) 474.

[4] S.Rines, Diagrams of phase equilibrium in metallurgy, Metallurgizdat, Moscow, (1960).

[5] M.Hansen and K.Anderko, Structure of binary alloys, Vol.2, Metallurgizdat, Moscow (1962).

[6] A.K.Pefford-Long, M.B.Stearns et al., J .Appl .Phys. 61, (1987) 1422.

[7] A.Traverse, J.Less, Common Metals, 145, (1988) 451.

[8] I.V.Kozhevnikov and A.V.Vinogradov, Physica Scripta,17, (1987) 134.

[9] S.S.Borisova, I.V.Kozhevnikov, V.V.Kondratenko et al., Journal of technical physics, 59, (1989) 78 (in Russian).

[10] A.V.Vinogradov, I.V.Kozhevnikov, V.E.Levashov et al. Proc.SPIE, 1140, (1989) 530.

PART V

EXCIMER LASER RESEARCH

THEORY OF HIGH-POWER WIDE-APERTURE EXCIMER LASERS

A.G. Molchanov
P.N. Lebedev Institute of Physics, Academy of Science, Leninski Prospect 53, Moscow 117924, USSR

Abstract

Theory of excimer amplifiers including effects of amplified spontaneous emission (ASE) in application to KrF-laser is presented. The problems of scaling to high output energies are discussed.

1. Introduction

The theory of excimer lasers includes three main sections: 1) calculation of the gain and absorption coefficients on the base of the model of kinetic processes in the active medium; 2)the calculation of the pumping rate of the active medium by using an electron beam or electric discharge; and 3) self- consistend calculation of the emission fields [1]. The present work is focused on the determination of the e-beam pumped KrF-laser parameters. The noted laser is considered to be a promising candidate for ICF drivers [2,3]. The active medium of KrF-laser has a high gain 0.1 cm^{-1}, broad bandwidth of 50 cm^{-1}, and a short radiative decay upper level of 6.5 ns. Hence, the medium has very powerful spontaneous emission which may be amplified with a sufficiently large medium in all directions to a quantaty greater than the saturating power. The KrF-laser is saturated uniformly for energy extraction times greater than a few nanoseconds and, consequently, the gain saturation is weakly dependent on the frequency band of the saturating field, and, hence, the ASE may significantly reduce the gain and the extracted energy. Increasing the output energy of KrF-amplifiers involves using wide outcoupling apertures, since the output energy density is limited by a comparatively low in UV breakdown threshold of the mirrors and the outcoupling windows of the amplifier. Moreover, wide-aperture KrF-amplifiers with a diameter/length ratio D/L = 1 are requered to reduce pulse duration by angular multiplexing in ICF [3].

The e-beam is now practically the only one effective method presently available to be used for wide-aperture KrF-laser pumping. Calculations of e-beam pumping rates are based on Monte-Carlo method are sufficiently complicated as a rule. Section 2 presents a sufficiently simple technique to calculate the pumping rates for a e-beam gas mixture excitation. Section 3 gives a simple method for numerical simulations of ASE effects in the amplifiers which have the form of a parallelepiped. The method has been applied to calculation of KrF-amplifiers of different dimensions.

2. Kinetics and e-beam pumping rates

In spite of the wide veriety of media and pumping technique, the basis of all kinetic models is related to the central role played by rare gases in converting the energy of external source and transfering this energy to other components in the active media. In modern theory of excimer lasers the following species of i-th rare-gas component should be included in kinetics: $R^+(i)$, $R_2^+(i)$, $R^*(i)$, $R_2^*(i)$, $R^{**}(i)$, $R_2^{**}(i)$, where R^* - rare atom with electron in first excited ns-hell, R^{**} - electron in the next np-shell, and R^* R^{**} -according excited states of molecules. In the mixture of two rare gases i anl j the following species of crossing components must be included: $R^+(i)R(j)$, $R(i)R^+(j)$, $R(i)R^*(j)$, $R^*(i)R(j)$. Thus the number of species in kinetics is mainly defined by standard blocks of rare gas components with rare-gas-halogens and may be other additives.

The following relation between the slow-electron production rate and fast-electron stopping power S is exist:

$$\frac{dn_e}{dt} = \frac{j_b}{e\,I} S \qquad (2.1)$$

where j_b- current density, e- electron charge, I- the mean energy needed to form an electron-ion pair. The dependence of stopping power on depth may be presented in the simplest gaussian form:

$$S(x) = \frac{E_0}{X_{eff}} \exp\left[- \left(\frac{x - \delta_\circ x_\circ}{\xi_\circ x_\circ} \right)^2 \right] \qquad (2.2)$$

where $\delta_\circ$, $\xi_\circ$ -parameters are founded from numerical results [4] and may be approximated as $\delta_\circ$=0.58-(0.125-0.003 E_0) ln(Z) $\xi_\circ$= (0.52+0.01 E_0)(1-0.12 ln(Z)) ,E_0-initial electron energy, $x_\circ$-maximum range of fast electrons. The analytic expression for last quantity have been founded in [5] and with logarithim correction for high electron energies may be presented in form:

$$x_\circ = \frac{\varepsilon_\circ^2}{\alpha \ln(\beta\varepsilon_\circ)(1 + \varepsilon_\circ)} \qquad (2.3)$$

where $\varepsilon_\circ = E_0/mc^2$, $\alpha = \eta\rho Z/A$, $\eta = 4\pi r_e^2 N_A = 0.60\ cm^{-1}$, $\beta = 1.16\ mc^2/2J$, $J \simeq 12.5\ Z$, $Z/A = \Sigma c_i Z_i/\Sigma c_i A_i$, Z_i, A_i, c_i - number, weigth, and relative concentration of i-th sort of atom, ρ-density. The formula (2.3) is valid for electron energies from ≃10 keV to 10 MeV. The effective depth of excitation of medium x_{eff} in (2.2) is calculated from energy conservation condition namely the integral of (2.2) should be equal E_0. Approximating of error integral $\Phi(x) \simeq th(2x/\sqrt{\pi})$ for all $x \geq 0$, it's may be obtain:

$$x_{eff} = \gamma_\circ x_\circ, \quad \gamma_\circ = \frac{\sqrt{\pi}}{2} \xi_\circ [1 + th(\frac{2\delta_\circ}{\sqrt{\pi}\,\xi_\circ})] \qquad (2.4)$$

The quantity $W_b = j_b E_0 / e\, x_{eff}$ defines the value of pumping power density. For active medium of KrF-laser having gas mixture Ar/Kr/F=95.5/4/0.5 with total pressure 1.75 atm, the e-beam with 0.6 MeV energy (behind foil) gives $x_{eff} = 0.5$ m, $x_\circ = 1$ m, and $E_0/x_\circ = 12$ keV/cm. The latter value is close to 13 keV/cm calculated in [6] by the Monte Carlo method at the same conditions. It should be noted that the analytical approach permits to use the single Monte Carlo computation to scale it to different electron energies and densities of active medium.

3. ASE-waves method and characteristics of KrF-amplifiers

In order to calculate the influence of ASE on the characteristics of wide-aperture optical amplifiers we will devide the entire angular space into a six of adjacent solid angles in which ASE is averaged [1]. This method is a natural generalization of a "back-forward" approximation in radiation transfer theory. As a result, we obtain a system of 8 differential equations with boundary conditions: 6 for the ASE waves and 2 for the signal ones (Fig.1). In the centre of amplifier these equations are self-conjugated. The comparison of ASE-wave method applied to amplifying sphere, where the exact formula for ASE flux is known [1], show excelent agreement.

In general case the characteristics of amplifier are defined by 5 dimensionless parameters: gain length product $g_0 L$, gain absorption ratio g_0/a_0, aperture form factor H/B, aspect ratio L/H and life - time ratio τ/τ_r, where τ_r-radiation and τ- total life-times of upper laser level. Besides that input I_{in}/I_s and output I_{out}/I_s parameters should be added, where I_s - saturated flux in active media. In order to regulate the calculations of wide-aperture amplifiers it is convinient to devide its into energy classes. The mean results of calculations for KrF-laser are presented in Table1. It is easy to note that these classes on the whole are corresponding to a stages of amplification in ICF drivers.

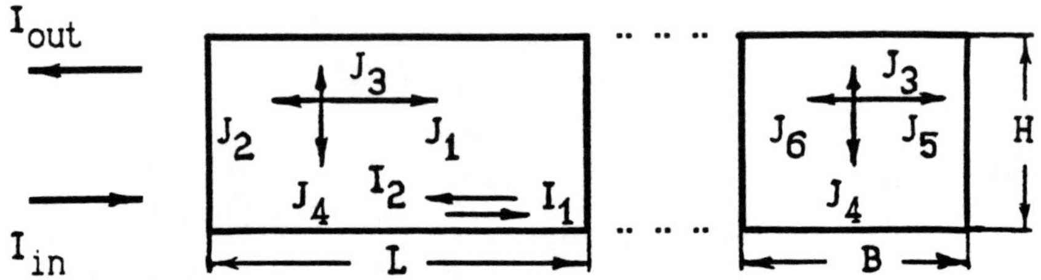

Fig. 1. Scheme for propagation of signal waves I_1, I_2 and ASE-waves $J_1, \ldots J_6$ in a two-pass amplifier having length L and output rectangle aperture HxB.

Table 1. Classes of KrF-amplifiers for ICF

PARAMETERS	CLASSES				
	A	B	C	D	E
Energy, Q(J)	50	10^3	10^4	10^5	10^6
Volume, V(l)	5	10^2	10^3	10^4	10^5
Aperture, H(m)	0.1	0.3	1	2	4.5
Length, L(m)	0.5	1.2	2	3	4.5
Aspect, A	5	4	2	1.5	1
Int. eff., (%)	12	11	10	9.5	8
Opt. load., (J/cm^2)	0.5	1	1.5	3	5

Fig. 2 gives results from self-consisted calculations of a system of kinetic and emission field equations for some characteristics of e-beam-pumped two-pass KrF-amplifier of D- class. The diffuse reflection coefficients of the side walls of the amplifiers were assumed to be 5 % . As we see from these data, in order to simultaneously achive a high efficiency as well as a high gain of $M = I_{out}/I_{in} \geq 10$ it is necessary to inject a rather intense signal to the input $I_{in} = (0.2\text{-}0.6)\ I_s$. Amplification of spontaneous emission transverse to the amplifier with large aperture serves to reduce the gain near the side walls. The magnitude of this reduction for weak input signals may comprise a noticible amount of the total gain and may influnce the angular divergence of laser emission (the complex "lens" effect). This effect may be partially compensated by varying the energy of electron beam but the complete means to obtain high quality divergency in wide-aperture KrF-amplifiers is phase conjugation method [7].

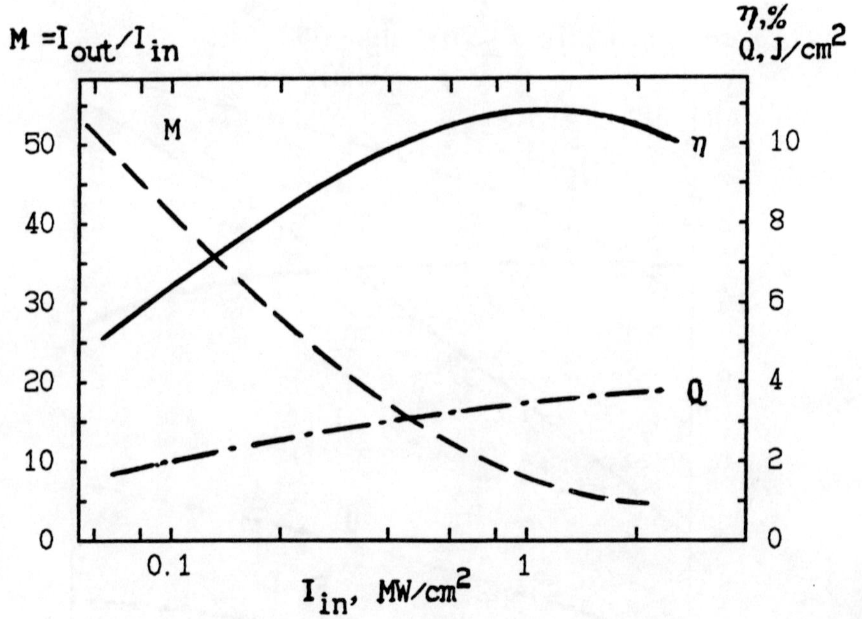

Fig.2. Gain characteristics of a two-pass KrF-amplifier of D - class: $L=3$ m, $S=2x2$ m, $Kr/F_2=99.7/0.3$, $P=0.8$ atm, $W_b=0.2$ MW/cm^3, $\tau_b=0.6$ mcs, $E_0=1.4$ MeV, $j_b=13$ A/cm^2, $g_0=0.024$ cm^{-1}, $a_0=0.002$ cm^{-1}, $I_s^0=2$ MW/cm^2, $\tau/\tau_r=0.2$.

The length dependence of the efficiency η, the maximum output density Q, the total gain M, the current density at the lateral walls j_b and the electron beam energy E_0 of two-pass KrF-amplifiers are given at Fig.3. The value of maximum laser output density Q was determinated by the product of the output intensity and the burnup time of the molecular fluorine. It should be noted that the direct scaling like presented at Fig.3 is not optimal for obtaining maximum efficiency and gain in a chain of amplifiers. For KrF-amplifiers any scaling should involve the strong restriction associated with comparatively low optical damage threshold of reflecting mirrors and output windows at lasing wavelength λ = 248 nm, moreover it must take into account restrictions of fluorine burning time and e-beam current density load. The KrF-amplifiers using Kr/F_2 mixtured (without argon component) have higher performance characteristics than the $Ar/Kr/F_2$ ternary mixture. The most important advantage is the possibility for efficient operation at low gas pressures of 0.8 atm which significantly reduces the requirement on foil.

It follows from these results that for KrF-amplifiers that may be used in ICF and having a total gain $M \geq 10$ the internal efficiency may reach values 9-10% if Kr/F_2 mixtures are used with a relative fluorine concentration of 0.2-0.5%, a total gas pressure of P=0.8 atm and pumping intensity of 0.2- 0.6 MW/cm^3. Here the input signal intensity must be high enough, not less than ≅60% of the saturating power.

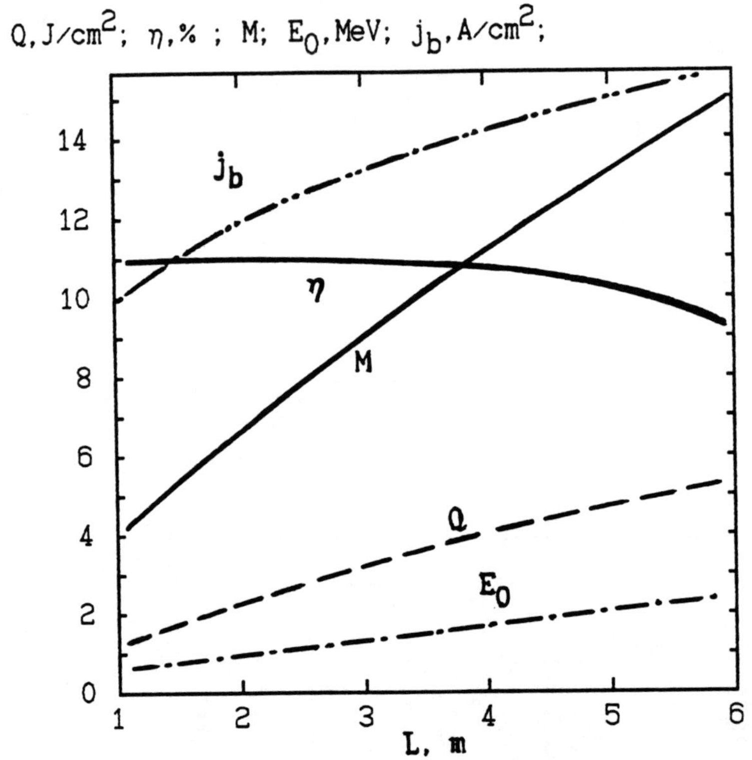

Fig. 3. Length scaling of a two-pass KrF-amplifier with fixed I_{in} = 0.8 MW/cm^2, L/H=1.5, Kr/F_2 =99.7/0.3, P=0.8 atm, W_b = 0.3 MW/cm^3.

References

1. A. G. Molchanov, The theory of active media in excimer lasers, In: Proc. Lebedev Phys. Inst. Ac. Sci., USSR, Nova Sci. Publ., vol. 171, 72-167 (1988).
2. Yu. V. Afanasiev, N. G. Basov, V. A. Danilychev, A. G. Molchanov, Laser thermonuclear fusion and the driver problem, Preprint FIAN, N 33, Moscow, 1984, 69 p.
3. R. J. Jensen, KrF for fusion: an overview of laser issues, Fusion Techn., vol. 11, 481-485 (1987).
4. L. V. Spencer, Energy dissipation by fast electrons, Nat. Bur. St., Monograph 1, 1959, 75 p.
5. A. G. Molchanov, Lasers in the vacuum ultraviolet and X-ray regions of the spectrum, UFN, vol. 106, 165-173 (1972).
6. L. A. Rosocha, J. A. Hanlon, J. McLeod, M. Kang, B. L. Kortegaard, M. D. Burrows, and P. S. Bowling, Aurora multikilojoule KrF laser system prototype for inertial confinement fusion, Fusion Techn., vol. 11, 497-531 (1987).
7. L. V. Tomov, R. Fedosejevs, D. C. D. McKen, C. Domier, A. A. Offenberger, Phase conjugation and pulse compression of KrF-laser by stimulated Raman scattering, Opt. Letts, vol. 8, 9-11 (1983).

COMPACT EXCIMER LASER PUMPED BY ELECTRIC DISCHARGES WITH WEAK ELECTRON BEAM PREIONIZATION

V.B. Karpov, I.N. Knyazev, V.V. Korobkin, and A.K. Naboichenko
General Physics Institute of the USSR Academy of Sciences, SU-117942 Moscow, Vavilov Street, 38, USSR

In view of the fact that discharge pumped excimer lasers have become a very popular instrument for scientific investigations in different laboratories now [1], it is obviously that creation of the new models of such lasers should be able to promote new scientific and applied fields development.

But some new applications of the excimer lasers concerning, for example, the amplification of short diffraction limited pulses for ultrahigh-intensity electromagnetic fields production require lasers with the high uniformity of their volume discharge. Also it is important to have excimer lasers with long optical pulse length because light pulses with a most long duration can be used for the transmission of maximum energy through lightguide optical systems without ones destruction.

It is well known that the preionization of a laser gas discharge volume is the very important factor mostly determining the possibilities of the laser [2,3]. Excimer lasers pumped by an electric discharge with UV photopreionization are presently the most commonly used due to their simplicity. But they have some important disadvantages. So, their discharge has a rather nonuniform structure [3]. That is why such lasers have the nonuniform optical properties of their active volumes and if be used as amplifiers will disturb an input high quality optical pulse. The optical pulse time length of such lasers is usually limited to values 20 - 50 ns because of their discharge instability [3,4]. These disadvantages concern largely the properties of the preionization [6].

From this point of view it is interesting to increase the start electron density of the discharge volumes of the lasers produced by preionization supplies to its values high enough for the discharge steady state obtaining for a simple one-step electric discharge circuit: Ne $\sim 10^{11}$ cm^{-3} [5,6]. It is known that all types of the UV photopreionization are not able to provide the start electron density greater than 10^{8} - 10^{10} cm^{-3} [4,7] which is some orders of magnitude less than the needed one. Moreover, there are some difficulties for obtaining a most uniform preionization of pure gases by UV radiation [2].

It is possible to produce the adequate value of "Ne" by such an external ionizer as a short electron beam of the appropriate electron current, duration and energy. The previous calculations [6] have shown that it is enough to have a rather weak electron beam with the current density about 0.1 A/cm^2 which is as short as some tens of ns. Certainly, the preionization must be uniform enough over the discharge volume. Besides, it is significantly to have as compact an installation as possible which should be able to operate at a pulse-repetition regime. Therefore it is clear that the great electron guns which are usually used for direct electron beam pumping are not suitable here. It is important also that some problems of the great electron guns such as the foil destruction by the electron beam itself or the beam pinching should not take place for as weak beams as required here.

The main aim of the work is creation of an excimer laser pumped by an electric discharge with the weak electron beam preionization and its possibilities studying.

The experimental installation setup is shown in Fig.1. It consists of the following main parts: an electron gun (1), laser chamber (2), discharge system (3) and optical cavity (4,5).

Our original compact coaxial electron gun allows to get uniform enough preionization of a gas over a discharge volume of 70 cm length and 3 cm width. The gun produces a "weak" electron beam with current density about 0.3 A/cm^2, electron energy 130 KeV and duration 50 ns (FWHM). A high-voltage pulse on a cold cathode (6) is produced by the resonance circuit containing a step-up pulse transformer (7), low inductance oil capacitor (8) and sharping spark gap (9) placed in a chamber filled by the elegas (SF_6) at a pressure 5 atm. A longitudinal thin foil electron beam window (10) with dimensions 80 * 2 cm^2 and original coaxial cold cathode provide the preionization of the active volume with a most uniform crossection. The gun is switched on by a pulse thyratron TGI1-1000/25 (11) and can operate at a frequency repetition rate up to 10 Hz. The cold cathode chamber of the gun was evacuated to a pressure about 10^{-4} torr.

We have made previous experiments with a discharge pumping. An electric discharge supply of the Blumlain scheme mounted from ceramic capacitors with full capacitance 35 nF was charged to 35 KV. The discharge circuit was switched on

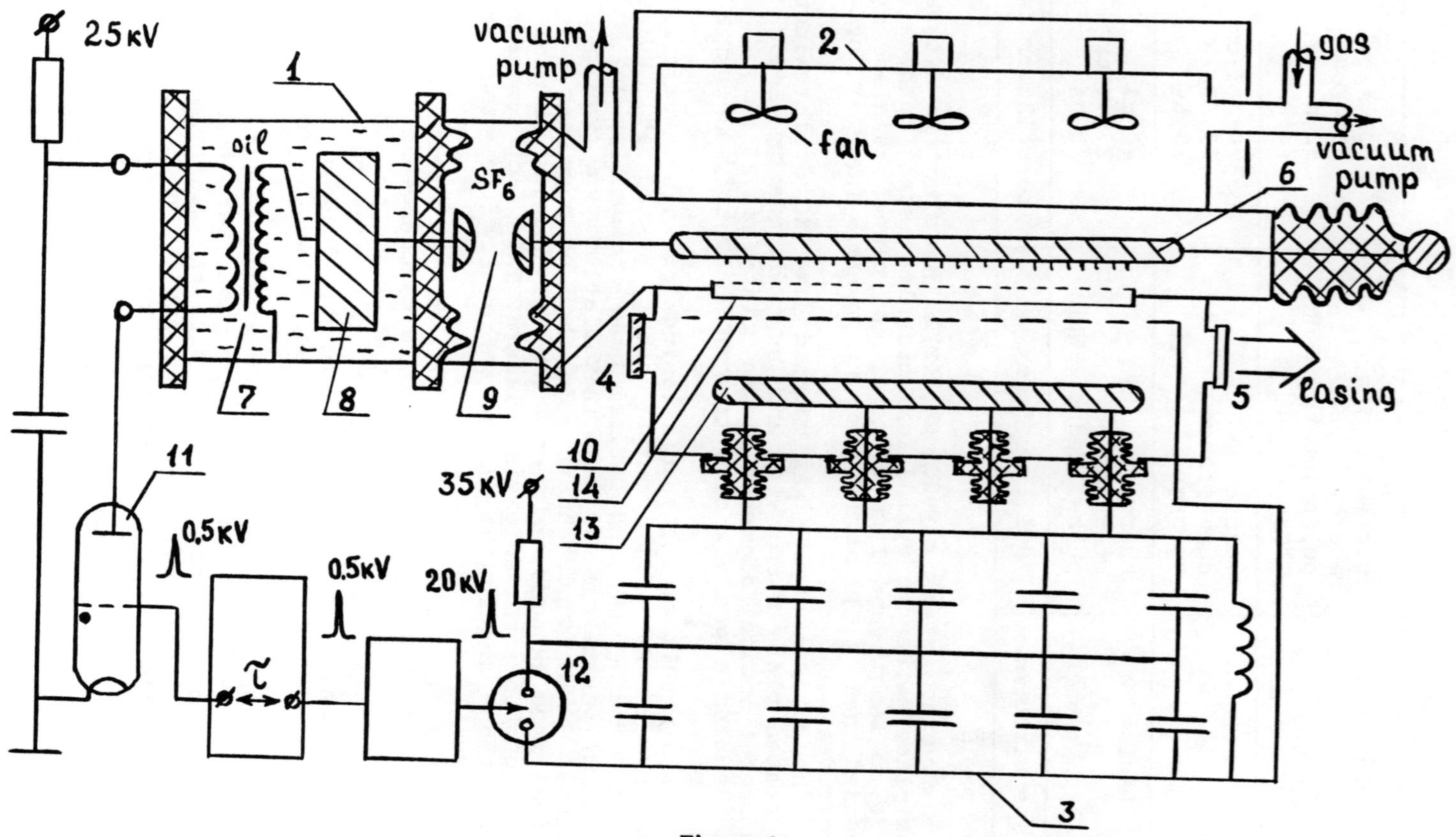
25кV
1
2
vacuum pump
gas
vacuum pump
oil
SF_6
fan
6
4
5
lasing
7
8
9
10
14
13
11
0,5кV
0.5кV
20кV
35кV
12
3

Figure 1

Table.1

Electron beam	current energy pulse length (FWHM) crossection	0.3 A/cm^2 130 KeV 50 ns 70*2 cm^2
Discharge	volume capacitance voltage	70*3*1 cm^3 35 nF 35 KV
Lasing	energy crossection pulse length (FWHM)	40 mJ 2.5*1 cm^2 30 ns
Gas mixture:	HCl : Xe : He	1torr:12torr:2atm

by a governed spark gap RU-65 (12). The electric discharge took place between a cylindrical stainless steel cathode (13) of 70 cm length and 3 cm diameter and plane stainless steel grounded net (14) served as an anode. The gap of the electrodes was 3 cm.

The optical cavity was mounted from a total reflection Al- coated plane mirror (4) and uncoated quartz window (5). The useful optical diameter of the cavity was 25 mm.

We used a gas mix of 1 torr HCl, 12 torr Xe and 2 atm He.

As a result we obtained a lasing for this first training discharge scheme with output energy 40 mJ at pulse length 30 ns (FWHM) and optical beam crossection 1*2.5 cm^2. The main parameters of our installation are presented in Table 1. It is important to note that we have not seen any optical beam nonuniformities which are usually obtained for the UV photopreionized excimer lasers. This fact confirms the proposition about the high uniformity of the volume discharge.

Since the time delay "τ" between the electron gun and discharge switching on moments could be changed simply with a step of 1 μs we have investigated how the output energy depends on that delay (see Fig.2). The value τ=0 corresponds to the moment of the gun switching on. One can see that the two times decrease of the output energy was achieved only in about 2 μs.

In conclusion we can say that these results of the previous experiments allow us to consider excimer lasers pumped by the electric discharge with the weak electron beam preionization prospect for high quality optical radiation obtaining. Also they may be usefull for the work with new and nontraditional excimer gas mixtures which demand the increased intensities of the preionization for their optimum

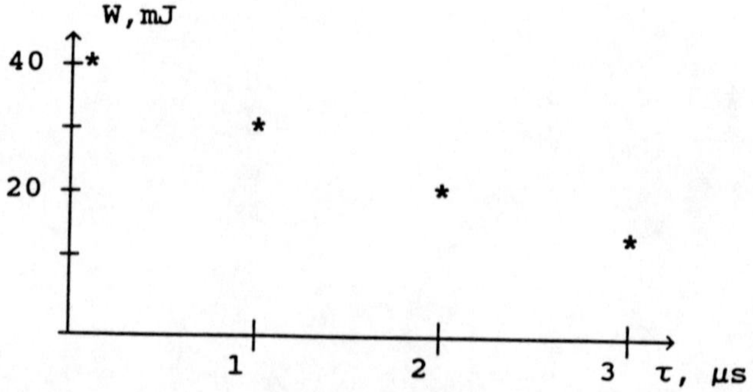

Figure 2

laser action [8]. And finally, there is a theoretical possibility [6] to obtain long output pulses (as long as 1 μs) at lower discharge voltages and simple one-step discharge circuit for the increased start electron density values.

REFERENCES

[1] Ch.K.Rhodes,In Excimer Lasers,(Springer-Verlag, Berlin, 1979).

[2] Baloshin et.al.,In *Application of computers for creation of lasers*,(Leningrad, Mashinostroenie, 1989)

[3] P.N.Mace, Optical Engineering, vol.20, No 6, (1981) 935

[4] Yu.I.Bychcov et. al., Soviet J. Quantum Electronics, vol.9, No 12,(1982) 2423.

[5] Yu.I.Bychcov et. al., Soviet J. Plasma Physics, vol.15, 3, (1989) 330.

[6] A.V.Borovskiy, V.B.Karpov, I.N.Knyazev, V.V.Korobkin, V.A.Yurovskiy, Preprint No 75 (in Russian) Gen. Phys. Inst.USSR Acad.Sci.,Moscow (1990) - (in press).

[7] V.U.Baranov, V.M.Borisov, U.U.Stepanov,In *Rare-gas halide excimer lasers pumped by electric discharge*, (Moscow, Energoatomizdat, 1988).

[8] A.N.Panchenco, V.F.Tarasenko. High-power laser action at λ=222 nm under pumping by self-sustained discharge. (International symposium "Short wavelength lasers and their applications", Samarkand, 1990, USSR).

MULTISTAGE EXCIMER SYSTEM "CACTUS" WITH SBS MIRROR AND PULSE LENGTH TRANSFORMERS

V.B. Karpov, I.N. Knyazev, V.V. Korobkin, and A.K. Naboichenko
General Physics Institute of the USSR Academy of Sciences, SU-117942 Moscow, Vavilov Street, 38, USSR

Excimer lasers have developed over the past few years into an efficient source of UV radiation. However certain important applications such as high-resolution spectroscopy, photochemistry, UV phase conjugation require high-brightness narrow-bandwidth lasers [1].

Spectral narrowing of rare-gas halide excimer lasers by using grazing-incidence grating configurations has already been reported [2-5]. However, in these schemes the spectral narrowing was limited to bandwidth values about 0.1 cm^{-1} and laser pulse energy to some μJ because of the high optical losses of gratings.

Our experimental installation we have named "CACTUS" is the multistage excimer laser system which was built on a principle of a master oscillator-preamplifiers-final amplifier (Fig.1). It is aimed for the high-quality UV radiation (λ=308 nm) (of small divergence and narrow bandwidth) production which should be enough for different nonlinear optical phenomena studying.

We have made a novel master oscillator with laser radiation of near diffraction limited divergence and the narrow spectral bandwidth approaching to the Fourier transformation limit of its optical pulse. It is important that the cavity of our master oscillator does not have any gratings and its spectral selective properties are based only on the phenomenon of interference. That is why the optical losses of our cavity are smaller and its output pulse energy is greater as compared with the previously reported master oscillators. It must be noted also that our master oscillator is rather simple in work and includes a little number of optical elements. Owing to this fact it is possible to work at the shortest length of the cavity.

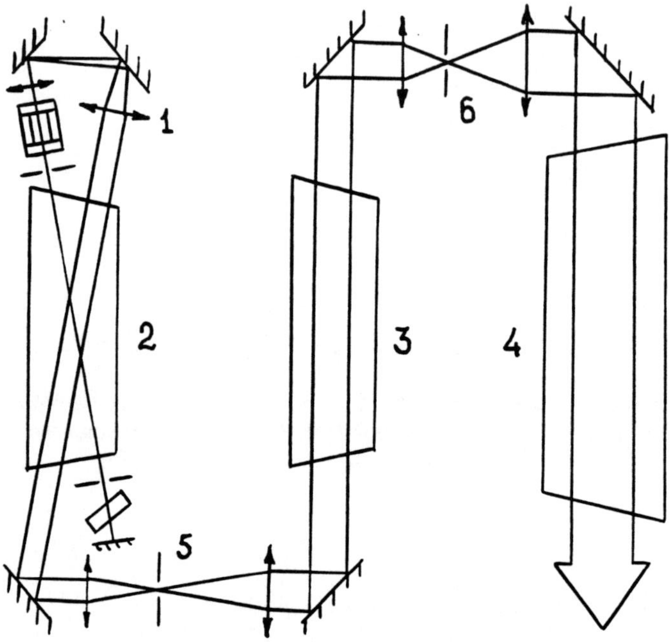

Figure 1

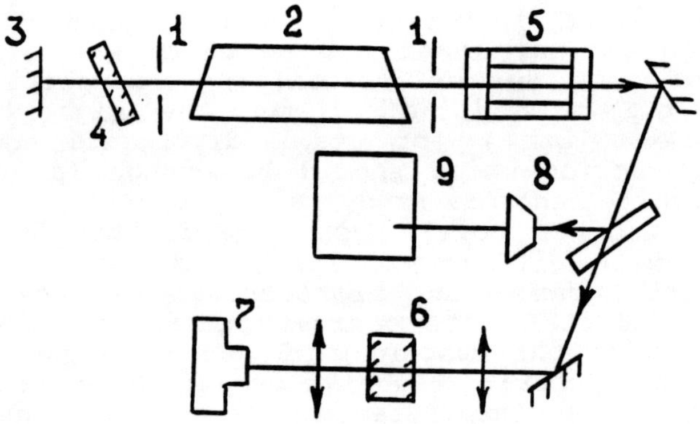

Figure 2

The master oscillator setup is shown in Fig.2. The diffraction limited divergence of its radiation is provided by two pinholes (1) of 1.6 mm diameter placed at the both sides of an active volume (2). So they extract the TEM_{00q} mode of the radiation.

The free nonselective regime spectrum densitogram of the industrial excimer laser ELI-91 which was used for the master oscillator creation is presented in Fig.3. It was obtained by a Fabry-Perot etalon with a free dispersion region (FDR) value 17 cm^{-1}. So the free regime bandwidth value is as large as 15 cm^{-1}.

The laser output spectrum is narrowed by a spectral selective cavity. Its total-reflection side includes a high-reflection plane mirror (3) and Fabry-Perot etalon (4) of properly chosen FDR and finess 15 tilted to the optical axis. There is an interferometric spectral selector (5) at the another side of the cavity. The selector is an indivisible hard quartz unit and it serves as an exit mirror. The properties of the selector are based on the phenomenon of interference and its reflectivity depends significantly on a frequency [5]. There are not any additional elements in our cavity. Spectrum bandwidth was controlled by Fabry-Perot etalons and interferometers (6) with different FDR values and camera (7). The densitogram of the output spectrum of our master oscillator is presented in Fig.4. It was obtained by a Fabry-Perot etalon with a FDR value 0.3 cm^{-1} and finess 15. One can see that this spectrum bandwidth is not greater than 0.02 cm^{-1} (FWHM). But this value is as large as the accuracy of our Fabry-Perot etalon. That is why we have made the additional control of the bandwidth by a fast photodiode (8) and 5 GHz-pass band oscillograph (9). The smooth shape of the output pulse of the master oscillator (Fig.5) allow us to suppose that only one longitudinal mode is there and its spectrum bandwidth value is about $3.5*10^{-3}$ cm^{-1} for our case. The main parameters of our master oscillator are presented in Table 1. The properties of the nonselective free regime operation of the laser are presented in the same Table for convenience.

The setup of the most simple excimer master oscillator which we invented and used for a narrow bandwidth lasing is presented in Fig.6. Its cavity consists of two pinholes of 1.6 mm diameter and besides only two optical elements. The first one is an interferometric selector (1) similar to that one in Fig.2. And the second is an uncoated quartz Fabry-Perot etalon (2) with FDR similar to that of the etalon (4) in Fig.2 used as an exit mirror. Only Fresnel's reflections from uncoated plane facets and the interference are used for this cavity spectral selective operation. It must be noted that the selector consists of different quartz etalons which have no coatings too. That is why this cavity not only very simple but also rather cheep because it does not require any coatings which may be problematic for the UV region. The output spectrum densitogram of such a master oscillator is presented in Fig.7. This spectrum with a bandwidth value about 0.5 cm^{-1}

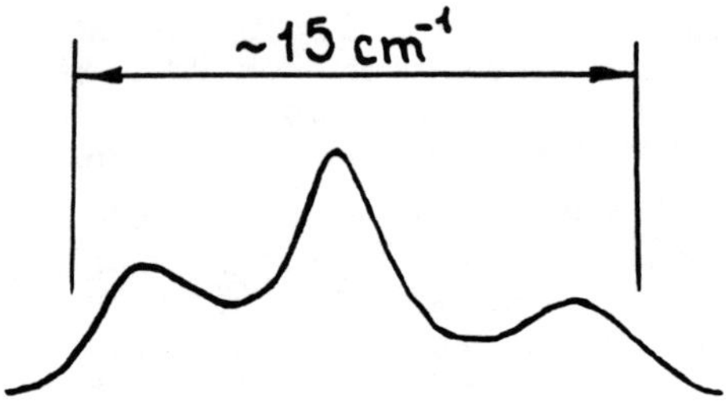

Figure 3

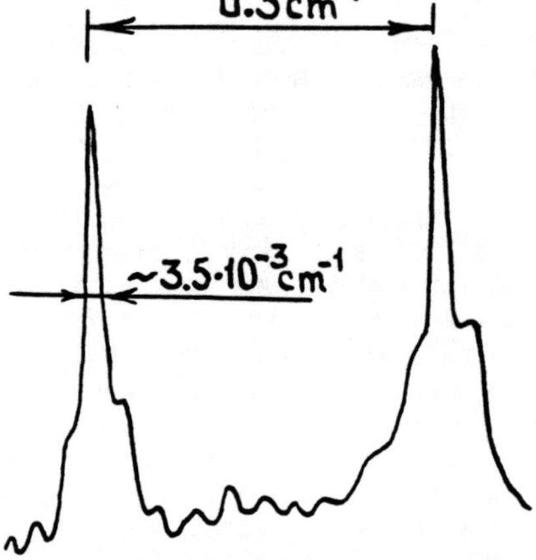

Figure 4

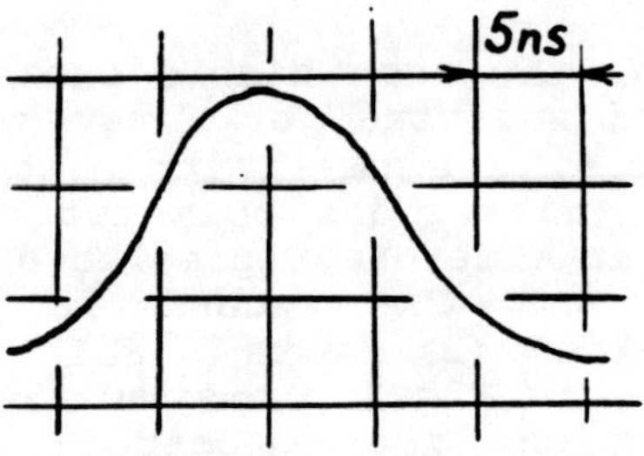

Figure 5

Table.1.Master oscillator

Regime	nonselective	selective
spectral bandwidth	15 cm^{-1}	$3.5*10^{-3}$ cm^{-1}
laser beam crossection	$10*20$ mm^2	ø 1.6 mm
divergence	10^{-2}rad	$6*10^{-4}$rad
energy	50 mJ	50 μJ
duration	20 ns	10 ns
wavelength	308 nm	308 nm

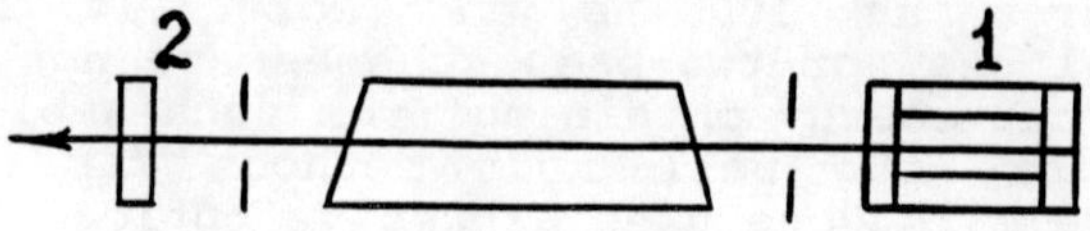

Figure 6

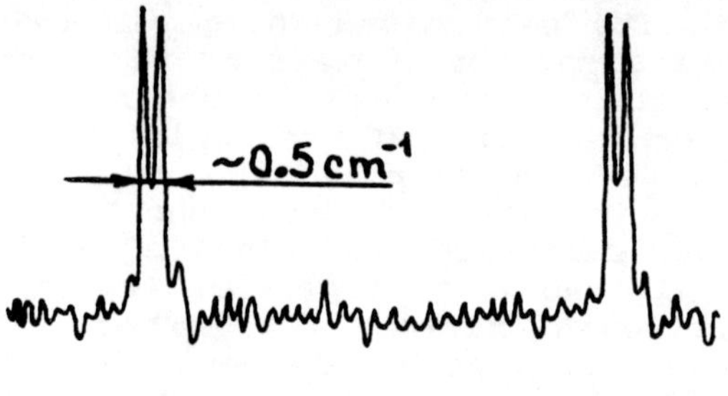

Figure 7

consists of two narrow lines with the bandwidth value <0.05 cm^{-1} for each one. Such a strange spectrum is produced on account of the interference of optical waves reflected from the different plane facets of the cavity. The output pulse energy of such a master oscillator was about 60 μJ at pulse length 10 ns (FWHM).

The setup of our multistage installation "CACTUS" is presented in Fig.1. Here the laser pulse of the master oscillator after its expansion by a lens telescope (1) is preamplified in the same active volume of the master oscillator (2). That is why its efficiency increasing is achieved. Besides, the automatic synchronization of this first preamplifier is obviously realized. After the next two one-pass amplifiers (3,4) and spatial filters (5,6) we had output energy 30 mJ for the narrow bandwidth (about $3.5*10^{-3}cm^{-1}$), pulse length 10 ns (FWHM) and near diffraction limited divergence. All the system can operate at a frequency repetition rate 0.5 Hz but its part that does not include the final amplifier (4) can operate at 50 Hz with pulse energy 5 mJ (other parameters do not significantly change).

There is such a problem of some multistage systems as the transformation of laser pulse duration. So, in contrast to the excimer master oscillators which usually have laser pulse duration about 10-20 ns these ones of powerfull amplifiers, for example, with direct electron beam pumping are as large as 100 ns and more. It is clear that conventional one- or two-pass schemes are not enough for the maximum output energy obtaining from such amplifiers.

There are two well-known methods for such a problem solving. The first one uses a passive optical time delay line (see Fig.8). But it has some disadvantages. First of all, it has very large dimensions and UV radiation may be absorbed over them. And besides, it has a large number of optical elements which can aberrate a high-quality input beam. But these aberrations cannot be compensated even by the phase conjugation because such a scheme is not reversible. The second method is based on a multipass cavity scheme (see,for example, Fig.9). Its main disadvantage is that for excimer lasers which have great pump powers at very small relaxation times it is necessary to compensate aberrations produced by an amplifier after each pass of the amplifier.

That is why we have proposed the next scheme presented in Fig.10. Let's suggest that a short input pulse of high-quality laser radiation from a master oscillator through a beam splitter (1) passes an amplifier (2) which has a phase conjugation Brillouin mirror (3) after its exit. The radiation with an aberrated by the amplifier wave front will be reflected from the Brillouin mirror and the back pass through the amplifier will be carried out by the phase conjugated beam. So we will have an inversely directed amplified beam of high quality again on the beam splitter. Such a phase conjugation mirror can be very effective for

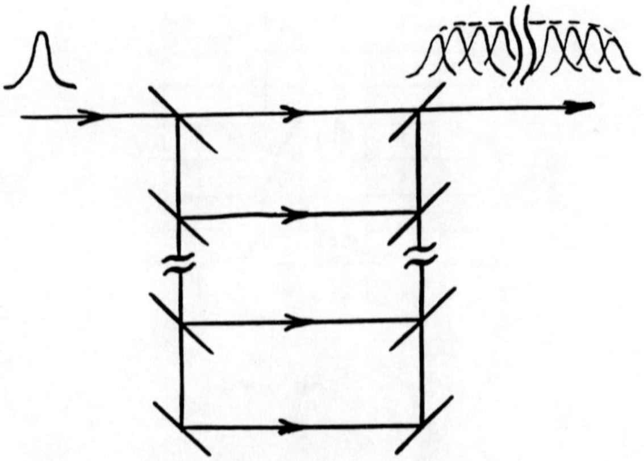

Figure 8

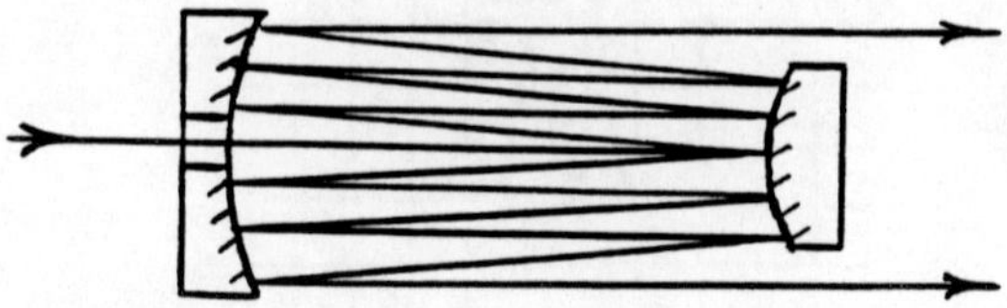

Figure 9

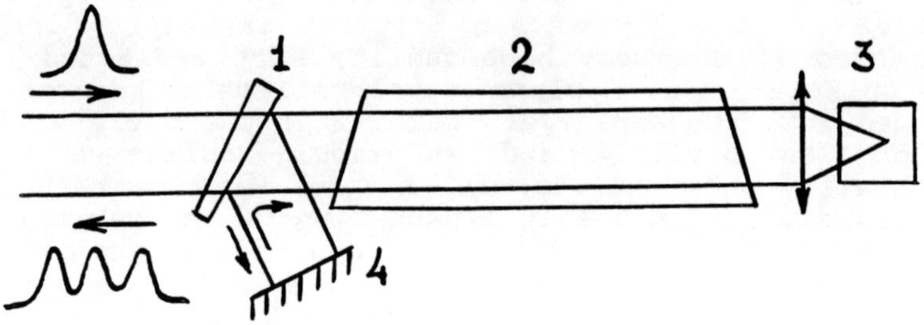

Figure 10

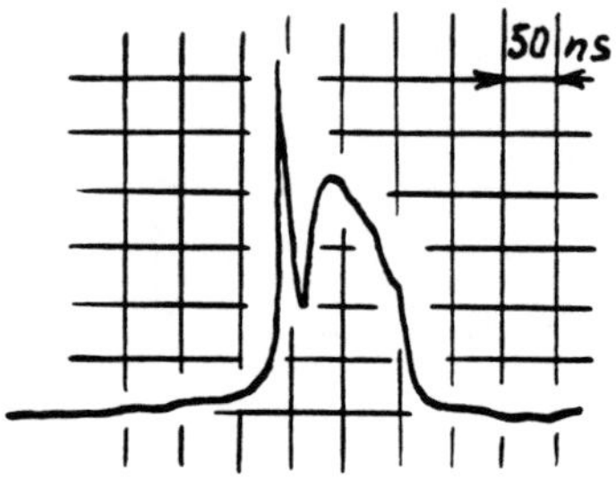

Figure 11

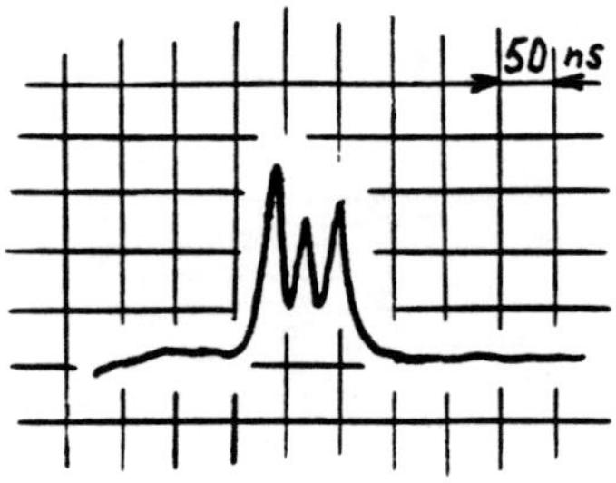

Figure 12

excimer lasers [4]. Then the most part of the amplified high-quality beam passed through the beam splitter is utilized (with help of some optical isolator which is not shown here). The rest high-quality part reflected from the beam splitter and a plane total-reflection mirror (4) is directed into the amplifier again. Then the cycle will repeat as long as possible and we should obtain a train of high-quality short pulses with a time distance between every two tops similar to the full pass time of the cavity.

For this idea checking we conducted an experiment with an optical scheme similar to that of Fig.10. We used liquid heptane as a nonlinear medium for the Brillouin mirror and a discharge pumped excimer laser "LIDA" as an amplifier with an enhancement time 50 ns (FWHM). The high-quality short pulse (10 ns, FWHM) was produced by the master oscillator described above. The oscillograms of that one and of the amplified spontaneous emission signal from "LIDA" are shown in Fig.11. Finally, the oscillogram of the useful amplified radiation from such a system is shown in Fig.12. One can see that it consists of three short (10 ns,FWHM) tops with the time delay 10 ns between each pair of them. The full length of the cavity for this case was 3 m and its full pass time 10 ns was in agreement with the time delay of the tops.

REFERENCES

[1] E.Armandillo et.al., Optics Letters, vol.9, No 8,(1984).
[2] M.C.Gower, Optics Letters, vol. 7, No 9,(1982) 423.
[3] M.C.Gower, Optics Letters, vol. 7, No 4,(1982) 162.
[4] E.Armandillo,D.Proch, Optics Letters,v.8,No 10,(1983)523.
[5] A.L.Egorov, V.V.Korobkin, R.V.Serov, Soviet J. Quant. Electron., vol.2, No 3, (1975) 513.

SEMICONDUCTOR SURFACE MODIFICATION BY KrF LASER RADIATION

V.L. Kantsyrev*, N.V. Morozov, P.B. Sergeyev, and M.A. Tyunina†
P.N. Lebedev Institute of Physics, Moscow, USSR
**Institute of Technical Glass, Moscow, USSR*
†Latvian State University, Riga, Estonia

Many problems in the technology, concerning surface characteristics modification, can be solved by irradiation with lasers, among which better ones are excimer lasers KrF and ArF with wavelengths 248nm and 193nm. Absorption of the majority of materialshere is about 10^6 cm^{-1}, providing materials heating up to melting and evaporation even with energy densities E = 1-2 J/cm^2 in pulse of 10-100 ns. At the same time high energy photons are able to initiate photochemical processes, permitting semiconductor etching and deposition of various materials. These reasons make it possible to use effectively the excimer lasers in microelectronics technology.

This work deals with the experiments on a high power electron beam KrF laser [1] irradiation of Si substrates with area of 1-2 cm^2. Laser pulse duration - 80 ns, pulse energy - up to 10 J. E on the substrates surfaces was varied from 0,5 to 5,0 J/cm^2. In our experiments we used monocrystalline silicon wafers of n- and p-type and (100) and (111) orientation, with 300 μm thickness and with polished surface. Also the layers of polycrystalline silicon (Si^*), phosphorous silicate glass (PSG), silicon dioxide (SiO_2), aluminium (Al) were formed on the substrates. Irradiation took place in vacuum (10^{-5} torr) under the room temperature. Then the surface was investigated with the help of optical and scanning electron microscopy, X-ray structure analysis, secondary ions mass spectroscopy, measurements of electrical parameters.

First of all, Si surface morphology in dependence of E was studied under the influence of focused radiation with divergence of 10^{-4} mrad. There were found periodical structures with d = 0.1-0.3 mm, where areas without any destructions alternated with areas of strong erosion in the form of separate craters of 20-70 μm diameter, or groups of overlapping craters, or continuous destructions. Such destruction alternated with areas of strong erosion in the form of separate craters of 20-70 μm diameter, or groups of overlapping craters, or continuous destructions. Such structures appeared under the $E > 0.5$ J/cm^2, became more clear with E increase and remained till 5 J/cm^2. The same experiments, but with more homogeneous irradiation, using laser radiation with divergence of 5 mrad and 2 m focus lense, have shown the lack of such structures under any values of E . Thus, one can relate them to the corresponding inhomogeneous distribution of E . All the other experiments were done under homogeneous radiation.

They have shown that under $E < 2.5$ J/cm^2 Si surface remained clean, without any defects. From $E = 2.5$ J/cm^2 in the periphery of the irradiated area separate cracks appeared, of several dozens microns length . The number of the cracks increased with E increase. Under $E = 3.0$ J/cm^2 we found groups of cracks, measuring 100 μm, either with common central point or in the form of closed line with cracks, running from it along crystallographical directions of the substrate. The cracks can be connected with local stress in crystal lattice, arising near defects or contaminations during cooling. Note, that melting point of Si, according to [2], is equal 0.7 J/cm^2.

When E was 3.5 J/cm^2 separate craters with several dozens microns diameter and approximately 2 μm depth arised. These craters were surrounded by the roller of the transported silicon melt and cracks, running from the crater through the roller of melt. The energy $E > 4,0$ J/cm^2 resulted in chaotically distributed groups of craters.

Under such values of E along with melting, evaporation of Si took place. Repelling pressure force acted on the substrate surface. Instability of laser plasma torch was accompanied by repelling pressure oscillations [3], that could result in local rise of mass transport and the formation of craters, which, in turn, became centers of mechanical stresses and cracking during cooling.

Si substrate surface melting determined the state of Si-SiO_2 system (thickness of SiO_2 - 0.2 μm) in dependence of E. Note that SiO_2 is transparent for the wavelength of 248 nm and not preventing substrate's absorption and heating up to melting (1412°C) without SiO_2 melting (1700°C). It was found out, that under $E < 1.5$ J/cm^2 defects and disturbances of SiO_2 on Si didn't appear.

Under $1.5 < E < 3.5$ J/cm^2 in the irradiated area chaotically oriented periodical structures measuring about 10 μm came into being. Emergence of the same structures measuring approximately wavelength was studied earlier [4]. We may suppose, that the origin of the waves in the system (Si-SiO_2) is also connected with diffraction of light on modulated relief (in our case - on the surface of melted Si). Increase of waves period may be connected with the presence of SiO_2 layer, but the value of E of their rise -

with SiO_2 softening.

Under $E > 3.5$ J/cm^2 the surface of the system Si-SiO_2 became covered with craters and areas of bared Si, where SiO_2 was destroyed by Si vapor pressing.

PSG film with thickness 0.4-0.6 µm appeared to be more efficient protective coating in comparison with SiO_2. (PSG-Si) system's surface didn't suffer any destructions up to $E = 4.5 - 5.0$ J/cm^2. Only under $E > 5.0$ J/cm^2 PSG film was partially destroyed and removed, with overlapping craters appearing in Si .

In the studies of surface morphology of Si^* layer with thickness 0.2 µm, deposited on monocrystalline Si, in dependence of E, we have observed the sequence of changes described earlier: separate cracks, groups of cracks, craters, groups of craters. Periodical structures, like these in the system Si-SiO_2, were also observed. Moreover, increase in E more than 4 J/cm^2 resulted not only in larger density of craters, but also in complete evaporation of Si^* layer. More intensive evaporation took place in the system Si-SiO_2-Si^*, where isolating SiO_2 with low heat conductivity limited heat diffusion from Si^*, providing more efficient Si^* heating and evaporation

Surface defects emergence and evaporation of material have determined the upper limit of E suitable for Si crystallization: 2.5 J/cm^2. Under $E > 1.0$ J/cm^2 the depth of the melting was greater than the thickness of Si^* layer and crystallization during substrate cooling began from the substrate, reproducing its structure. For Si(100) disorientation of lattice of about 3°-4° was found, connected, probably, with high velocity of crystallization, equal approximately to 10 m/s .

Surface changes of Al films with the thickness up to 1.3 µm, deposited on Si, were also investigated in dependence of E . Under $E > 3$J/cm^2 there appeared cracks in the substrate without disturbances of Al film. Further rise of E resulted in craters, evaporation of the film and the substrate. Defects dimensions and periodical structures were analogous to the earlier described ones in Si and SiO . The results of measurements of electrical parameters allowed to suppose that under $E > 1.5$ J/cm^2 the process of Al and Si melting took place. The resistance of Al-Si contact decreased considerably. The current-voltage characteristics (CVC) have shown, that both ohmic and rectifying contacts could be obtained. The diodes, formed in Si, demonstrated the ideal CVC with $n = 1.4 - 1.6$. The degradation of parameters under $E > 2.5 - 3.0$ J/cm^2 was related to the defects origin, already discussed here. SIMS analysis of Al-Si alloy, obtained under $E = 4$ J/cm^2, has shown, that the relation Al:Si was changing from 3:1 near surface to 1:3 in the depth 0.2 µm. Varying Al film thickness and the value of E permitted to form alloys of different composition.

The results, presented here, can ground the prospects of using high power KrF lasers in microtechnology and provide the development of integrated circuits technology 'in situ ' on the basis of these lasers.

REFERENCES

[1] V.M.Buchnev et. al., Kvantovaya elektronika, 8 (1981) 1235; 10 (1983) 2048.

[2] V.A.Puziryov, A.S.Gerasimov, Zarubejnaya radioelektronika, №4, (1989) 56.

[3] A.A.Samokhin, Trudi IOFAN, 13 (1988) 3.

[4] V.Y.Panchenko et. al.,Itogi nauki i techniky, VINITI Moscow, Vol.3,(1989) 93.

EXCIMER LASER OPTICS

N.V. Morozov, S.I. Sagitov, and P.B. Sergeyev
P.N. Lebedev Institute of Physics, Moscow, USSR

INTRODUCTION

Starting of the first excimer quantum Xe_2 [1] generator, realised in 1970 by the group of the scientists of the P.N.Lebedev Physical Institute, USSR, lead by N.G.Basov, and further researches in this field proved the possibilities of getting high power coherent radiation both in UV and VUV spectrum regions up to 100nm and shorter with different excimer lasers. The method of pumping the corresponding active mediums exists now. The progress in the field of creating reliable lasers with the shorter wavelength of the generation is practically fully determined by the success in deciding laser optics problem. First of all it is a resonator optics: mirrors with different reflection coefficients as well as laser chamber output windows. The aim of this article is to reflect the state of things with resonator optics of UV diapason for excimer XeCl, KrF, ArF lasers with λ=308, 248 and 193 nm.

Researches of the excimer lasers, undertaken in the laboratory and now in the department of Quantum Radiophysics, since the moment of starting the first Xe generator have been touching the questions of optics. Realizing the importance of the latter, a special group, investigating technological problems of the creation of mirrors coatings with high laser damage resistance at wavelengths of the lasers,is formed. This group works together with the laser engineers.

Since the beginning of the 80's an experimental e-beam pumped gas medium laser set up EL-1 [2] has been working. Specifics of the work of e-beam pumped XeCl, KrF and ArF lasers at high specific power and energy pumping were examined. The lasers have the following parameters: active volume - 0.45 l, length 23 cm and light beam diameter 5 cm. Resonator mirrors can be fixed just to the laser chamber. An outside resonator can also be used at setting the laser chamber windows. Generation energy at the EL1 set up is : for XeCl-laser - up to 5 J, for KrF- laser - up to 14 J, for ArF-laser - up to 6J under pulse duration ∿ 60-80 ns. Radiation divergence, depending on the resonator type can be changed from 5 to 0.1 mrad.

The data obtained at EL-1 on laser damage resistance of optical elements of the named lasers has become the basis of the given work.

I.DEMANDS TO THE ACCURACY OF THE RESONATOR OPTICS PRODUCING

The most important laser parameter determining laser consumer possibilities is radiation divergence - θ. In the reality this parameter depends on many factors, but the minimum divergence of radiation is determined by the optical active medium inhomogeneity. In excimer lasers gas medium is used. In nonexcited condition the homogeneity of the gas density is high,and under the pump duration ∿ 100ns there is no time for dynamic processes in gas to disturb it. However, optical inhomogeneity of the active medium can appear because of the specific power pumping (W) inhomogeneity. Laser with the transversal e-beam excitation or electric discharge excitation have such an inhomogeneity along all active medium length - L.

Let the pumping profile along one of the transversals coordinates (x) relatively to lasers axis (z) have the form shown on the figure 1. Here D is a light laser aperture.

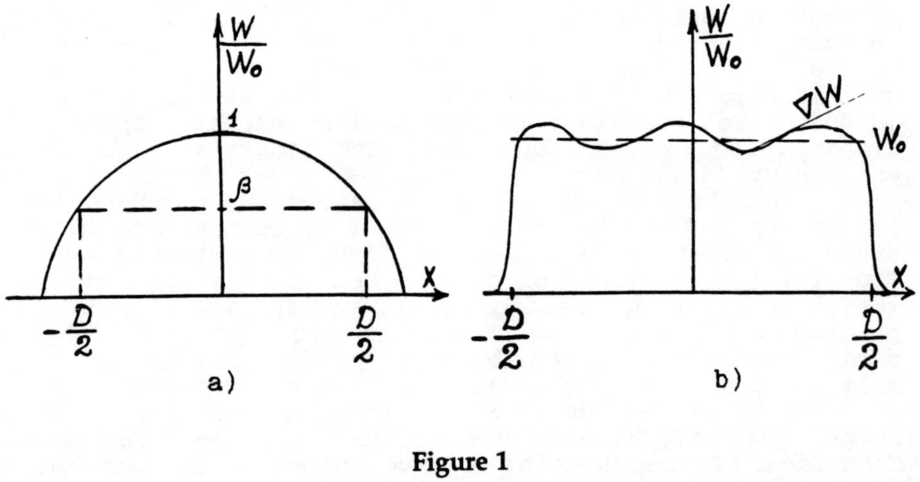

Figure 1

As it was cleared up [3], dependence of the active medium refractive index n on generation wavelength can be presented in the following way:

$$n = n_o + K \cdot W \quad \text{or} \quad \Delta n = (n - n_o) = K \cdot W \qquad (1)$$

where K is a proportion coefficient. For the mixtures $Ar/Kr/F_2$ and Kr/F_2 in e-beam pumped KrF laser $K=3 \cdot 10^{-7}$ (cm^3/MW) [3].

Resulting from (1), profile n will repeat profile W. Then optical inhomogeneity along x for two cases can be characterised by value ∇n:

$$\text{a)} \quad \nabla n = K \cdot (W_o - \beta W)/(D/2) = 2(1-\beta)KW_o/D$$

$$\text{b)} \quad \nabla n = K \cdot \nabla W = (K \cdot W_o) \cdot (\nabla W/W_o) \qquad (2)$$

For one way amplifier limit of θ_a will be, according to [4] and (2),

$$\text{a)} \quad \theta_a = 2\, \nabla n \cdot L = 4(1-\beta)KW_oL/D$$

$$\text{b)} \quad \theta_a = 2\, K(W_oL) \cdot (\nabla W/W_o) \qquad (3)$$

Relatively to diffractional divergence $\theta_d = 2.44 \cdot \lambda/D$ we get:

$$\theta_a/\theta_d = 4K(W_oL) \cdot (1-\beta)/2.44\lambda = 2(1-\beta) \qquad (4),$$

because for KrF e-beam pumped laser (KrF-EPL) $\lambda=2.48 \cdot 10^{-5}$cm and optimal $(W_o \cdot L)=100$ MW/cm^2.

So we can resume that,when the pumping inhomogeneity with the profile shown at figure 1 isn't more than 50% on laser aperture ($\beta>0.5$), it's possible to attain the diffractional divergence radiation with such a medium.

Distortions of the output wave front can appear on optical elements, in the way of laser beam. This fact is well known. It's easy to show, that in order to get θ close to diffractional divergence, local optical elements surface distortions of resonator optics must not exceed $\lambda/20$. It puts severe demands upon mechanical solidity of excimer lasers optical materials.

2. MATERIALS FOR EXCIMER LASER WINDOWS

With moving on in UV and VUV spectrum region the quantity of transparent materials steadfastly reduces. A list of materials fit for producing excimer laser windows counts just a number of units, because of the high demands to the solidity, chemical damage resistance, minimum hydroscopics, price and extremely high homogeneity.

For XeCl and XeF lasers with generation wavelength 308 and 353 nm the ideal window material is fused silica, of course. It's relatively cheap. It has good optical characteristics, and a technology for polishing and producing is worked out. The problem of increasing chemical damage surface resistance to Cl and fluorides is solved by choosing the corresponding coatings. In any way, it's necessary to use these materials in antireflection surfaces.

Moving on to the shorter waves region the possibilities of using fused silica are restricted. The problem arises in choosing optical materials. In Table 1 some parameters of optical materials , used most widely in UV region, are represented. The main data are taken from [5].

Basing on the data, presented in the table, it's easy to conclude, that as it follows from the solidity, transparency and window price for 250nm (KrF-laser), the between them remains right down to 190nm. For shorter waves lasers, MgF_2 becomes, obviously, more preferable.

In usage of these materials for making raregas-halides laser windows the chemical surface resistance to F_2 influence becomes an important factor. According to [6], MgF_2 is considerably better than fused silica.But in practice the necessity of antireflection coatings usage levels this difference.

Table I.CHARACTERISTICS OF OPTICAL MATERIALS FOR UV AND VUV SPECTRUM REGIONS

№	Material	Price, rubles per kg	Microsolidity, kG/mm^2	Absorption in cm^{-1} on λ 250 nm	220 nm	I90 nm	I70 nm	I60 nm	E_g, eV
1	LiF	1500	96	.02	.03	.05	.08	.011	13
2	CaF_2	300	168	.026	.035	.065	.14	.22	10
3	MgF_2	700	450	.007	.01	.02	.05	.08	11
4	SiO_2	300	740	.01	.02	.05	.19	.95	7.7
5	Al_2O_3	6000	2000	.01	.03	.18	.70	.90	8.3

The most important parameters of the optical material, suitable for work with laser beam, are it's volume and surface damage resistance as well as two-photon absorption. The studies of the dependence of these values on different factors for the materials presented in Table I for the wavelengths of KrF and ArF lasers have been realizing at EL-1 for a long time [7-8]. The main results of this work is summed in Tables 2, 3, 4.

As it is clear from the data table, MgF_2, working with the radiation in the region less than 250nm, according to the main parameters, has obvious favours. It is also supported by the results of the resource experiments on windows of fused silica, CaF_2 and MgF_2. Experiments were held at 200 W periodical-pulse KrF laser [11].

Table 2. SURFACE DAMAGE THRESHOLDS OF SAMPLES IN AIR FOR λ=248nm AND 193nm AND IRRADIATED AREA ABOUT I mm^2

№	Sample material	Damage threshold, J/cm^2 λ = 248 nm	λ = I93 nm
I	LiF	20	–
2	CaF_2	70	6
3	MgF_2	500	25
4	SiO_2	200	I5
5	Al_2O_3	70	–
		τ = 80 ns	τ = 60 ns

Table 3. VOLUME DAMAGE RESISTANCE OF THE OPTICAL MATERIALS FOR WAVELENGTHS OF KrF AND ArF LASERS AT IRRADIATION REGION ∿I50μm

№	Material	E and I on λ = 248 nm		E and I on λ=I93 nm	
		J/cm^2	GW/cm^2 (70ns)	J/cm^2	GW/cm^2 (60ns)
I	LiF	350	5	I20	2
2	CaF_2	70	0.9	6	0.I
3	MgF_2	I400	20	480	8
4	SiO_2	700	I0	I20	2

Table 4. TWO-PHOTON ABSORPTION COEFFICIENT β (cm/GW) FOR UV OPTICAL MATERIALS ON 248 nm

№	Material	τ=70ns	τ=0.7 ps [9]	τ= 8ps [10]
I	LiF	< 0.I	< 0.00I3	-
2	CaF_2	< 0.I	< 0.0083	$2\cdot 10^{-2I} cm^3/W^2$
3	MgF_2	0.05	< 0.00I3	-
4	SiO_2	0.5	0.045	0.08

But when planning the practical usage of MgF_2 it's nessessary to mind that now MgF_2 monocrystal samples are produced with the dimensions equal to 160mm or lower, and they often have large inhomogeneity. It's necessary to remember, that this material possesses birefringence.

3. MIRROR AND ANTIREFLECTION COATINGS

The technology of producing mirrors and antireflection coatings forwork with UV and VUV regions excimer lasers is being intensively developed now. The technology includes many stages. They are: choosing the appropriate material for the substrate and the way of preparing it for coating fabrication, choosing many-layer coating construction and the way of putting it. The ranks of the short article make it impossible to describe these questions in details, so we'll represent only the consumer parameters for a number of samples with record characteristics.

Determination of the laser damage resistance of the coatingswas made on the EL-1 set-up according to the methodics described in [7]. It's necessary to point out that the area of irradiated surface in experiments was no less than 3-5 mm^2.

3.1. Aluminium Mirrors

The first excimer laser and most of the next ones were started on the resonators with aluminium mirrors usage. It is conditioned by the simplicity of their producing and the reflection coefficient reaching 85-90% in a broad spectrum region from the infrared region to 120nm. But their laser damage resistance is often low.

Results of the experiments with aluminium mirrors produced by the thermal evaporation of the substance with the following precipiaton of it on the substrates, made of glass K-8 and fused silica KU-1, were presented in the article [7]. On the KrF laser wavelength the damage threshold is $\sim$ 0.25 J/cm^2, and on 193nm - 0.15 J/cm^2. It's practically impossible to work with the mirrors with such parametres.

Currently aluminium mirrors with much higher laser damage resistance became to appear. Not so long ago [12] we reported about the experiments on some samples, produced with the help of the pulse-plasma method of coatings fabrication. On the wavelengths 355, 248 and 193nm the damage thresholds were 2, 1.5 and 1 J/cm^2.

The usage of these mirrors on EL-1 set-up made it possible to get generation with the energy 4J at ArF laser, and after the change of the gas mixture, without changing the resonator optics, - 9J at KrF-EPL. On laser chamber in direct contact with gas mixture one of such mirrors worked 30 hours. During this time 200 KrF laser pulses with the energy density on the mirror equal to $\sim$ 0.1J/cm^2 and 70 ArF laser pulses with the energy density ∿ 0.1-0.15 J/cm^2 were made.

After finishing work there was an obvious blue thin layer on the mirror. It will be noted, that these aluminium mirrors were without safety coating from F_2. In spite of such a condition, the mirror was quite workable.

After that a number of mirrors, wich were received from different organizations, were tested. But we couldn't attain better results so far, though in some cases the damage thresholds on 248nm were about 0.7-1 J/cm^2. It's necessety to notice that aluminium diamond turned mirrors had the threshold of 1,6 J/cm^2.

These results show that given aluminium mirrors can be used as resonator optics for excimer laser in UV region, and in VUV region they may be just irreplaceable [13]. It proves that it's resonable to continue the work on the development of mirror producing technology.

3.2. Dielectric Mirrors and Antireflection Coatings

At set up EL-1 a lot of samples, produced according to the improved technology, have been experienced since the time of publishing article [7]. Some of obtained results are presented in the Table 5.

As one can see from the table, laser damage resistance of mirrors and antireflection coatings on 248nm during the last 3 years increased twice and now is 10 and 7 J/cm^2, respectively. It corresponds to the best world's results [14]. Laser damage resistance on 193 nm has also increased from 0.6 to 3 J/cm^2.

In lasers based on rare gas-halides, antireflecting window coatings work under more severe condition of simultaneous influence both of radiation and of agressive gas mixture. It reduces significantly resources possibilities of these elements.

In this connection the tests of fused silica windows with antireflection coatings (Sc_2O_3/AlF_3 in MgF_2) are very demonstrative. The damage threshold of the windows in the air was 3 J/cm^2. After 40 hours of work in contact with F_2, when 370 impulses with laser energy density about 0.2 J/cm^2 were made, window trasparency fell from 97.5 to 91.5% . Scattering on the window sides been contacted with laser gas mixture and air were accordingly 0.9 and 0.6%. This example is demonstrative for the evaluation of the resource abilities of low damage threshold dielectric coatings for the excimer laser optics.

Table 5. LASER DAMAGE THRESHOLDS (E) OF THE MIRRORS AND ANTIREFLECTION COATINGS ON WAVELENGTH 308, 248 AND 193 nm

Substrate materials	Sistem of coatings	R, %	λ, nm	E, J/cm²	Notes
	ALUMINIUM MIRRORS				
Al	Al	90	248	1.6	Diamond turning
fused silica	Al	90	248	1.5	pulse-plasma
/-/	-/-/-/-/-/-/-/-/-/-/	/-/	350	2.0	method
/-/	-/-/-/-/-/-/-/-/-/-/	/-/	193	1.0	coatings
	DIELECTRIC MIRRORS				
fused silica	$(Sc_2O_3/MgF_2)^9Sc_2O_3$	98	308	5.5	/7/
	$(HfO_2/MgF_2)^{12}MgF_2$	99	308	7*	Made in KBTEM, Minsk
/-/	$(HfO_2/MgF_2)^7HfO_2MgF_2$	98	248	5.5	/7/.
/-/	$(ScO_3/MgF_2)^{12}MgF_2$	99	248	9	Made in KBTEM, Minsk
/-/	$(NdF_3/Na_3AlF_6)^6NdF_3$	72	248	9	
/-/	$(-/-/-/-/-/-)^9NdF_3$	88	248	9.2	
/-/	-/- + UV irradiation	88	248	10	
/-/	$(Al_2O_3/MgF_2)^9Al_2O_3$	90	193	0.4	/7/
MgF_2	$(-/-/-/-/-)^7Al_2O_3$	70	193	0.6	/7/
fused silica	$(Al_2O_3/MgF_2)^{19}MgF_2$	91	193	3.2	Made in KBTEM, Minsk
	$(LaF_3/MgF_2)^{19}MgF_2$	93	193	2.7	/-/
	ANTIREFLECTION COATINGS				
/-/	$(Sc_2O_3/AlF_3$ in $MgF_2)$	0.5	248	3	Made in KBTEM, Minsk
/-/	$(MgF_2/Al_2O_3/MgF_2)$	0.4	248	6	
/-/	$Na_3AlF_6/NdF_3/Na_3AlF_6$	0.6	248	7	Made in KBTEM, Minsk
/-/	$(MgF_2/Al_2O_3/MgF_2)$	0.2	193	0.4	

* -Theshold at 10 times irradiation.

Thus, presented results show that for KrF and longer wavelength excimer lasers the basic possibilities can be realised with the usage of the fused silica optics and traditional technology of dielectric coatings. The providing of stable work of the excimer laser in a shorter waves region demands working out new optical materials and mirror technology, and in particular metall-dielectric mirrors.

The authors would like to express their gratitude to V.I.Pashkevich and S.V.Kalinnikov for information on mirrors, made in KBTEM and also to B.M.Reiterov for some data on fluoride crystals.

REFERENCES

[1] N.G.Basov, V.A.Danilitchev, J.M.Popov, D.D.Hodkevitch, Pis'ma v JETF, 12, (1970)473.
[2] V.M.Buchnev, A.D.Klementov, P.B.Sergeyev, Kvantovaya electronika, 8, (1981)1235; 10, (1983) 2048.
[3] A.D.Klementov, N.V.Morozov, P.B.Sergeyev, Kratkie soobsheniya po fizike, № 5, (1987) 49.
[4] G.A.Kirillov et.al., Kvantovaya elektronika, 2,(1975)666.
[5] "Akustitsheskie kristally", Moscow,"Nauka",(1982).
[6] B.D.Vvedenskiy, L.D.Micheyev, S.I.Sagitov, Izvestiya AN SSSR Neorganitcheskie materialy, № 10,(1987) 1704.
[7] A.D.Klementov, N.V.Morozov, S.I.Sagitov, P.B.Sergeyev, Kvantovaya elektronika, 13,(1986) 2141.
[8] A.D.Klementov, N.V.Morozov, B.M.Reiterov, P.B.Sergeyev, Kvantovaya elektronika, 16, (1989) 1238.
[9] A.J.Taylor, R.B.Gibson, J.B.Roberts,Optics letters, 13, (1988) 814.
[10] T.Tomie, I.Okuda, M.Yano, Appl.Phys.Lett.,55,(1985) 325.
[11] T.S.Fahlen, IEEE J.Quant.Electr., QE-16, (1980) 1260.
[12] N.N.Antipov, O.A.Logunov, N.V.Morozov et.al.- Proceedings 7 conf. 'Interaction optical radiation with matter', Leningrad, (1988), p.107
[13] E.Spiller, Laser Techn. Extrime Ultrav. Meet., Boulder, Colo, March,1984, New York, (1984) p.312
[14] K.Ueda, Laser and Particle Beams, 7,(1989) p.3-375

WAVEFRONT CORRECTOR FOR EXCIMER LASERS

A.V. Ikramov, S.V. Romanov, I.V, Safronov, and A.O. Sulimov
NPO "Kompozit" 141070, Moscow d, Kaliningrad

The successful application of the adaptive optics methods to the visible and infrared ranges of spectrum has shown an effectiveness of the intracavity correction of radiation with the aim of improvement of laser characteristics. Recently the experiments have appeared which show that the usage of adaptive mirror in the eximer laser resonator leads to the possibility to control the light spot configuration in the far field and to stabilize the parameters of laser itself [1]. However,the known wavefront correctors [2,3] didn't allow to solve those problems effectively enough.

On the other hand,modern eximer lasers have a divergence of 0.15 to 3 mrad,that in tens and even in hundreds times exceeds the diffraction limit. Since the divergence is determined by the beam wavefront quality,the wavefront correctors with the reflective surface deformations of tens of wavelength (λ=0.15 - 0.35μm) are needed for it's sufficient improvement.

In spite of this average power of modern eximer laser can reach of loow,so it's desired to use cooled adaptive mirror to prevent thermal distortions of the reflective surface.

Present paper describes the cooled wavefront corrector based on bimorph structure that was intended for the shortwave optical systems.

The constructive design of adaptive mirror consists of housing,flexible cooled plate and two piezoceramic plates. The housing is the metallic cylinder with the diameter of 76 mm and the face height of 1.5 mm. Throughout the face of housing pits are made in such a way that it's central part - the flexible cooled plate - is suspended by eight lintels. The flexible cooled plate is the structure consisting of reflective plate and cooled support structure. On the support structure with the diameter of 50 mm the "wafer" type cooling system is formed which is represented by the grid with channels.For the increase of heat remove

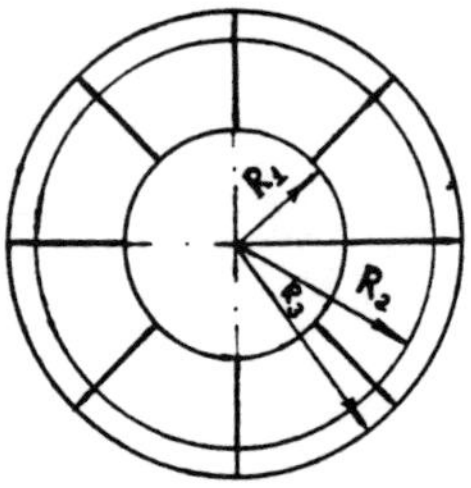

Fig.1 The control electrodes of the lower piezoceramic plates

the channels are oriented at 45° with respect to the line connecting inlet and outlet of coolant inside the housing (see [4]). Also the cooling channels are formed inside all lintels which connect the wafer structure of the flexible plate with the collectors of coolant remove and supply inside the housing. Two piezoceramic plates with the diameters of 48 mm and 43 mm are sequentally attached to the cooled flexible plate,and besides on the lower one the system of 17 sectionalized electrodes are formed (fig.1). Thus the total number of the adaptive mirror's electrodes is 19.

For the numerical simulation of mirror construction the finite element method was chosen. During the calculation following problems were solved:

a) optimization of cooling system parameters (i.e. depth, width and step of channels) for providing maximum heat remove;

b) optimization of adaptive mirror geometric parameters (i.e.reflective plate's,cooled support's and both piezoceramic plates thicknesses) for maximum deformations of the reflective surface;

c) optimization of the electrodes system (i.e.radii of rings) for the best approximation of defocus and spherical aberration with the adaptive mirror's surface;

d) calculation of the influence functions form for individual electrodes of phase corrector and of the construction's free oscillations lowest frequency.

Following parameters of cooling system were chosen according to the calculations:

- step of channels is 1.1 mm;
- width - 0.6 mm;
- depth - 0.5 mm.

It was determined for given system of cooling that if the environmental temperature change is 1° C the reflective plate's deformation is 0.44μm (peak value) because of substantial discrepancy between thermal expansion coefficients of the piezoceramics and the material of cooled plate (cooper).Besides the influence of laser beam with power density of 1 kW/sm^2 and diameter of 40 mm on reflective surface causes it's heating on 2.5°C and insignificant heating of the piezoceramics on 0.15° C. Such temperature distribution was obtained with the coolant expenditure of 1 l/ min. In

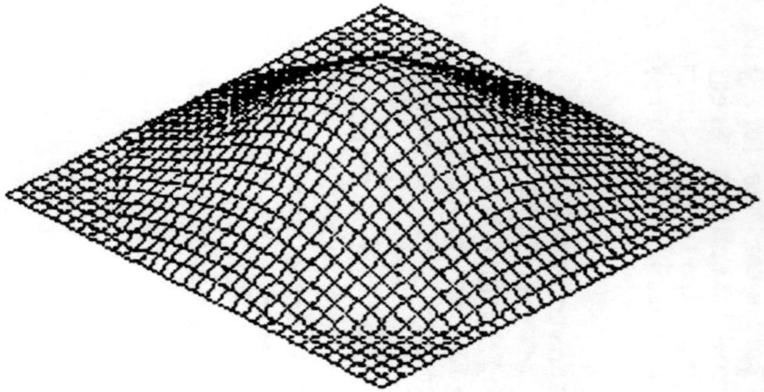

Fig.2 The central electrode influence function.

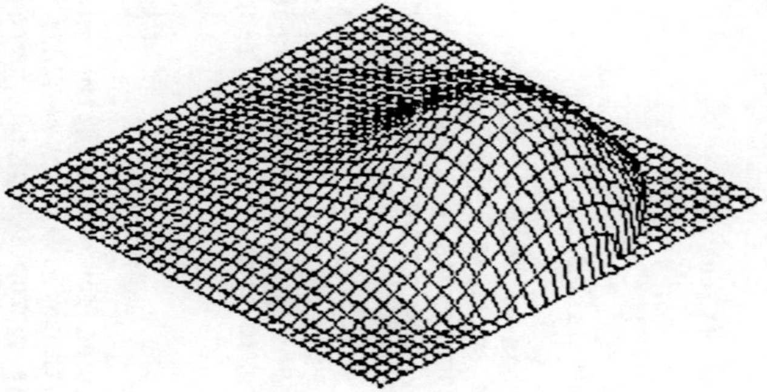

Fig.3 The middle ring electrode influence function.

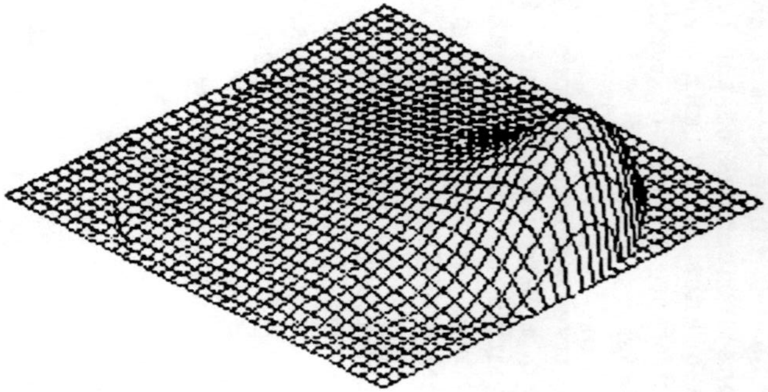

Fig.4 The outer ring electrode influence function.

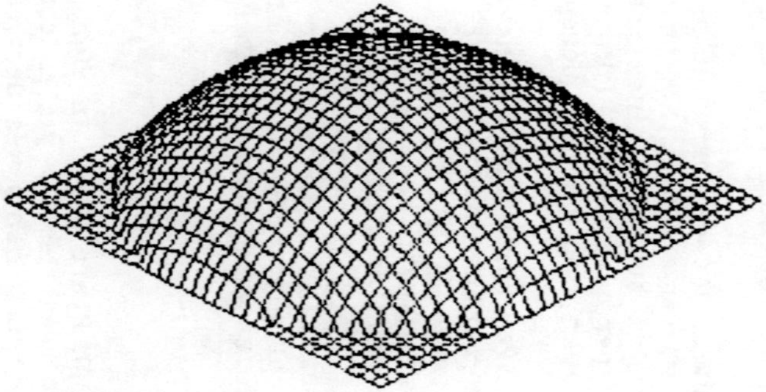

Fig.5 The upper piezoplate's influence function.

this case the maximum reflective surface displacement not exceeds 2.4μm.

Optimization of corrector geometry gives next results:

- reflective plate thickness is 0.5 mm;
- cooled support's thickness is 1 mm;
- upper piezoplate thickness is 0.3 mm;
- lower (sectionized) piezoplate thickness is 0.3 mm.

At given parameters the maximum deformation of adaptive mirror is +/-15μm.For further increase of the adaptive mirror's reflective surface deformations it is nesessary to use piezoceramics plates of lesser thickness,however their making is limited by the technological abilities of piezoceramics production.

According to optimization of the lower piezoelectric plate's system of control electrodes it was obtained following relationship in case of best defocus and spherical aberration's approximation by the corrector's surface:

$$R_1:R_2:R_3=5:9:10, \quad (1)$$

where R_1, R_2, R_3 are shown in fig.1. Result obtained for the adaptive mirror with two piezoelectric plates coincides with those given in ref.[5] where the same relationship was derived for a single piezoplate phase corrector.

On the 3-D finite-element model of phase corrector the form and free oscillations frequencies of the construction were determined. Frequency of first resonance is 2.7 kHz. Fig. 2-5 represent influence functions of phase corrector in case of maximum control action on the corresponding electrodes.

Adaptive bimorph mirrors of described construction were made from cooper and molybdenum. Figure 6 is the view of the mirror. To the back cover of mirror connector assembly are attached to provide voltage supply to electrodes. Piezoelectric plates were made of theЦТС-19 ceramics (lower one) andЦТСНВ-1 (upper). The maximum control voltage was +/-300 V.

In described construction of adaptive mirror the additional (upper) piezoplate has following functions:

1) to provide fast compensation of thermal defocus that appears at the intensive laser beam propogation inside gases at the expanse of high sphericity of influence function;

2) to increase adaptive mirror's reflective surface deformations;

Fig.6 The adaptive mirror

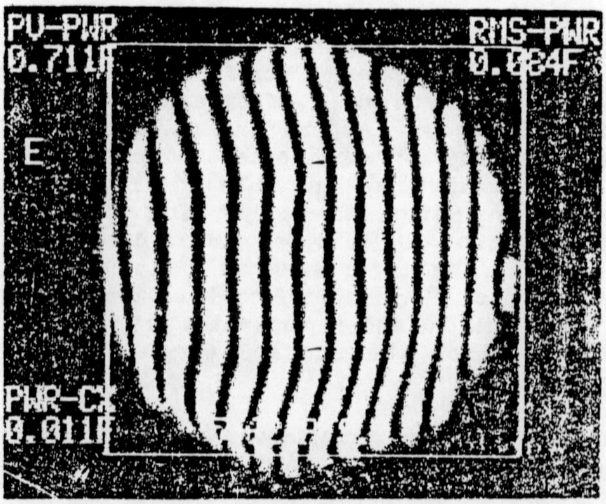

Fig.7 The initial surface of the adaptive mirror

3) to remove hysteresis of the lower (sectionized) control plate by the applying the voltage taking away residual deformations of the reflective surface.

Fig.7 shows interferogram of the cooper adaptive mirror's initial shape. All the quantities on the interferogram are given in terms of fringes at λ=632.8 nm. As it seen from the interferogram the mirror's local error is less then $\lambda/2$ (without any correction). Fig.8 represents

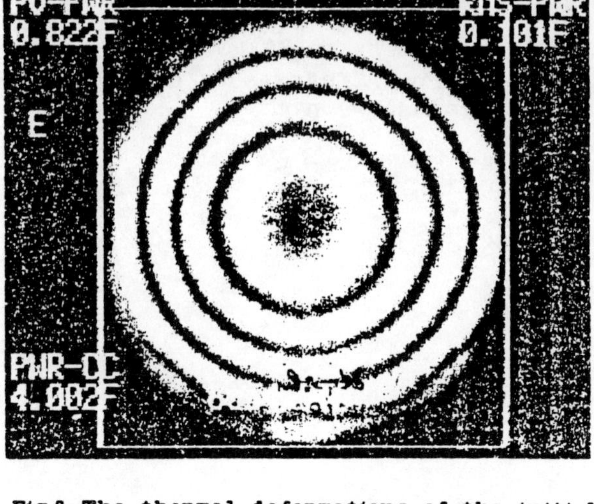

Fig.8 The thermal deformations of the initial surface

interferogram of the same surface but at higher temperature (a few centigrade heating). It is seen that mirror's surface deformation because of difference in thermal expansion coefficients is of 4 fringes (1.2μm) and it posseses high sphericity. If to take into consideration the value of mirror's deformation at temperature changes of 1°C of 0.4μm (1.3 fringe),one can state the correspondence of thermal mirror's deformation with results of calculations. In addition if to assume the adaptive mirror's surface deformation to be not due to difference in thermal expension coefficients but due to upper plate's deformations at applying a voltage to it, then the interferogram can be interpreted as the interference function of mirror's surface. On comparing fig.5 and 8 a good agreement of the calculated and experimental interference function can be seen. Interferograms of the surface were obtained with "MARK-III" interferometer.

On the base of obtained results one can conclude that real bimorph adaptive mirror due to their high surface deformations (+/-15μm) are able to work effectively in shortwave controlled optical systems (e.g.,in the eximer laser's resonator) as the correctors of the optical aberrations of great amplitude (~70λ,λ=193 nm). More succesfully phase corrector works with axial symmetric aberrations like defocus and spherical aberrations. Frequency response of the mirror (f=2.7 kHz) allow it's application to systems with typical frequency range up to 2 kHz.

REFERENCES

[1] A.S.Akhmanov, M.A.Vorontsov,A.V.Kudryashov et al., Quantum electronics,15,(1988) 1525.
[2] V.G.Taranenko, Yu.G.Gorokhov,and N.S.Romanyuk,Zarubezhnaya radioelectronika,No 8,(1982) 19.
[3] P.V.Nikolaev and A.V.Smirnov,Optiko-mekhanicheskaya promyshlennost,No 11,(1987) 47.
[4] A.V.Ikramov, A.V.Kudryashov, S.V.Romanov et al.,SU Pat.,No (199), Int.Cl. H01S 3/02, filed Dec.7,1989, No 4766709/24-25.
[5] M.A.Vorontsov,S.A.Gnedoy,A.V.Kudryashov et al.,Prepr.SRCTL USSR AS 87-29,Shatura,(1987) p.14.

FLUORESCENCE Xe_2^* MODEL IN PULSE SELF-SUSTAINED DISCHARGES

K.S. Gochelashvily, I.V. Kochetov, and L.R. Yangurazova
General Physics Institute, Moscow, Vavilova 38, USSR

Power vacuum ultraviolet (VUV) sources with high efficiency are the subject of great interest in many applications in science and technique [1]. Xe_2^*, Ar_2^*, Ne_2^* dimerouses transitions wave-lengths correspond to 172 nm, 126 nm, 90 nm; energies of these wavelengths can directly break chemical bonds of different chemical compounds, and it makes VUV sourses applicable in the chemical technology; due to the short wavelenghts these sourses are perspective in microelectronics and photolitography [2].

A quasi-continuous KrF (B → X) fluorescence was observed in a microwave discharge-pumped KrF lamp, and besides a power efficiency was of ~12% [2].

In works [3,4] the fluorescence with the high efficiency of 50% was obtained by the fast electron beam pumping. But the technical realization of such pumping is rather difficult, demands special protection and does not permit to obtain the long chain of pulses.

In work [5] the uniform steady independent pulse discharge in inert gases was obtained, but problems of the fluorescence efficiency were not examined.

In this work the computational method and numerical theoretical studies of the fluorescence in the self-sustained pulse discharge in Xe_2^* are proposed. Mathematical model, formulated for a spatial homogeneous medium, includes the balance equations for electron levels and ion-molecular kinetics, the Boltzmann equation for spherically-symmetric part of electron energy distribution function and the equations for external discharge circuit [6]. Our kinetic scheme was based on the reactions from [3], but we did not make difference between singlet $Xe_2^*(^3\Sigma)$ eximers and, as a result, did not take into account the direct mixing processes between them. We considered the optically thin active medium and therefore the processes of photoabsorbtion are not taken into consideration unlike [3]. All reactions used in this work are shown in Table 1. The reactions with electron participation are shown in the upper part of Table 1, constants of this elementary processes were calculated with the help of electron energy distribution function with the cross sections taken from [6,7].

TABLE 1.

Reactiors with electrons						
1	Xe	+	e	$\rightleftarrows$	Xe^*	+ e
2	Xe	+	e	$\rightleftarrows$	Xe^{**}	+ e
3	Xe	+	e	$\rightleftarrows$	Xe^+	+ 2e
4	Xe_2^+	+	e	$\longrightarrow$	Xe^{**}	+ Xe

5	Xe_3^+	+ e	$\longrightarrow$	Xe^{**}	+ 2Xe	
6	Xe^*	+ e	$\longrightarrow$	Xe^+	+ 2e	
7	Xe_2^*	+ e	$\longrightarrow$	Xe_2^+	+ 2e	
8	Xe_2^*	+ e	$\longrightarrow$	2Xe	+ e	
9	Xe_2^*	+ e	$\longrightarrow$	Xe^*	+ Xe	+ e
10	Xe_2^*	+ e	$\rightleftarrows$	Xe_2^{**}	+ e	
11	Xe^*	+ e	$\rightleftarrows$	Xe^{**}	+ e	
12	Xe^{**}	+ e	$\longrightarrow$	Xe^+	+ 2e	
13	Xe_2^+	+ e	$\longrightarrow$	Xe^+	+ Xe	+ e
14	Xe_2^{**}	+ e	$\longrightarrow$	Xe_2^+	+ 2e	

	Reactions with heavy particles						Rates($cm^6/s, cm^3/s, 1/s$)
1	Xe^+	+ 2Xe	$\longrightarrow$	Xe_2^+	+ Xe		$2 \cdot 10^{-31}$
2	Xe_2^+	+ 2Xe	$\longrightarrow$	Xe_3^{**}	+ Xe		$6 \cdot 10^{-32}$
3	Xe^{**}	+ 2Xe	$\longrightarrow$	Xe_2^{**}	+ Xe		$5 \cdot 10^{-31}$
4	Xe_2^{**}		$\longrightarrow$	Xe^*	+ Xe		$1 \cdot 10^{8}$
5	Xe_2^{**}	+ Xe	$\longrightarrow$	Xe^*	+ 2Xe		$1 \cdot 10^{-11}$
6	Xe^*	+ 2Xe	$\longrightarrow$	Xe_2^*	+ Xe		$4.4 \cdot 10^{-32}$
7	$2Xe_2^*$		$\longrightarrow$	Xe_2^+	+ 2Xe	+ e	$8 \cdot 10^{-11}$
8	$2Xe^*$		$\longrightarrow$	Xe^+	+ Xe	+ e	$8 \cdot 10^{-11}$
9	$2Xe^{**}$		$\longrightarrow$	Xe^+	+ Xe	+ e	$1 \cdot 10^{-11}$
10	$2Xe_2^{**}$		$\longrightarrow$	Xe_2^+	+ 2Xe	+ e	$8 \cdot 10^{-11}$
11	Xe_2^*		$\longrightarrow$	2Xe	+ $h\nu_1$		$2 \cdot 10^{8}$
12	Xe^{**}		$\longrightarrow$	Xe^*	+ $h\nu_2$		$6 \cdot 10^{6}$

Energy levels of Xe and Xe_2 used in our kinetic model are s own in Fig.1. The radiation at 172 nm from the first exited s ate of Xe_2^* to the ground state is considered useful.

The test calculations with fast electron beam pumping were made to examine our kinetic model. Results of our calculations are in good agriment with experimental and theoretical results of Xe_2^* fluorescence dynamic [3], curves of the fluorescence rate at 0.5 atm and 3atm are shown in Fig.2.

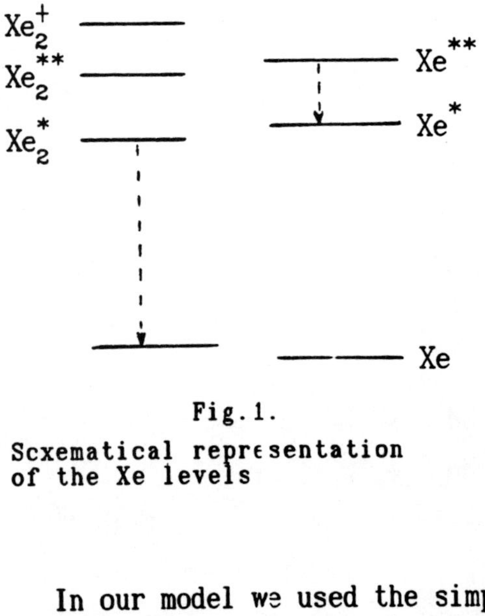

Fig.1.
Scxematical representation of the Xe levels

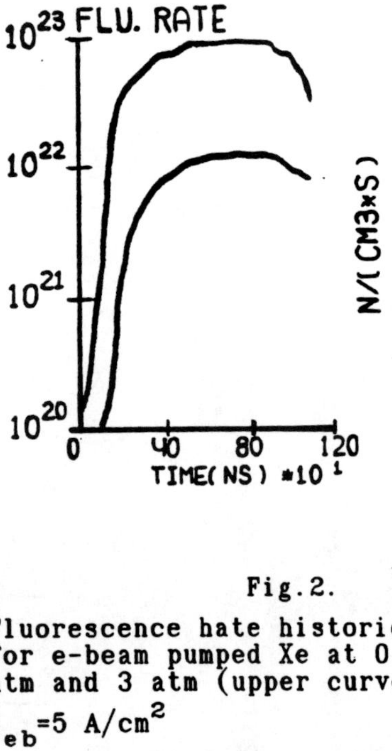

Fig.2.
Fluorescence hate histories for e-beam pumped Xe at 0.3 atm and 3 atm (upper curve); j_{eb}=5 A/cm^2

In our model we used the simple external driving circuit, consisting of LC contour; the choice of the simple circuit was determined by studing the main-points of the processe, without detalization. Computations were made for different parameters of electrical circuit, experimentally realized for eximer discharge lasers in [5]. For initial parameters 50 kV, inductance L=110 nH, capacitance C=50 nF, separation between anode and cathode d=3 cm, area of discharge A=240 cm^2 and pressure P=1 atm, efficiency was of 32%.

The fluorescence efficiency was determined as a relation of all spontaneous radiation energy to energy, contributed in active volume of discharge. The determined efficiency varied from 13% to 42% for different initial conditions.

The Fig.3 shows the typical discharge voltage history and current behaviour through discharge. The strong electron multiplication was observed on maximum carrent phase due to step ionization processe. At this point voltage presents the typical horisontal line coresponding to quasi-stational stage of discharge burning, is sufficiently high, approximately 10 in comparison with eximer lasers, where this value is equal 3-5.

As in the case of eximer lasers the initial power of preionization practically did not influence on efficiency, but affected only the time of reaching the current's maximum.

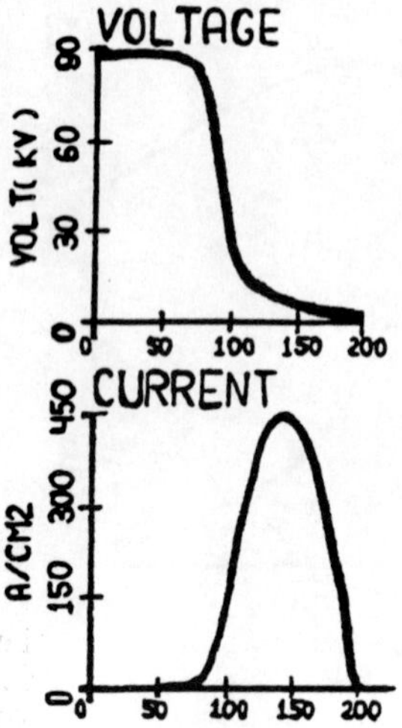

Fig.3.
Typical voltage and current histories for discharge

We studied the efficiency in dependence of pressure, initial discharge voltage, initial electron concentration and other paramerters. The Fig.4 shows the dependence of the energy contributed into discharge on the initial discharge voltage. The next Fig.5 shows the dependence of the efficiency and output energy also on the initial discharge voltage. (U > 80) leads to decreasing of efficiency and output energy. Results shown in Fig 3-5 are obtained for P=1 atm, C=50 nF, L=40 nH, A=240 cm^2, d=8 cm.

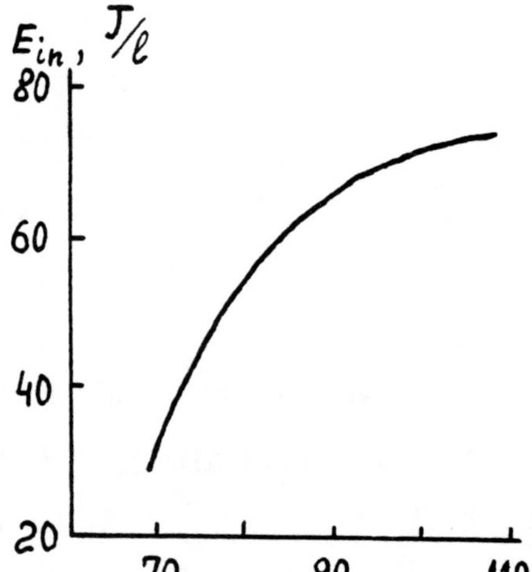

Fig.4
Dependence of the energy contributed into discharge on the initial discharge voltage

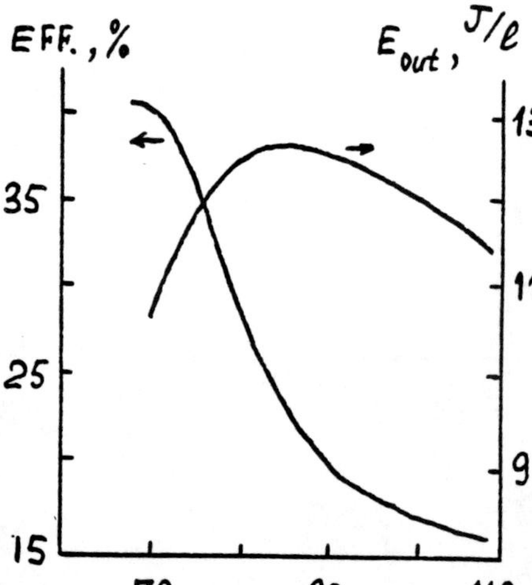

Fig.5
Dependence of the effeciency and output energy on the initial discharge voltage

We made the calculations for more high pressures, 2 atm and 4 atm, in order to investigate the possibility of creation the more compact installation. We did not change the initial voltage but decreased the distance between electrods. In Table 2 values of the input energy and the efficiency are given for pressures 1 atm, 2 atm and 4 atm.

TABLE 2.

P (atm)	1	2	4
E_{in} (j/l)	26	41	213
eff (%)	41	39	37

Thus we theoretically show the possibility of reaching the high Xe_2^* fluorescence efficiency in the pulse self-sustained discharge supposing the existence of uniform burning.

REFERENCES

[1] J.R.Murray and C.K.Rhodes, J.Appl.Phys., 47, (1976)5041.

[2] Kumagai Hiroshy and Obara Minoru, Appl.Phys.Lett., 54, (1989)2619.

[3] D.J.Eckstrom, H.H.Nakano, D.C.Lorents, J.Appl.Phys., 64, (1988)1679.

[4] D.J.Eckstrom, H.H.Nakano, D.C.Lorents, J.Appl.Phys., 64, (1988)1691.

[5] V.Yu. Barauov, V.M. Borisov, Yu.Yu. Stepanov, Discharge Experimental Lasers Based on Halide Inert Gases, Moscow. Energoatomizetal, 1988.

[6] A.V. Dem'yanov, A.V. Kochetov, A.P. Napurtovich, et al. Prepreint, IAE-4099/12 Moscow, 1985.

[7] V.Yu. Baranov, V.M. Borisov, F.I. Vysikaylo, et al. Preprint, IAE-3080, Moscow, 1979.

Authors index

Subject Index